Medicinal Plants and Malaria

Applications, Trends, and Prospects

Traditional Herbal Medicines for Modern Times

Each volume in this series provides academia, health sciences, and the herbal medicines industry with in-depth coverage of the herbal remedies for infectious diseases, certain medical conditions, or the plant medicines of a particular country.

Series Editor: Dr. Roland Hardman

Volume 1
Shengmai San, edited by Kam-Ming Ko

Volume 2
Rasayana: Ayurvedic Herbs for Rejuvenation and Longevity, by H.S. Puri

Volume 3
Sho-Saiko-To: (Xiao-Chai-Hu-Tang) Scientific Evaluation and Clinical Applications, by Yukio Ogihara and Masaki Aburada

Volume 4
Traditional Medicinal Plants and Malaria, edited by Merlin Willcox, Gerard Bodeker, and Philippe Rasoanaivo

Volume 5
Juzen-taiho-to (Shi-Quan-Da-Bu-Tang): Scientific Evaluation and Clinical Applications, edited by Haruki Yamada and Ikuo Saiki

Volume 6
Traditional Medicines for Modern Times: Antidiabetic Plants, edited by Amala Soumyanath

Volume 7
Bupleurum *Species: Scientific Evaluation and Clinical Applications,* edited by Sheng-Li Pan

Volume 8
Herbal Principles in Cosmetics: Properties and Mechanisms of Action, by Bruno Burlando, Luisella Verotta, Laura Cornara, and Elisa Bottini-Massa

Volume 9
Figs: The Genus Ficus, by Ephraim Philip Lansky and Helena Maaria Paavilainen

Volume 10
Phyllanthus Species: Scientific Evaluation and Medicinal Applications, edited by Ramadasan Kuttan and K. B. Harikumar

Volume 11
Honey in Traditional and Modern Medicine, edited by Laïd Boukraâ

Volume 12
Caper: The Genus Capparis, by Ephraim Philip Lansky, Helena Maaria Paavilainen, and Shifra Lansky

Volume 13
Chamomile: Medicinal, Biochemical, and Agricultural Aspects, by Moumita Das

Volume 14
Rhodiola rosea, by Alain Cuerrier and Kwesi Ampong-Nyarko

Volume 15
Connecting Indian Wisdom and Western Science: Plant Usage for Nutrition and Health, by Luisella Verotta, Maria Pia Macchi, and Padma Venkatasubramanian

Traditional Herbal Medicines for Modern Times

Medicinal Plants and Malaria

Applications, Trends, and Prospects

Woon-Chien Teng
Ho Han Kiat
Rossarin Suwanarusk
Hwee-Ling Koh

CRC Press
Taylor & Francis Group
Boca Raton London New York

CRC Press is an imprint of the
Taylor & Francis Group, an **informa** business

CRC Press
Taylor & Francis Group
6000 Broken Sound Parkway NW, Suite 300
Boca Raton, FL 33487-2742

First issued in paperback 2021

ISBN-13: 978-1-03-209811-1 (pbk)
ISBN-13: 978-1-4987-4467-6 (hbk)

Publisher's Note
The publisher has gone to great lengths to ensure the quality of this reprint but points out that some imperfections in the original copies may be apparent.

Library of Congress Cataloging-in-Publication Data

Teng, Woon-Chien, author.
Medicinal plants and malaria : applications, trends, and prospects / Woon-Chien Teng, Ho Han Kiat, Rossarin Suwanarusk, and Hwee-Ling Koh.
p. ; cm. -- (Traditional herbal medicines for modern times ; v. 16)
Includes bibliographical references and index.
ISBN 978-1-4987-4467-6 (hardcover : alk. paper)
I. Kiat, Ho Han, author. II. Suwanarusk, Rossarin, author. III. Koh, Hwee Ling, author. IV. Title. V. Series: Traditional herbal medicines for modern times ; v. 16.
[DNLM: 1. Antimalarials. 2. Malaria--drug therapy. 3. Phytotherapy--methods. 4. Plants, Medicinal. QV 256]

RA644.M2
614.5'32--dc23 2015028512

Visit the Taylor & Francis Web site at
http://www.taylorandfrancis.com

and the CRC Press Web site at
http://www.crcpress.com

To our families,

for their unwavering support and love

Contents

List of Figures

List of Tables

List of Tables

Foreword

The search for efficacious solutions to a devastating disease plaguing some 207 million people worldwide especially in Africa and Southeast Asia is never-ending and must be pursued relentlessly. This is the central message of this informative book that recognizes that the fight against malaria has to be multifaceted and plants play an essential role in the research.

Plants are prolific production houses for innumerable active compounds against malaria. Many books have been written on plants and their therapeutic potential against malaria, but I have yet to read one such as this work that provides a comprehensive overview of the disease, the worldwide measures for its prevention and treatment, the very long list of plants reportedly used in malaria control, and recommendations of plant species for further research. The authors specifically describe four species of traditional plants with known antimalarial properties and highlight four others that hold great promise as sources of novel therapeutic leads. However, they warn that even where "it is clearly evident that medicinal plants have a vital role to play in malaria control… very few are further developed" into drugs and vaccines for various reasons.

I commend the authors for this ethnobotanical approach and survey of the 1800 medicinal plants reportedly used in the fight against malaria. This book is a most up-to-date guide for researchers, students, and the general public on the potential uses of plants in the search for the elusive cure against a scourge of mankind.

Leo W. H. Tan
Professor
Fellow of the Singapore National Academy of Science

Preface

Malaria is a potentially life-threatening disease that affects millions worldwide, especially in Africa. More recently, the emergence and spread of multidrug resistance in parts of Southeast Asia prompts the urgent need for novel and effective therapy against the disease. Of particular note, there has not been a recent book on plants used in malaria in the past decade.

This book was written to highlight the therapeutic potential of plants for treating malaria. Useful information on malaria, current prevention and treatment, and scientific research carried out are collated for easy reference. With insights into the plant species used for malaria and how they are prepared, this is the first book detailing medicinal plants used for treating malaria. The plant parts used, methods of preparation, and doses are included where available. The scope of the book covers the uses and effects at the individual, population, environment, vector, and parasite levels of medicinal plants in combating malaria. It begins with a brief introduction to malaria, its epidemiology and implications on public health, the life cycle of *Plasmodium* parasites, clinical presentation of the disease, currently available antimalarial drugs and their roles in therapy, and medicinal plants used for malaria including those that are currently in clinical use in various countries. It includes the latest findings from ethnobotanical research, the challenges and isolation of antiplasmodial phytochemicals from medicinal plants, as well as results from clinical trials and public health interventions using medicinal plants. A list of plant species to be prioritized for further research is also proposed. As Singapore is a thriving research hub located in Southeast Asia where some countries have malaria drug resistance, plants used for malaria that can be found in Singapore are also highlighted.

This book will be a vital resource for students, teachers, healthcare professionals, academics, and researchers from multidisciplinary fields interested in ethnopharmacology, drug discovery from medicinal plants, and promising botanical antimalarial preparations to meet unmet medical needs. It will also inspire the general public and researchers who are interested in herbal medicine and medicinal plants, and who would like to know more about the science behind how plants may potentially be used to prevent and treat malaria. To cater to the wide audience of readers, the authors have made every effort to present the information in a manner as comprehensive and yet simple as possible.

It is hoped that this book will stimulate general interest and awareness of the potential of medicinal plants in the prevention and treatment of malaria, as well as serve as a useful resource for research in natural products, drug discovery, biodiversity, pharmacology, medicinal chemistry, and other related fields.

Acknowledgments

The authors express their heartfelt gratitude to:

World Health Organization (WHO), Centers for Disease Control and Prevention (CDC), and Public Health England (PHE), for permission to use their data and images (link to PHE content license: http://www.nationalarchives.gov.uk/doc/open-government-licence/version/3/)

Dreamstime.com and Mr. Sebastian Kaulitzki, Mr. James Gathany, Mr. Andrew Kator, and Ms. Jennifer Legaz, for permission to use their images and/or photographs in Figure 1.1

Ms. Ria Tan (wildsingapore.com) and Mr. Kwan-Han Ong (http://www.natureloveyou.sg), for permission to use their photographs in Figure 3.1

Leeward Pacific Pte. Ltd. and the National University of Singapore, for financial support of a collaborative research grant (R-148-000-172-592 to Associate Professor Hwee-Ling Koh) and National University of Singapore Industrial PhD Programme (NUS IPP) scholarship (to Ms. Woon-Chien Teng)

Photographers: Ms. Emily Woon-Hui Teng, Mr. Hai-Ning Wee, Ms. Woon-Chien Teng, and Associate Professor Hwee-Ling Koh

Mr. Cheng-Shoong Chong, for his contributions to Section 3.5 on the conservation of medicinal plants

The staff of CRC Press and Taylor & Francis Group, for their professionalism and support

Colleagues and friends, for their interesting deliberations on medicinal plants

Our families, for their unwavering support and love

Guide to Using This Book

This book is a concise compilation of up-to-date information on malaria and the use of medicinal plants in treating it.

Chapter 1 gives an overview of malaria, namely, the epidemiology and implications on public health; the life cycle of the *Plasmodium* parasite; and the signs, symptoms, and pathophysiology of the disease.

Chapter 2 presents a comprehensive guide to current antimalarial therapy, namely, drugs for treatment, prevention, and emergency standard treatment. The information is adapted from the World Health Organization (WHO), the Centers for Disease Control and Prevention (CDC) in the United States, and Public Health England (PHE).

Chapter 3 compiles published literature of research conducted on medicinal plants used for treating malaria. The need for medicinal plants, ethnobotanical research, general information on bioassay-guided fractionation for drug discovery (including extraction and bioassay screening), chemical classes of compounds isolated, and conservation of medicinal plants are presented.

Chapter 4 highlights two well-established medicinal plants (*Artemisia annua* and *Cinchona* bark), two promising medicinal plants (*Dichroa febrifuga* and *Vernonia amygdalina*), and four herbal preparations. Three of the herbal preparations are in clinical use and they are Sumafoura Tiemoko Bengaly (Mali), N'Dribala (Burkina Faso), and Phyto-laria (Ghana), whereas the fourth preparation (PR 259 CT1) has shown promising results in clinical trials in the Congo.

For a listing of medicinal plants reported to be used to treat malaria, see Table A.1 in the Appendix. The corresponding family name is also shown. For a listing of medicinal plants used for malaria from ethnobotanical surveys that provide details on the preparation of the plants, see Table A.2 in the Appendix. Medicinal plant species listed by family appear at the end of the Appendix.

You may start with Chapter 1 to understand the devastating disease, or go straight to any topic of your choice without affecting your reading pleasure.

This book has something for everyone, whether you are a traveler to a malaria-affected country wondering what prophylaxis measure to take; or you are interested to know which plant in your garden may have antimalarial properties; or you are already involved in the field of malaria research and control, wondering what medicinal plants can offer. The list goes on. In addition to increasing public awareness, we hope that the information in this book will inspire more people to appreciate what Mother Nature can offer, as well as to inspire more people to take up the challenge to better understand medicinal plants.

The contents of this book serve to provide both general and scientific information about medicinal plants used in treating malaria and are not intended as a guide to self-medication by consumers or as a substitute for the medical advice by healthcare professionals. The general public is advised to discuss the information contained herein with a physician, pharmacist, or other healthcare professional. Neither the authors nor the publisher can be held responsible for the accuracy of the information itself or the consequences from the use or misuse of the information in this book.

The resources are not vetted and it is the responsibility of the reader to ensure the accuracy of the cited information. Readers are reminded that the information presented is subject to change as research is ongoing and there may be interindividual variation. Although every effort has been made to minimize errors, there may be inadvertent omissions or human error in the compilation of information.

Authors

Woon-Chien Teng is a PhD candidate in the Department of Pharmacy, National University of Singapore (NUS). She graduated with a BSc (Pharmacy) (Honors) from NUS. Teng is also a registered pharmacist with the Singapore Pharmacy Council. She has previously worked as a pharmacist at Changi General Hospital and as a preregistration pharmacist at Kandang Kerbau Women's and Children's Hospital.

Ho Han Kiat is an associate professor in the Department of Pharmacy, National University of Singapore (NUS). He received his BSc (Pharmacy) with first class honors from NUS in 2000. Subsequently, he obtained his PhD in medicinal chemistry from the University of Washington in 2005, investigating the molecular mechanisms of specific drug-induced liver toxicity (under the mentorship of Professor Sidney Nelson). He then returned to Singapore for a 3-year postdoctoral fellowship with Professor Axel Ullrich, investigating the roles of tyrosine kinases for various malignancies. Since 2009, Dr. Kiat started building his own research program focusing on drug-induced liver toxicity, as well as exploring new drug targets for liver cancer. He directs a toxicology division within the Drug Development Unit at NUS. He has published more than 40 papers in internationally recognized journals and has won multiple university-level teaching excellence awards.

Rossarin Suwanarusk is a research scientist with the Malaria Immunobiology Group at the Singapore Immunology Network (SIgN), A*STAR. She graduated with a PhD from Mahidol University (Medical Technology), Thailand in 2003. Suwanarusk is well known for her research on the pathobiology and drug-resistance markers of *Plasmodium vivax*. She has published more than 30 articles in international peer-reviewed journals, including *Nature*, *Science*, *Blood*, and *Journal of Infectious Diseases.*

Hwee-Ling Koh is an associate professor in the Department of Pharmacy, National University of Singapore (NUS). She graduated with a PhD from the University of Cambridge (United Kingdom), as well as a BSc (Pharmacy) (Honors) and MSc (Pharmacy) from NUS. She is a registered pharmacist with the Singapore Pharmacy Council. She has been teaching and conducting research on traditional Chinese medicine and medicinal plants at NUS for more than 15 years. Her research areas include quality control and safety of botanical products, and the study of natural products and medicinal plants as potential sources of lead compounds of novel therapeutics. Dr. Koh is a technical/expert assessor with the Singapore Accreditation Council–Singapore Laboratory Accreditation Scheme (SAC-SINGLAS); a member of the Agri-Food and Veterinary Authority (AVA) Advisory Committee on Evaluation of Health Claims; and a member of the Complementary Health Product Advisory Committee, Ministry of Health, Singapore. She is also a member of the United States Pharmacopoeial Expert Panel. Dr. Koh has written various articles published in international peer-reviewed journals, including the *Journal of Ethnopharmacology*, *Drug Discovery Today*, *Journal of Chromatography A*, *Drug Safety*, *Journal of Pharmaceutical and Biomedical Analysis*, *Journal of Agricultural and Food Chemistry*, and *Food Additives and Contaminants*. She has also coauthored a book titled *A Guide to Medicinal Plants: An Illustrated, Scientific and Medicinal Approach* (Singapore: World Scientific).

List of Abbreviations

ACT	artemisinin-based combination therapy
BD	twice daily
CDC	Centers for Disease Control and Prevention
CITES	Convention on International Trade in Endangered Species of Wild Fauna and Flora
CYP450	cytochrome P450
DR Congo	Democratic Republic of the Congo
G6PD	glucose-6-phosphate dehydrogenase
GC-MS	gas chromatography-mass spectrometry
GPI	glycosylphosphatidylinositol
HEPES	4-(2-hydroxyethyl)piperazine-1-ethanesulfonic acid
HPLC	high-performance liquid chromatography
IC_{50}	concentration at which there is 50% growth inhibition
IUCN	International Union for Conservation of Nature
IV_{mal}	importance value for the treatment of malaria
LC-MS	liquid chromatography-mass spectrometry
NA	not applicable
$NaHCO_3$	sodium bicarbonate
NO_3^-	nitrate
OD	once daily
P-gp	P-glycoprotein
PHE	Public Health England
PO	per os (orally)
QDS	four times daily
RPMI	Roswell Park Memorial Institute
SERCA	sarcoplasmic-endoplasmic reticulum calcium transport adenosine triphosphatase
TDS	three times daily
TLR	toll-like receptor
TNF	tumor necrosis factor
WHO	World Health Organization
WWF	World Wide Fund for Nature

CHAPTER 1

Malaria

An Overview

1.1 EPIDEMIOLOGY AND IMPLICATIONS ON PUBLIC HEALTH

Malaria is a life-threatening tropical disease affecting about 207 million people worldwide with about 627,000 fatalities in 2012. About 90% of deaths occur in Africa, with 77% being children less than 5 years of age (World Health Organization [WHO] 2013a). Estimated numbers of cases and deaths from different parts of the world in 2012 are shown in Table 1.1.

About half the world's population is at risk of malaria. Table 1.2 shows the malarious areas worldwide. Malaria is, however, preventable and curable. Malaria is closely associated with poverty, and this leads to a vicious cycle, as interventions would be required to be affordable and accessible to this high-risk group (WHO 2013a). Malaria is one of the top priorities of the Bill and Melinda Gates Foundation (2015), which has channeled approximately $2 billion in malaria control thus far. International funding for control of malaria has sharply increased from less than $100 million in 2000 to $1.71 billion in 2010 and is estimated to be $1.84 billion in 2012. However, this is still inadequate in achieving worldwide malaria targets. Millions of people still do not have access to long-lasting insecticidal nets, preventive interventions, rapid diagnostic testing, and quality-assured therapy, of which substandard and counterfeit drugs pose additional problems that remain to be addressed (Nayyar et al. 2012; WHO 2013a).

Several antimalarial drugs are in place, but the world today faces the urgent problem of the emergence of multidrug-resistant *Plasmodium falciparum*, which leads to treatment failure. Resistance to all categories of antimalarials has been documented for *P. falciparum*, *P. vivax*, and *P. malariae* (WHO 2010). Resistance to artemisinins is spreading in Southeast Asia. It was initially detected in west Cambodia, east Myanmar, west Thailand, and south Vietnam (WHO 2013a), but resistance has also emerged in south Laos and northeast Cambodia (Ashley et al. 2014). Resistance to chloroquine, the cheapest and most commonly used antimalarial, and sulfadoxine-pyrimethamine is spreading in the endemic areas (WHO 2013b). In addition, existing chemotherapy comes with potential adverse effects and toxicity.

The most advanced malaria vaccine candidate, "RTS, S/AS01E" (Mosquirix), a recombinant protein-based vaccine with a protein of *P. falciparum* circumsporozoite

Table 1.1 Estimated Malaria Cases and Deaths in 2012

Region	Estimated Number of Cases	Estimated Number of Deaths
Global	207 million	627,000
Africa	165 million	562,000
Americas	800,000	800
Eastern Mediterranean	13 million	18,000
Europe	30	0
Southeast Asia	27 million	42,000
Western Pacific	1 million	3500

Source: World Health Organization (WHO), 2013, "World Malaria Report 2013," accessed August 20, 2014, http://www.who.int/malaria/publications/world_malaria_report_2013/en/. With permission.

Table 1.2 Areas with Malaria

Region	Country
Central Africa	Algeria, Benin, Burkina Faso, Cabo Verde, Côte d'Ivoire, Gambia, Ghana, Guinea, Guinea-Bissau, Liberia, Mali, Mauritania, Niger, Nigeria, Sao Tome and Principe, Senegal, Sierra Leone, Togo
East and South Africa	Comoros, Eritrea, Ethiopia, Kenya, Madagascar, Malawi, Mozambique, Rwanda, Uganda, United Republic of Tanzania (Mainland), United Republic of Tanzania (Zanzibar), Zambia
West Africa	Angola, Burundi, Cameroon, Central African Republic, Chad, Congo, Democratic Republic of the Congo, Equatorial Guinea, Gabon
Region of the Americas	Argentina, Belize, Bolivia, Brazil, Colombia, Costa Rica, Dominican Republic, Ecuador, El Salvador, French Guiana, France, Guatemala, Guyana, Haiti, Honduras, Mexico, Nicaragua, Panama, Paraguay, Peru, Suriname, Venezuela
Eastern Mediterranean Region	Afghanistan, Djibouti, Iran, Iraq, Pakistan, Saudi Arabia, Somalia, South Sudan, Sudan, Yemen
European Region	Azerbaijan, Georgia, Kyrgyzstan, Tajikistan, Turkey, Uzbekistan
Southeast Asia Region	Bangladesh, Bhutan, Democratic Republic of Korea, India, Indonesia, Myanmar, Nepal, Sri Lanka, Thailand, Timor-Leste
Western Pacific Region	Cambodia, China, Laos, Malaysia, Papua New Guinea, Philippines, Republic of Korea, Solomon Islands, Vanuatu, Vietnam

Source: World Health Organization (WHO), 2013, "World Malaria Report 2013," accessed August 20, 2014, http://www.who.int/malaria/publications/world_malaria_report_2013/en/. With permission.

fused with a hepatitis B surface antigen (rts), and combined with hepatitis B surface antigen (s), was recently given a "positive scientific opinion" by the Committee for Medicinal Products for Human Use (European Medicines Agency). With that, WHO will evaluate and give guidance for its use by November 2015, following approval by respective national regulatory authorities (GlaxoSmithKline 2015). The vaccine will be assessed not to replace but to add on to existing measures for prevention, diagnosis, and treatment. It has undergone a Phase III trial in seven countries in Africa on infants (6–12 weeks old) and young children (5–17 months old). Although its efficacy is found to be reduced over the 4-year follow-up period (Olotu et al. 2013) and during the extended period of follow-up (RTS,S Clinical Trials Partnership 2015), the vaccine, which was administered as a schedule of three monthly doses, demonstrated 28% and 18% efficacy in the absence of a booster dose against clinical malaria in children and infants, respectively, bringing about significant reduction in clinical malaria cases. Having an 18-month booster dose was found to lengthen the period of protection in both age groups. It has shown a good safety profile, although there were some cases of meningitis in older children who were vaccinated that remain to be investigated. Nevertheless, the benefits of this vaccine outweigh the risks in the age groups studied, especially for pediatrics in high-transmission areas with high mortality (European Medicines Agency 2015). Due to its composition, it also protects against hepatitis B but should not be used in cases where malaria prevention is not warranted (European Medicines Agency 2015). Twenty vaccine projects are still undergoing clinical trials, but their development is approximately 5 to 10 years behind that of RTS,S/AS01E. For example, there is a vaccine based on the antimerozoite surface protein of blood-stage parasites and another that inhibits transmission of malaria (Aguiar et al. 2012). The ideal drug, which has also been termed a "single exposure, radical cure and prophylaxis," should also be affordable, as many antimalarial drugs are sponsored by nonprofit organizations and governments (Flannery, Chatterjee, and Winzeler 2013). However, despite the multipronged approach in drug discovery, which includes but is not limited to medicinal chemistry-, whole cell-, and target-based methods, the "single exposure, radical cure and prophylaxis" drug remains elusive.

1.2 THE *PLASMODIUM* LIFE CYCLE

Malaria is transmitted by the female *Anopheles* mosquito, which feeds between dusk and dawn. It is caused by *Plasmodium* parasites, four of which are commonly known to infect humans: *P. falciparum* (the most deadly), *P. vivax*, *P. ovale*, and *P. malariae*. *P. knowlesi*, a malaria parasite in monkeys from Southeast Asia, has been established as the fifth species to infect humans, after reports of human infections were described in Malaysia, Thailand, China, Myanmar, Singapore and the Philippines (Centers for Disease Control and Prevention [CDC] 2010).

Following a bite from an infected *Anopheles*, the *Plasmodium* sporozoites enter the bloodstream to reach the liver, where they invade hepatocytes and mature into schizonts, or hypnozoites (in *P. vivax* and *P. ovale*) (Figure 1.1). Hypnozoites are responsible for malaria relapses, as they can remain dormant in the liver for months to years. The schizonts then rupture after a week to release merozoites into the blood, where they invade erythrocytes and continue asexual replication. Some differentiate into gametocytes (sexual erythrocytic stage), which when ingested by another *Anopheles* mosquito will transform into male and female gametes that then join to form diploid zygotes. Zygotes then invade the midgut of the mosquito where they become oocysts, which then rupture to release sporozoites. The latter then travel to the salivary glands of the mosquito, and the cycle repeats with each new blood meal.

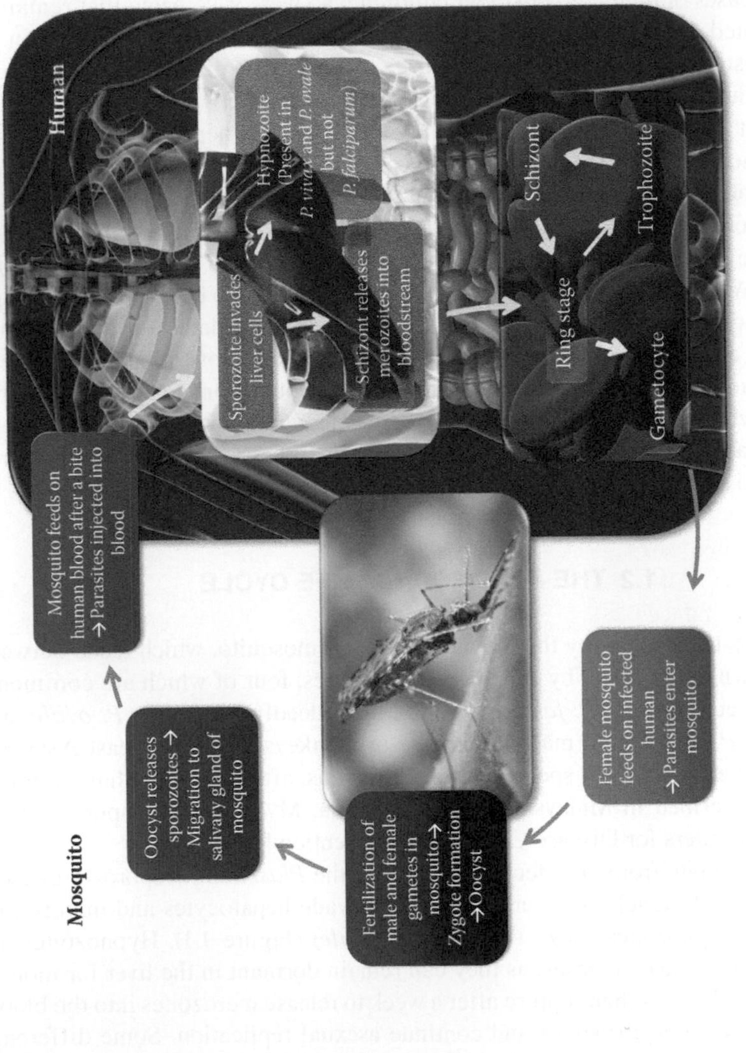

Figure 1.1 Life cycle of *Plasmodium*. (Photo of *Anopheles albimanus* from CDC, James Gathany; image of human anatomy from Mr. Sebastian Kaulitzki, Dreamstime.com; diagram of red blood cells from Andrew Kator and Jennifer Legaz, © 2004–2015. With permission.)

1.3 THE DISEASE

After infection from a carrier-mosquito bite, symptoms appear after at least a week. In uncomplicated malaria, patients present with nonspecific signs and symptoms (Table 1.3) every 1, 2, or 3 days (CDC 2010). This is related to the duration of the blood stage cycle of the parasite; *P. malariae* has the longest cycle of 72 hours, followed by about 48 hours for *P. falciparum*, *P. vivax*, and *P. ovale*. *P. knowlesi* has the shortest cycle of 24 hours. Infection caused by *P. knowlesi* follows a similar course as that in falciparum and vivax malaria (Singh and Daneshvar 2013).

Clinical symptoms manifest as a result of release of waste substances (e.g., hemozoin), which stimulate macrophages and other immune cells to produce cytokines and other factors (CDC 2010). The severity of the condition is also influenced by factors, such as genetic makeup of the host, and extent of immunity against the parasite, the relationships between which are not well understood (Oakley et al. 2011).

Relapses may occur weeks to months later in malaria caused by *P. vivax* and *P. ovale*.

Severe malaria occurs when there is organ failure, or metabolic or hematologic complications, and requires emergency treatment. It can manifest as cerebral malaria, severe anemia, hemoglobinuria, acute respiratory distress syndrome, acute renal failure, hypotension due to cardiovascular collapse, hyperparasitemia, metabolic acidosis, and hypoglycemia.

The pathophysiology of malarial fever is still poorly understood. It was observed in the 19th century that fever in malaria is synchronous with the intermittent release of malarial parasites during rupture of schizonts (Golgi 1886). A minimum parasite density (pyrogenic threshold) is needed before a febrile response can occur in malaria. It has also been recognized that pyrogenic toxins released from schizonts upon rupture induces fever in malaria. Malaria hemozoin and glycosylphosphatidylinositol (GPI), the latter of which is a glycoprotein that anchors proteins to cell membranes, are two such substances. These are *Plasmodium* pathogen-associated molecular patterns that interact with toll-like receptors (TLRs) that are produced by immune cells, and these receptors act to produce a proinflammatory state upon activation by nonself ligands (Gowda 2007). Hemozoin, an insoluble crystal formed in the acidic food vacuole from heme detoxification in the parasite, was found to stimulate the host by presenting DNA to TLR9 in the cell (Parroche et al. 2007). Interaction between malaria GPI with TLR2 on the surface of cells leads to macrophage production of MyD88-dependent TNF-α (Krishnegowda et al. 2005). Administration of monoclonal antitumor necrosis factor (anti-TNF) antibody brought about remission of malarial fever, suggesting that TNF may be the primary cytokine mediating malarial fever.

Table 1.3 Signs and Symptoms of Uncomplicated Malaria

Signs	Symptoms
Elevated temperature, diaphoresis, hepatosplenomegaly, jaundice, increased respiratory rate	Fever, chills, sweating, headache, nausea, vomiting, body ache, general malaise

REFERENCES

Aguiar, A.C.C., E.M.M.D. Rocha, N.B.D. Souza, T.C.C. França, and A.U. Krettli. 2012. "New approaches in antimalarial drug discovery and development: A review." *Memórias do Instituto Oswaldo Cruz* 107 (7):831–845.

Ashley, E.A., M. Dhorda, R.M. Fairhust, C. Amaratunga, P. Lim, S. Suon, S. Sreng et al. 2014. "Spread of artemisinin resistance in *Plasmodium falciparum* malaria." *New England Journal of Medicine* 371:411–423.

Bill and Melinda Gates Foundation. 2015. "Malaria: Strategy overview." Accessed March 8, 2015. http://www.gatesfoundation.org/What-We-Do/Global-Health/Malaria.

Centers for Disease Control and Prevention (CDC). 2010. "Malaria." Accessed June 3, 2015. http://www.cdc.gov/malaria/about/disease.html.

European Medicines Agency. 2015. "First malaria vaccine receives positive scientific opinion from EMA." Accessed August 10, 2015. http://www.ema.europa.eu/ema/index.jsp?curl=pages/news_and_events/news/2015/07/news_detail_002376.jsp&mid=WC0b01ac058004d5c1.

Flannery, E.L., A.K. Chatterjee, and E.A. Winzeler. 2013. "Antimalarial drug discovery—Approaches and progress towards new medicines." *Nature Reviews. Microbiology* 11:849–862.

GlaxoSmithKline. 2015. "GSK's malaria candidate vaccine, Mosquirix™ (RTS,S), receives positive opinion from European regulators for the prevention of malaria in young children in sub-Saharan Africa." Accessed August 10, 2015. https://www.gsk.com/en-gb/media/press-releases/2015/gsk-s-malaria-candidate-vaccine-mosquirix-rtss-receives-positive-opinion-from-european-regulators-for-the-prevention-of-malaria-in-young-children-in-sub-saharan-africa/.

Golgi, C. 1886. "Sull'infezione malarica." *Archivio per le Scienze Mediche (Torino)*. 10:109–135.

Gowda, D.C. 2007. "TLR-mediated cell signaling by malaria GPIs." *Trends in Parasitology* 23:596–604.

Krishnegowda, G., A.M. Hajjar, J. Zhu, E.J. Douglass, S. Uematsu, S. Akira, A.S. Woods, and D.C. Gowda. 2005. "Induction of proinflammatory responses in macrophages by the glycosylphosphatidylinositols of *Plasmodium falciparum*: Cell signaling receptors, glycosylphosphatidylinositol (GPI) structural requirement, and regulation of GPI activity." *Journal of Biological Chemistry* 280:8606–8616.

Nayyar, G.M., J.G. Breman, P.N. Newton, and J. Herrington. 2012. "Poor-quality antimalarial drugs in southeast Asia and sub-Saharan Africa." *Lancet Infectious Diseases* 12 (6):488–496.

Oakley, M.S., N. Gerald, T.F. McCutchan, L. Aravind, and S. Kumar. 2011. "Clinical and molecular aspects of malaria fever." *Trends in Parasitology* 27:442–449.

Olotu, A., G. Fegan, J. Wambua, G. Nyangweso, K. Awuondo, A. Leach, M. Lievens et al. 2013. "Four-year efficacy of RTS, S/AS01E and its interaction with malaria exposure." *New England Journal of Medicine* 368:1111–1120.

Parroche, P., F.N. Lauw, N. Goutagny, E. Latz, B.G. Monks, A. Visintin, K.A. Halmen et al. 2007. "Malaria hemozoin is immunologically inert but radically enhances innate responses by presenting malaria DNA to toll-like receptor 9." *Proceedings of the National Academy of Sciences of the United States of America* 104:1919–1924.

RTS, S Clinical Trials Partnership. 2015. "Efficacy and safety of RTS,S/AS01 malaria vaccine with or without a booster dose in infants and children in Africa: Final results of a phase 3, individually randomised, controlled trial." *The Lancet*. doi: 10.1016/S0140-6736(15)60721-8.

Singh, B., and C. Daneshvar. 2013. "Human infections and detection of *Plasmodium knowlesi.*" *Clinical Microbiology Reviews* 26 (2):165–184.
World Health Organization (WHO). 2010. "Guidelines for the treatment of malaria. Second edition." Accessed March 4, 2015. http://whqlibdoc.who.int/publications/2010/9789241547925_eng.pdf?ua=1.
World Health Organization (WHO). 2013a. "World Malaria Report 2013." Accessed August 20, 2014. http://www.who.int/malaria/publications/world_malaria_report_2013/en/.
World Health Organization (WHO). 2013b. "Drug resistance: Malaria." Accessed February 4, 2014. http://www.who.int/drugresistance/malaria/en/.

CHAPTER 2

Current Antimalarial Drugs

This chapter presents the role of commercially available antimalarial agents in treatment and prevention, with country-specific recommendations of preventive measures for travelers, dosing regimens, and precautions for specific antimalarial medications. Most antimalarial agents target the blood stage of the infection. For targeting the liver forms of the parasite, terminal prophylaxis with primaquine is required. Table 2.1 shows the four main classes of drugs used for malaria prevention and treatment to date (Travassos and Laufer 2012).

2.1 DRUGS FOR TREATMENT OF MALARIA

Artemisinin-based combination therapy (ACT) forms the first-line treatment in uncomplicated falciparum malaria in adults, children, and lactating women, whereas artesunate is the first-line recommendation for severe malaria in adults and children, and in the second and third trimesters of pregnancy (World Health Organization [WHO] 2010). Quinine is used as first-line treatment of falciparum malaria in the first trimester of pregnancy. Primaquine is the current sole agent that provides radical cure (defined as elimination of symptoms and asexual erythrocytic stages of parasites, and prevention of relapses by killing hypnozoites) (WHO 2010). Chloroquine is still used against *Plasmodium vivax*, *Plasmodium ovale*, and *Plasmodium malariae*.

2.2 DRUGS FOR PREVENTION OF MALARIA

For malaria prevention, the Centers for Disease Control and Prevention (CDC) website offers reliable updated country-specific information with regard to areas with malaria, prevalent *Plasmodium* species, drug resistance, and estimated relative risk of malaria for U.S. travelers, as well as recommendations on antimalarial prevention (e.g., medications and mosquito avoidance) (CDC 2014). Table 2.2 summarizes country-specific recommendations by the CDC on antimalarial preventive medications; and Table 2.3 shows the antimalarial preventive medications for nonimmune travelers classified based on the prevalence of chloroquine- or mefloquine-resistant

Table 2.1 Current Antimalarial Drugs

Class	Modes of Action	Drugs and Combination Therapies
	Artemisinins	
(a) Artemisinin 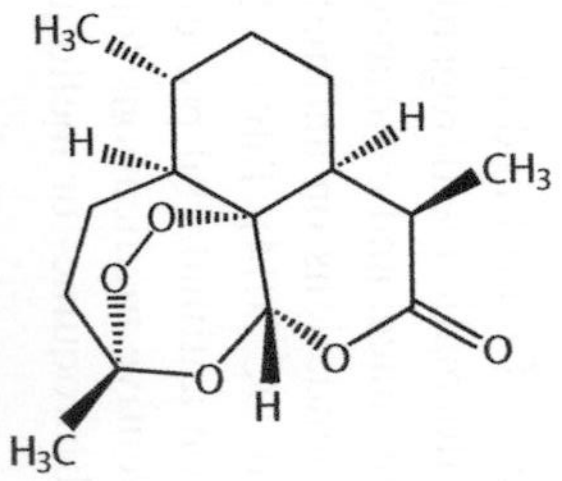	Still not fully understood. Some mechanisms elucidated thus far are: • Inhibition of sarcoplasmic-endoplasmic reticulum calcium transport adenosine triphosphatase (SERCA), PfATP6 (Eckstein-Ludwig et al. 2003; Jambou et al. 2005). • Cleavage of Fe^{2+} in endoperoxide bridge results in generation of free radicals, damaging parasite proteins (Meshnick 2002; O'Neill, Barton, and Ward 2010).	Artemisinin-based combination therapies: artemether-lumefantrine, artesunate-amodiaquine, artesunate-mefloquine, artesunate-sulfadoxine-pyrimethamine, dihydroartemisinin-piperaquine
	Quinolines	
(b) Chloroquine 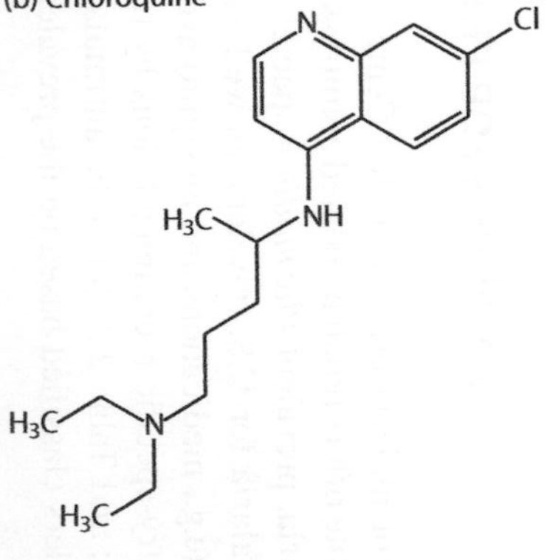	Accumulation in parasitic food vacuole and complexation with heme, which prevents crystallization, resulting in inhibition of heme polymerase and build-up of free heme, which is cytotoxic (Müller and Hyde 2010). Primaquine is also active against intrahepatic forms of all *Plasmodium* species and *P. falciparum* gametocytes.	4-aminoquinolines: chloroquine, amodiaquine 4-methanolquinolines: quinine, quinidine and mefloquine 8-aminoquinolines: primaquine Others: lumefantrine and halofantrine

(Continued)

Table 2.1 (Continued) Current Antimalarial Drugs

Class	Modes of Action	Drugs and Combination Therapies
	Antifolates	
(c) Proguanil 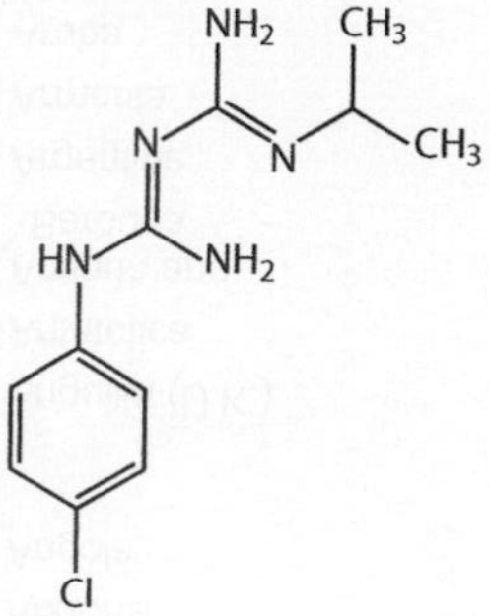	Sulfadoxine, pyrimethamine, and proguanil: Inhibition of enzymes involved in synthesis of folate, resulting in disruption of DNA synthesis in parasite (Curd, Davey, and Rose 1945; Falco et al. 1951). Atovaquone: Disruption of mitochondrial electron transport chain (Fry and Pudney 1992).	Sulfadoxine-pyrimethamine, atovaquone-proguanil
	Antimicrobials	
(d) Doxycycline 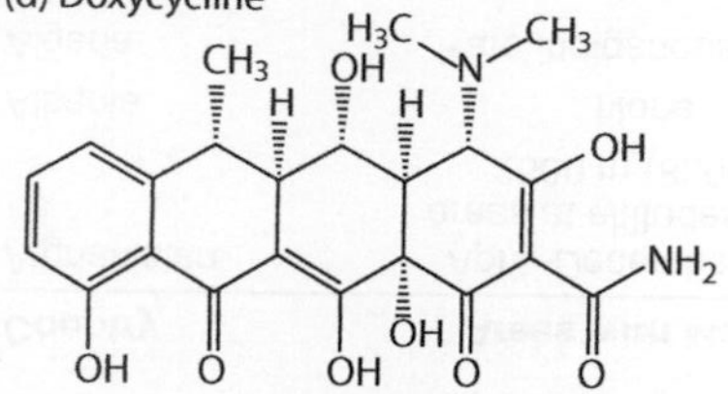	Target the apicoplast and interfere with parasitic protein synthesis (Fidock et al. 2004).	Doxycycline, tetracycline, clindamycin

Table 2.2 Country-Specific Antimalarial Prevention Recommendations

Country	Areas with Malaria	Estimated Relative Risk of Malaria for U.S. Travelers	Drug Resistance	Malaria Species	Recommended Chemoprophylaxis
Afghanistan	April–December in all areas at altitudes below 2500 m (8202 ft)	High	Chloroquine	*P. vivax* 80%–90%; *P. falciparum* 10%–20%	Atovaquone-proguanil, doxycycline, or mefloquine
Albania	None	None	Not applicable	Not applicable	Not applicable
Algeria	Rare indigenous cases	No information	Chloroquine	*P. falciparum*, *P. vivax*	Mosquito avoidance only
American Samoa (U.S.)	None	None	Not applicable	Not applicable	Not applicable
Andorra	None	None	Not applicable	Not applicable	Not applicable
Angola	All	Moderate	Chloroquine	*P. falciparum* 90%; *P. ovale* 5%; *P. vivax* 5%	Atovaquone-proguanil, doxycycline, or mefloquine
Anguilla (U.K.)	None	None	Not applicable	Not applicable	Not applicable
Antarctica	None	None	Not applicable	Not applicable	Not applicable
Antigua and Barbuda	None	None	Not applicable	Not applicable	Not applicable
Argentina	None	None	Not applicable	Not applicable	Not applicable
Armenia	None	None	Not applicable	Not applicable	Not applicable
Aruba	None	None	Not applicable	Not applicable	Not applicable
Australia	None	None	Not applicable	Not applicable	Not applicable
Austria	None	None	Not applicable	Not applicable	Not applicable
Azerbaijan	May–October in rural areas below 1500 m (4921 ft); none in Baku	Very low	None	*P. vivax* 100%	Mosquito avoidance only
Azores (Portugal)	None	None	Not applicable	Not applicable	Not applicable

(*Continued*)

Table 2.2 (Continued) Country-Specific Antimalarial Prevention Recommendations

Country	Areas with Malaria	Estimated Relative Risk of Malaria for U.S. Travelers	Drug Resistance	Malaria Species	Recommended Chemoprophylaxis
Bahamas, The	None	None	Not applicable	Not applicable	Not applicable
Bahrain	None	None	Not applicable	Not applicable	Not applicable
Bangladesh	All areas, except in the city of Dhaka	Low	Chloroquine	More than half *P. falciparum*, remainder *P. vivax*	Atovaquone-proguanil, doxycycline, or mefloquine
Barbados	None	None	Not applicable	Not applicable	Not applicable
Belarus	None	None	Not applicable	Not applicable	Not applicable
Belgium	None	None	Not applicable	Not applicable	Not applicable
Belize	All areas, especially the districts of Cayo, Stann Creek, and Toledo; none in Belize City and islands frequented by tourists	Low	None	*P. vivax* 95%; *P. falciparum* 5%	Districts of Cayo, Stann Creek, and Toledo: Atovaquone-proguanil, chloroquine, doxycycline, mefloquine, or primaquine; all other areas with malaria: Mosquito avoidance
Benin	All	High	Chloroquine	*P. falciparum* 85%; *P. ovale* 5%–10%; *P. vivax* rare	Atovaquone-proguanil, doxycycline, or mefloquine
Bermuda (U.K.)	None	None	Not applicable	Not applicable	Not applicable
Bhutan	Rural areas below 1700 m (5577 ft) especially the southern belt districts along the border with India: Chirang, Geylegphug, Samchi, Samdrup, Jongkhar, and Shemgang	Very low	Chloroquine	*P. falciparum* 60%; *P. vivax* 40%	Atovaquone-proguanil, doxycycline, or mefloquine

(*Continued*)

Table 2.2 (Continued) Country-Specific Antimalarial Prevention Recommendations

Country	Areas with Malaria	Estimated Relative Risk of Malaria for U.S. Travelers	Drug Resistance	Malaria Species	Recommended Chemoprophylaxis
Bolivia	All areas below 2500 m (8202 ft); none in the city of La Paz	Low	Chloroquine	*P. vivax* 93%; *P. falciparum* 7%	Atovaquone-proguanil, doxycycline, mefloquine, or primaquine
Bosnia and Herzegovina	None	None	Not applicable	Not applicable	Not applicable
Botswana	Present in the following districts: Central and North West (including Chobe National Park); none in the cities of Francistown and Gabarone	Very low	Chloroquine	*P. falciparum* 90%; *P. vivax* 5%; *P. ovale* 5%	Atovaquone-proguanil, doxycycline, or mefloquine
Brazil	States of Acre, Amapa, Amazonas, Maranhão, Mato Grosso, Para, Rondonia, Roraima, and Toncantins; also present in urban areas, including cities such as Belem, Boa Vista, Macapa, Manaus, Maraba, Porto Velho, and Santarem; rare cases in Cuiaba City; no transmission at Iguassu Falls	Low	Chloroquine	*P. vivax* 85%; *P. falciparum* 15%	Areas with malaria except Cuiaba City: atovaquone-proguanil, doxycycline, or mefloquine; Cuiaba City: mosquito avoidance only
British Indian Ocean Territory; includes Diego Garcia (U.K.)	None	None	Not applicable	Not applicable	Not applicable
Brunei	None	None	Not applicable	Not applicable	Not applicable

(Continued)

Table 2.2 (Continued) Country-Specific Antimalarial Prevention Recommendations

Country	Areas with Malaria	Estimated Relative Risk of Malaria for U.S. Travelers	Drug Resistance	Malaria Species	Recommended Chemoprophylaxis
Bulgaria	None	None	Not applicable	Not applicable	Not applicable
Burkina Faso	All	High	Chloroquine	*P. falciparum* 80%; *P. ovale* 5%–10%; *P. vivax* rare	Atovaquone-proguanil, doxycycline, or mefloquine
Burma (Myanmar)	Present at altitudes below 1000 m (3281 ft); none in the cities of Mandalay and Rangoon (Yangoon)	Moderate	Chloroquine, mefloquine	*P. falciparum* 90%; remainder *P. malariae*, *P. ovale*, and *P. vivax*	In the provinces of Bago, Kachin, Kayah, Kayin, Shan, and Tanintharyi: atovaquone-proguanil or doxycycline; all other areas with malaria: atovaquone-proguanil, doxycycline, or mefloquine
Burundi	All	Moderate	Chloroquine	*P. falciparum* 86%; remainder *P. malariae*, *P. ovale*, and *P. vivax*	Atovaquone-proguanil, doxycycline, or mefloquine
Cambodia	Present throughout the country including Siem Reap city; rare cases in Phnom Penh; none at the temple complex at Angkor Wat, and around Lake Tonle Sap	Low	Chloroquine, Mefloquine	*P. falciparum* 86%; *P. vivax* 12%; *P. malariae* 2%	In the provinces of Banteay Meanchey, Battambang, Kampot, Koh Kong, Odder Meanchey, Pailin, Preah Vihear, Pursat, and Siem Reap bordering Thailand: atovaquone-proguanil or doxycycline; all other areas with malaria: atovaquone-proguanil, doxycycline, or mefloquine; Phnom Penh: mosquito avoidance only

(*Continued*)

Table 2.2 (Continued) Country-Specific Antimalarial Prevention Recommendations

Country	Areas with Malaria	Estimated Relative Risk of Malaria for U.S. Travelers	Drug Resistance	Malaria Species	Recommended Chemoprophylaxis
Cameroon	All	High	Chloroquine	*P. falciparum* 85%; *P. ovale* 5%–10%; *P. vivax* rare	Atovaquone-proguanil, doxycycline, or mefloquine
Canada	None	None	Not applicable	Not applicable	Not applicable
Canary Islands (Spain)	None	None	Not applicable	Not applicable	Not applicable
Cape Verde	Limited cases in Sao Tiago Island	Very low	Chloroquine	Primarily *P. falciparum*	Mosquito avoidance only
Cayman Islands (U.K.)	None	None	Not applicable	Not applicable	Not applicable
Central African Republic	All	High	Chloroquine	*P. falciparum* 85%; *P. malariae*, *P. ovale*, and *P. vivax* 15% combined	Atovaquone-proguanil, doxycycline, or mefloquine
Chad	All	High	Chloroquine	*P. falciparum* 85%; *P. malariae*, *P. ovale*, and *P. vivax* 15% combined	Atovaquone-proguanil, doxycycline, or mefloquine
Chile	None	None	Not applicable	Not applicable	Not applicable

(Continued)

Table 2.2 (Continued) Country-Specific Antimalarial Prevention Recommendations

Country	Areas with Malaria	Estimated Relative Risk of Malaria for U.S. Travelers	Drug Resistance	Malaria Species	Recommended Chemoprophylaxis
China	Present year round in rural parts of Anhui, Guizhou, Hainan, Henan, Hubei, and Yunnan Provinces; rare cases occur in other rural parts of the country below 1500 m (4921 ft) May–December; none in urban areas; some major river cruises may go through malaria endemic areas in Anhui and Hubei Provinces	Low	Chloroquine, mefloquine	Primarily *P. vivax*; *P. falciparum* in select locations	Along China-Burma (Myanmar) border in the western part of Yunnan province: atovaquone-proguanil or doxycycline; Hainan and other parts of Yunnan province: atovaquone-proguanil, doxycycline, or mefloquine; Anhui, Guizhou, Henan, and Hubei provinces: atovaquone-proguanil, chloroquine, doxycyline, mefloquine, or primaquine; all other areas with malaria including river cruises that pass through malaria-endemic provinces: mosquito avoidance only
Christmas Island (Australia)	None	None	Not applicable	Not applicable	Not applicable
Cocos (Keeling) Islands (Australia)	None	None	Not applicable	Not applicable	Not applicable
Colombia	All areas below 1700 m (5577 ft); none in Bogota and Cartagena	Low	Chloroquine	*P. falciparum* 35%–40%; *P. vivax* 60%–65%	Atovaquone-proguanil, doxycycline, or mefloquine
Comoros	All	No data	Chloroquine	Primarily *P. falciparum*	Atovaquone-proguanil, doxycycline, or mefloquine

(*Continued*)

Table 2.2 (Continued) Country-Specific Antimalarial Prevention Recommendations

Country	Areas with Malaria	Estimated Relative Risk of Malaria for U.S. Travelers	Drug Resistance	Malaria Species	Recommended Chemoprophylaxis
Congo, Republic of the (Congo-Brazzaville)	All	High	Chloroquine	*P. falciparum* 90%; *P. ovale* 5%–10%; *P. vivax* rare	Atovaquone-proguanil, doxycycline, or mefloquine
Cook Islands (New Zealand)	None	None	Not applicable	Not applicable	Not applicable
Costa Rica	None	None	Not applicable	Not applicable	Not applicable
Côte d'Ivoire	All	High	Chloroquine	*P. falciparum* 85%; *P. ovale* 5%–10%; *P. vivax* rare	Atovaquone-proguanil, doxycycline, or mefloquine
Croatia	None	None	Not applicable	Not applicable	Not applicable
Cuba	None	None	Not applicable	Not applicable	Not applicable
Cyprus	None	None	Not applicable	Not applicable	Not applicable
Czech Republic	None	None	Not applicable	Not applicable	Not applicable
Congo, Democratic Republic of the (Congo-Kinshasa)	All	Moderate	Chloroquine	*P. falciparum* 90%; *P. ovale* 5%; *P. vivax* rare	Atovaquone-proguanil, doxycycline, or mefloquine
Denmark	None	None	Not applicable	Not applicable	Not applicable
Djibouti	All	No data	Chloroquine	*P. falciparum* 90%; *P. vivax* 5%–10%	Atovaquone-proguanil, doxycycline, or mefloquine
Dominica	None	None	Not applicable	Not applicable	Not applicable

(*Continued*)

Table 2.2 (Continued) Country-Specific Antimalarial Prevention Recommendations

Country	Areas with Malaria	Estimated Relative Risk of Malaria for U.S. Travelers	Drug Resistance	Malaria Species	Recommended Chemoprophylaxis
Dominican Republic	All areas (including resort areas), except none in the cities of Santiago and Santo Domingo	Low	None	*P. falciparum* 100%	Atovaquone-proguanil, chloroquine, doxycycline, or mefloquine
Easter Island (Chile)	None	None	Not applicable	Not applicable	Not applicable
Ecuador, including the Galápagos Islands	All areas at altitudes below 1500 m (4921 ft); not present in the cities of Guayaquil and Quito, or the Galápagos Islands	Low	Chloroquine	*P. vivax* 90%; *P. falciparum* 10%	Guayas, Esmeraldas, and Canar provinces: atovaquone-proguanil, doxycycline, or mefloquine; all other areas with malaria: atovaquone-proguanil, doxycycline, mefloquine, or primaquine
Egypt	None	None	Not applicable	Not applicable	Not applicable
El Salvador	Rare cases along the Guatemalan border	Very low	None	*P. vivax* 99%; *P. falciparum* <1%	Mosquito avoidance only
Equatorial Guinea	All	High	Chloroquine	*P. falciparum* 85%; *P. malariae*, *P. ovale*, and *P. vivax* 15% combined	Atovaquone-proguanil, doxycycline, or mefloquine
Eritrea	All areas at altitudes below 2200 m (7218 ft); none in Asmara	No data	Chloroquine	*P. falciparum* 85%; *P. vivax* 10%–15%; *P. ovale* rare	Atovaquone-proguanil, doxycycline, or mefloquine
Estonia	None	None	Not applicable	Not applicable	Not applicable

(Continued)

Table 2.2 (Continued) Country-Specific Antimalarial Prevention Recommendations

Country	Areas with Malaria	Estimated Relative Risk of Malaria for U.S. Travelers	Drug Resistance	Malaria Species	Recommended Chemoprophylaxis
Ethiopia	All areas at altitudes below 2500 m (8202 ft) except none in Addis Ababa	Moderate	Chloroquine	*P. falciparum* 60%–70%; *P. vivax* 30%–40%; *P. malariae*, *P. ovale* rare	Atovaquone-proguanil, doxycycline, or mefloquine
Falkland (Las Islas Malvinas), South Georgia, and South Sandwich Islands (U.K.)	None	None	Not applicable	Not applicable	Not applicable
Faroe Islands (Denmark)	None	None	Not applicable	Not applicable	Not applicable
Fiji	None	None	Not applicable	Not applicable	Not applicable
Finland	None	None	Not applicable	Not applicable	Not applicable
France	None	None	Not applicable	Not applicable	Not applicable
French Guiana	All areas except none in city of Cayenne or Devil's Island (Ile du Diable)	Moderate	Chloroquine	*P. falciparum* <50%; remainder *P. vivax*; *P. malariae* rare	Atovaquone-proguanil, doxycycline, or mefloquine

(Continued)

Table 2.2 (Continued) Country-Specific Antimalarial Prevention Recommendations

Country	Areas with Malaria	Estimated Relative Risk of Malaria for U.S. Travelers	Drug Resistance	Malaria Species	Recommended Chemoprophylaxis
French Polynesia, includes the island groups of Society Islands (Tahiti, Moorea, and Bora-Bora); Marquesas Islands (Hiva Oa and Ua Huka); and Austral Islands (Tubuai and Rurutu)	None	None	Not applicable	Not applicable	Not applicable
Gabon	All	Moderate	Chloroquine	*P. falciparum* 90%; remainder *P. malariae*, *P. ovale*, *P. vivax*	Atovaquone-proguanil, doxycycline, or mefloquine
Gambia, The	All	High	Chloroquine	*P. falciparum* ≥85%; *P. ovale* 5%–10%; *P. malariae*, *P. vivax* rare	Atovaquone-proguanil, doxycycline, or mefloquine
Georgia	None	None	Not applicable	Not applicable	Not applicable
Germany	None	None	Not applicable	Not applicable	Not applicable
Ghana	All	High	Chloroquine	*P. falciparum* 90%; *P. ovale* 5%–10%; *P. vivax* rare	Atovaquone-proguanil, doxycycline, or mefloquine
Gibraltar (U.K.)	None	None	Not applicable	Not applicable	Not applicable
Greece	None	None	None	None	Not applicable

(*Continued*)

Table 2.2 (Continued) Country-Specific Antimalarial Prevention Recommendations

Country	Areas with Malaria	Estimated Relative Risk of Malaria for U.S. Travelers	Drug Resistance	Malaria Species	Recommended Chemoprophylaxis
Greenland (Denmark)	None	None	Not applicable	Not applicable	Not applicable
Grenada	None	None	Not applicable	Not applicable	Not applicable
Guadeloupe, including Saint Barthélemy and Saint Martin (France)	None	None	Not applicable	Not applicable	Not applicable
Guam (U.S.)	None	None	Not applicable	Not applicable	Not applicable
Guatemala	Rural areas only at altitudes below 1500 m (4921 ft); none in Antigua, Guatemala City, or Lake Atitlán	Low	None	*P. vivax* 97%; *P. falciparum* 3%	Escuintla Province: atovaquone-proguanil, chloroqouine, doxycycline, or mefloquine; all other areas with malaria: atovaquone-proguanil, chloroquine, doxycycline, mefloquine, or primaquine
Guinea	All	High	Chloroquine	*P. falciparum* 85%; *P. ovale* 5%–10%; *P. vivax* rare	Atovaquone-proguanil, doxycycline, or mefloquine
Guinea-Bissau	All	No data	Chloroquine	*P. falciparum* 85%; *P. ovale* 5%–10%; *P. vivax* rare	Atovaquone-proguanil, doxycycline, or mefloquine

(Continued)

Table 2.2 (Continued) Country-Specific Antimalarial Prevention Recommendations

Country	Areas with Malaria	Estimated Relative Risk of Malaria for U.S. Travelers	Drug Resistance	Malaria Species	Recommended Chemoprophylaxis
Guyana	All areas below 900 m (2953 ft); rare cases in the cities of Amsterdam and Georgetown	Moderate	Chloroquine	*P. falciparum* 50%; *P. vivax* 50%	Areas with malaria except the cities of Amsterdam and Georgetown: atovaquone-proguanil, doxycycline, or mefloquine; cities of Amsterdam and Georgetown: mosquito avoidance only
Haiti	All (including Port Labadee)	High	None	*P. falciparum* 99%; *P. malariae* rare	Atovaquone-proguanil, chloroquine, doxycycline, or mefloquine
Honduras	Present throughout the country and in Roatán and other Bay Islands; none in San Pedro, Sula, and Tegucigalpa	Moderate	None	*P. vivax* 93%; *P. falciparum* 7%	Atovaquone-proguanil, chloroquine, doxycycline, mefloquine, or primaquine
Hong Kong (China)	None	None	Not applicable	Not applicable	Not applicable
Hungary	None	None	Not applicable	Not applicable	Not applicable
Iceland	None	None	Not applicable	Not applicable	Not applicable
India	All areas throughout country including cities of Bombay (Mumbai) and Delhi, except none in areas above 2000 m (6562 ft) in Himachal Pradesh, Jammu and Kashmir, and Sikkim	Moderate	Chloroquine	*P. vivax* 50%; *P. falciparum* >40%; *P. malariae* and *P. ovale* rare	Atovaquone-proguanil, doxycycline, or mefloquine

(*Continued*)

Table 2.2 (Continued) Country-Specific Antimalarial Prevention Recommendations

Country	Areas with Malaria	Estimated Relative Risk of Malaria for U.S. Travelers	Drug Resistance	Malaria Species	Recommended Chemoprophylaxis
Indonesia	Rural areas of Kalimantan (Borneo), Nusa Tenggara Barat (includes the island of Lombok), Sulawesi, and Sumatra; all areas of eastern Indonesia (provinces of Maluku, Maluku Utara, Nusa Tenggara Timur, Papua, and Papua Barat); none in the cities of Jakarta, Ubud, or resort areas of Bali and Java; low transmission in rural areas of Java, including Ujung Kulong, Sukalumi, and Pangadaran	Moderate	Chloroquine (*P. falciparum* and *P. vivax*)	*P. falciparum* 66%; remainder primarily *P. vivax*	Atovaquone-proguanil, doxycycline, or mefloquine
Iran	Rural areas of Fars Province, Sistan-Baluchestan Province, and southern, tropical parts of Hormozgan and Kerman Provinces	Very low	Chloroquine	*P. vivax* 88%; *P. falciparum* 12%	Atovaquone-proguanil, doxycycline, or mefloquine
Iraq	None	None	Not applicable	Not applicable	Not applicable
Ireland	None	None	Not applicable	Not applicable	Not applicable
Israel	None	None	Not applicable	Not applicable	Not applicable
Italy including Holy See (Vatican City)	None	None	Not applicable	Not applicable	Not applicable
Jamaica	None	None	Not applicable	Not applicable	Not applicable

(Continued)

Table 2.2 (Continued) Country-Specific Antimalarial Prevention Recommendations

Country	Areas with Malaria	Estimated Relative Risk of Malaria for U.S. Travelers	Drug Resistance	Malaria Species	Recommended Chemoprophylaxis
Japan	None	None	Not applicable	Not applicable	Not applicable
Jordan	None	None	Not applicable	Not applicable	Not applicable
Kazakhstan	None	None	Not applicable	Not applicable	Not applicable
Kenya	Present in all areas (including game parks) at altitudes below 2500 m (8202 ft); none in the highly urbanized, central part of the city of Nairobi	Moderate	Chloroquine	*P. falciparum* 85%; *P. vivax* 5%–10%; *P. ovale* up to 5%	Atovaquone-proguanil, doxycycline, or mefloquine
Kiribati (formerly Gilbert Islands), includes Tarawa, Tabuaeran (Fanning Island), and Banaba (Ocean Island)	None	None	Not applicable	Not applicable	Not applicable
Korea, North	Present in southern provinces	No data	None	Presumed to be *P. vivax* 100%	Atovaquone-proguanil, chloroquine, doxycycline, mefloquine, or primaquine
Korea, South	Limited to the months of March–December in rural areas in the northern parts of Incheon, Kangwon-do, and Kyonggi-do provinces including the demilitarized zone (DMZ)	Low	None	*P. vivax* 100%	Atovaquone-proguanil, chloroquine, doxycycline, mefloquine, or primaquine
Kosovo	None	None	Not applicable	Not applicable	Not applicable
Kuwait	None	None	Not applicable	Not applicable	Not applicable

(*Continued*)

Table 2.2 (Continued) Country-Specific Antimalarial Prevention Recommendations

Country	Areas with Malaria	Estimated Relative Risk of Malaria for U.S. Travelers	Drug Resistance	Malaria Species	Recommended Chemoprophylaxis
Kyrgyzstan	None	None	Not applicable	Not applicable	Not applicable
Laos	All, except none in the city of Vientiane	Very low	Chloroquine and mefloquine	*P. falciparum* 95%; *P. vivax* 4%; *P. malariae* and *P. ovale* 1% combined	Along the Laos-Burma (Myanmar) border in the provinces of Bokeo and Louang Namtha and along the Laos–Thailand border in the province of Champasack and Saravan: atovaquone-proguanil or doxycycline; all other areas with malaria: atovaquone-proguanil, doxycycline, or mefloquine
Latvia	None	None	Not applicable	Not applicable	Not applicable
Lebanon	None	None	Not applicable	Not applicable	Not applicable
Lesotho	None	None	Not applicable	Not applicable	Not applicable
Liberia	All	High	Chloroquine	*P. falciparum* 85%; *P. ovale* 5%–10%; *P. vivax* rare	Atovaquone-proguanil, doxycycline, or mefloquine
Libya	None	None	Not applicable	Not applicable	Not applicable
Liechtenstein	None	None	Not applicable	Not applicable	Not applicable
Lithuania	None	None	Not applicable	Not applicable	Not applicable
Luxembourg	None	None	Not applicable	Not applicable	Not applicable
Macau SAR (China)	None	None	Not applicable	Not applicable	Not applicable
Macedonia	None	None	Not applicable	Not applicable	Not applicable

(*Continued*)

Table 2.2 (Continued) Country-Specific Antimalarial Prevention Recommendations

Country	Areas with Malaria	Estimated Relative Risk of Malaria for U.S. Travelers	Drug Resistance	Malaria Species	Recommended Chemoprophylaxis
Madagascar	All	Moderate	Chloroquine	*P. falciparum* 85%; *P. vivax* 5%–10%; *P. ovale* 5%	Atovaquone-proguanil, doxycycline, or mefloquine
Madeira Islands (Portugal)	None	None	Not applicable	Not applicable	Not applicable
Malawi	All	Moderate	Chloroquine	*P. falciparum* 90%; *P. malariae*, *P. ovale*, and *P. vivax* 10% combined	Atovaquone-proguanil, doxycycline, or mefloquine
Malaysia	Present in rural areas of Malaysian Borneo (Sabah and Sarawak Provinces), and to a lesser extent in rural areas of Peninsular Malaysia	Low	Chloroquine	*P. falciparum* 40%; *P. vivax* 50%; remainder *P. malariae*, *P. knowlesi*, and *P. ovale*; *P. knowlesi* reported to cause 28% of cases in Sarawak, and known to cause cases in both Malaysia Borneo and Peninsular Malaysia	Atovaquone-proguanil, doxycycline, or mefloquine
Maldives	None	None	Not applicable	Not applicable	Not applicable

(*Continued*)

Table 2.2 (Continued) Country-Specific Antimalarial Prevention Recommendations

Country	Areas with Malaria	Estimated Relative Risk of Malaria for U.S. Travelers	Drug Resistance	Malaria Species	Recommended Chemoprophylaxis
Mali	All	High	Chloroquine	*P. falciparum* 85%; *P. ovale* 5%–10%; *P. vivax* rare	Atovaquone-proguanil, doxycycline, or mefloquine
Malta	None	None	Not applicable	Not applicable	Not applicable
Marshall Islands	None	None	Not applicable	Not applicable	Not applicable
Martinique (France)	None	None	Not applicable	Not applicable	Not applicable
Mauritania	Present in southern provinces, including the city of Nouakchott	High	Chloroquine	*P. falciparum* 85%; *P. ovale* 5%–10%; *P. vivax* rare	Atovaquone-proguanil, doxycycline, or mefloquine
Mauritius	None	None	Not applicable	Not applicable	Not applicable
Mayotte (France)	All	No data	Chloroquine	*P. falciparum* 40%–50%; *P. vivax* 35%–40%; *P. ovale* <1%	Atovaquone-proguanil, doxycycline, or mefloquine
Mexico	Present in Chiapas, Chihuahua, Durango, Nayarit, and Sinaloa; rare cases in Campeche, Jalisco, Oaxaca, Sonora, and Tabasco; rare cases in the municipality of Othon P. Blanco in the southern part of Quintana Roo bordering Belize; no malaria along the United States–Mexico border	Very low	None	*P. vivax* 100%	States of Chiapas, Chihuahua, Durango, Nayarit, and Sinaloa: atovaquone-proguanil, chloroquine, doxycycline, mefloquine, or primaquine; states of Campeche, Jalisco, Oaxaca, Sonora, Tabasco, and Othon P. Blanco municipality of Quintana Roo: mosquito avoidance only

(*Continued*)

Table 2.2 (Continued) Country-Specific Antimalarial Prevention Recommendations

Country	Areas with Malaria	Estimated Relative Risk of Malaria for U.S. Travelers	Drug Resistance	Malaria Species	Recommended Chemoprophylaxis
Micronesia, Federated States of; includes Yap Islands, Pohnpei, Chuuk, and Kosrae	None	None	Not applicable	Not applicable	Not applicable
Moldova	None	None	Not applicable	Not applicable	Not applicable
Monaco	None	None	Not applicable	Not applicable	Not applicable
Mongolia	None	None	Not applicable	Not applicable	Not applicable
Montenegro	None	None	Not applicable	Not applicable	Not applicable
Montserrat (U.K.)	None	None	Not applicable	Not applicable	Not applicable
Morocco	None	None	Not applicable	Not applicable	Not applicable
Mozambique	All	Moderate	Chloroquine	*P. falciparum* 90%; *P. malariae*, *P. ovale*, and *P. vivax* rare	Atovaquone-proguanil, doxycycline, or mefloquine
Namibia	Present in the provinces of Kunene, Ohangwena, Okavango (Kavango), Omaheke, Omusati, Oshana, Oshikoto, and Otjozondjupa, and in the Caprivi Strip	Low	Chloroquine	*P. falciparum* 90%; *P. malariae*, *P. ovale*, and *P. vivax* 10% combined	Atovaquone-proguanil, doxycycline, or mefloquine

(*Continued*)

Table 2.2 (Continued) Country-Specific Antimalarial Prevention Recommendations

Country	Areas with Malaria	Estimated Relative Risk of Malaria for U.S. Travelers	Drug Resistance	Malaria Species	Recommended Chemoprophylaxis
Nauru	None	None	Not applicable	Not applicable	Not applicable
Nepal	Present throughout country at altitudes below 2000 m (6562 ft); none in Kathmandu and on typical Himalayan treks	No data	Chloroquine	*P. vivax* 85%; *P. falciparum* 15%	Atovaquone-proguanil, doxycycline, or mefloquine
Netherlands	None	None	Not applicable	Not applicable	Not applicable
Netherlands Antilles (Bonaire, Curaçao, Saba, Saint Eustasius, and Saint Martin)	None	None	Not applicable	Not applicable	Not applicable
New Caledonia (France)	None	None	Not applicable	Not applicable	Not applicable
New Zealand	None	None	Not applicable	Not applicable	Not applicable
Nicaragua	Present in districts of Chinandega, Leon, Managua, Matagalpa, Región Autónoma Atlántico Norte (RAAN), and Región Autónoma Atlántico Sur (RAAS)	Low	None	*P. vivax* 90%; *P. falciparum* 10%	Atovaquone-proguanil, chloroquine, doxycycline, mefloquine, or primaquine

(Continued)

Table 2.2 (Continued) Country-Specific Antimalarial Prevention Recommendations

Country	Areas with Malaria	Estimated Relative Risk of Malaria for U.S. Travelers	Drug Resistance	Malaria Species	Recommended Chemoprophylaxis
Niger	All	High	Chloroquine	*P. falciparum* 85%; *P. ovale* 5%–10%; *P. vivax* rare	Atovaquone-proguanil, doxycycline, or mefloquine
Nigeria	All	High	Chloroquine	*P. falciparum* 85%; *P. ovale* 5%–10%; *P. vivax* rare	Atovaquone-proguanil, doxycycline, or mefloquine
Niue (New Zealand)	None	None	Not applicable	Not applicable	Not applicable
Norfolk Island (Australia)	None	None	Not applicable	Not applicable	Not applicable
North Korea	None	None	Not applicable	Not applicable	Not applicable
Northern Mariana Islands (U.S.), includes Saipan, Tinian, and Rota Island	None	None	Not applicable	Not applicable	Not applicable
Norway	None	None	Not applicable	Not applicable	Not applicable
Oman	Sporadic transmission in Ad Dakhliyah, North Al Batinah, and North and South Ash Sharqiyah; none in the city of Muscat	Very low	Chloroquine	*P. falciparum* and *P. vivax*	Mosquito avoidance only
Pakistan	All areas (including all cities) at altitudes below 2500 m (8202 ft)	Moderate	Chloroquine	*P. vivax* 70%; *P. falciparum* 30%	Atovaquone-proguanil, doxycycline, or mefloquine
Palau	None	None	Not applicable	Not applicable	Not applicable

(*Continued*)

Table 2.2 (Continued) Country-Specific Antimalarial Prevention Recommendations

Country	Areas with Malaria	Estimated Relative Risk of Malaria for U.S. Travelers	Drug Resistance	Malaria Species	Recommended Chemoprophylaxis
Panama	Transmission throughout the country; none in urban areas of Panama City or in the former Canal Zone	Low	Chloroquine (east of the Panama Canal)	*P. vivax* 99%; *P. falciparum* 1%	Provinces east of the Panama Canal: atovaquone-proguanil, doxycycline, mefloquine, or primaquine; other areas with malaria: mosquito avoidance only
Papua New Guinea	Present throughout country at altitudes below 2000 m (6562 ft)	High	Chloroquine (both *P. falciparum* and *P. vivax*)	*P. falciparum* 65%–80%; *P. vivax* 10%–30%; remainder *P. malariae* and *P. ovale*	Atovaquone-proguanil, doxycycline, or mefloquine
Paraguay	Present in the departments of Alto Paraná, Caaguazú, and Canendiyú	Very low	None	*P. vivax* 95%; *P. falciparum* 5%	Atovaquone-proguanil, chloroquine, doxycycline, mefloquine, or primaquine
Peru	All departments below 2000 m (6562 ft) including cities of Iquitos and Puerto Maldonado; none in Lima province and coast south of Lima, and none in the cities of Ica and Nazca; none in the highland tourist areas (Cuzco, Machu Picchu, and Lake Titicaca) and southern cities of Arequipa, Moquegua, Puno, and Tacna	Low	Chloroquine	*P. vivax* 85%; *P. falciparum* 15%	Atovaquone-proguanil, doxycycline, or mefloquine

(Continued)

Table 2.2 (Continued) Country-Specific Antimalarial Prevention Recommendations

Country	Areas with Malaria	Estimated Relative Risk of Malaria for U.S. Travelers	Drug Resistance	Malaria Species	Recommended Chemoprophylaxis
Philippines	Present in rural areas below 600 m (1969 ft) on islands of Basilan, Luzon, Mindanao, Mindoro, Palawan, Sulu (Jolo), and Tawi-Tawi; none in urban areas	Low	Chloroquine	*P. falciparum* 70%–80%; *P. vivax* 20%–30%	Atovaquone-proguanil, doxycycline, or mefloquine
Pitcairn Islands (U.K.)	None	None	Not applicable	Not applicable	Not applicable
Poland	None	None	Not applicable	Not applicable	Not applicable
Portugal	None	None	Not applicable	Not applicable	Not applicable
Puerto Rico (U.S.)	None	None	Not applicable	Not applicable	Not applicable
Qatar	None	None	Not applicable	Not applicable	Not applicable
Réunion (France)	None	None	Not applicable	Not applicable	Not applicable
Romania	None	None	Not applicable	Not applicable	Not applicable
Russia	None	None	Not applicable	Not applicable	Not applicable
Rwanda	All	Moderate	Chloroquine	*P. falciparum* 90%; *P. vivax* 5%; *P. ovale* 5%	Atovaquone-proguanil, doxycycline, or mefloquine
Saint Helena (U.K.)	None	None	Not applicable	Not applicable	Not applicable
Saint Kitts (Saint Christopher) and Nevis (U.K.)	None	None	Not applicable	Not applicable	Not applicable
Saint Lucia	None	None	Not applicable	Not applicable	Not applicable

(Continued)

Table 2.2 (Continued) Country-Specific Antimalarial Prevention Recommendations

Country	Areas with Malaria	Estimated Relative Risk of Malaria for U.S. Travelers	Drug Resistance	Malaria Species	Recommended Chemoprophylaxis
Saint Pierre and Miquelon (France)	None	None	Not applicable	Not applicable	Not applicable
Saint Vincent and the Grenadines	None	None	Not applicable	Not applicable	Not applicable
Samoa (formerly Western Samoa)	None	None	Not applicable	Not applicable	Not applicable
San Marino	None	None	Not applicable	Not applicable	Not applicable
São Tomé and Príncipe	All	Very low	Chloroquine	*P. falciparum* 85%; remainder *P. malariae*, *P. ovale*, and *P. vivax*	Atovaquone-proguanil, doxycycline, or mefloquine
Saudi Arabia	Present in emirates by border with Yemen, specifically Asir and Jizan; none in cities of Jeddah, Mecca, Medina, Riyadh, and Ta'if	Low	Chloroquine	*P. falciparum* predominantly; remainder *P. vivax*	Atovaquone-proguanil, doxycycline, or mefloquine
Senegal	All	High	Chloroquine	*P. falciparum* >85%; *P. ovale* 5%–10%; *P. vivax* rare	Atovaquone-proguanil, doxycycline, or mefloquine
Serbia	None	None	Not applicable	Not applicable	Not applicable
Seychelles	None	None	Not applicable	Not applicable	Not applicable

(Continued)

Table 2.2 (Continued) Country-Specific Antimalarial Prevention Recommendations

Country	Areas with Malaria	Estimated Relative Risk of Malaria for U.S. Travelers	Drug Resistance	Malaria Species	Recommended Chemoprophylaxis
Sierra Leone	All	High	Chloroquine	*P. falciparum* 85%; *P. ovale* potentially 5%–10%; *P. malariae* and *P. vivax* rare	Atovaquone-proguanil, doxycycline, or mefloquine
Singapore	None	None	Not applicable	Not applicable	Not applicable
Slovakia	None	None	Not applicable	Not applicable	Not applicable
Slovenia	None	None	Not applicable	Not applicable	Not applicable
Solomon Islands	All	High	Chloroquine	*P. falciparum* 60%; *P. vivax* 35%–40%; *P. ovale* <1%	Atovaquone-proguanil, doxycycline, or mefloquine
Somalia	All	High	Chloroquine	*P. falciparum* 90%; *P. vivax* 5%–10%; *P. malariae* and *P. ovale* rare	Atovaquone-proguanil, doxycycline, or mefloquine
South Africa	Present in northeastern KwaZulu-Natal Province as far south as the Tugela River, Limpopo (Northern) Province, and the Mpumalanga Province; present in Kruger National Park	Low	Chloroquine	*P. falciparum* 90%; *P. vivax* 5%; *P. ovale* 5%	Atovaquone-proguanil, doxycycline, or mefloquine

(*Continued*)

Table 2.2 (Continued) Country-Specific Antimalarial Prevention Recommendations

Country	Areas with Malaria	Estimated Relative Risk of Malaria for U.S. Travelers	Drug Resistance	Malaria Species	Recommended Chemoprophylaxis
South Georgia and the South Sandwich Islands	None	None	Not applicable	Not applicable	Not applicable
South Sudan, Republic of	All	High	Chloroquine	*P. falciparum* 90%; *P. vivax* 5%–10%; *P. malariae* and *P. ovale* rare	Atovaquone-proguanil, doxycycline, or mefloquine
Spain	None	None	Not applicable	Not applicable	Not applicable
Sri Lanka	Last locally transmitted case in October 2012	None	Not applicable	Not applicable	Not applicable
Sudan	All	High	Chloroquine	*P. falciparum* 90%; *P. vivax* 5%–10%; *P. malariae* and *P. ovale* rare	Atovaquone-proguanil, doxycycline, or mefloquine
Suriname	Present in provinces of Brokopondo and Sipaliwini; rare cases in Paramaribo	Moderate	Chloroquine	*P. falciparum* 70%; *P. vivax* 15%–20%	All areas except Paramaribo: atovaquone-proguanil, doxycycline, or mefloquine; Paramaribo: mosquito avoidance only
Swaziland	Present in eastern areas bordering Mozambique and South Africa, including all of Lubombo district and the eastern half of Hhohho, Manzini, and Shiselweni districts	Very low	Chloroquine	*P. falciparum* 90%; *P. vivax* 5%; *P. ovale* 5%	Atovaquone-proguanil, doxycycline, or mefloquine

(*Continued*)

Table 2.2 (Continued) Country-Specific Antimalarial Prevention Recommendations

Country	Areas with Malaria	Estimated Relative Risk of Malaria for U.S. Travelers	Drug Resistance	Malaria Species	Recommended Chemoprophylaxis
Sweden	None	None	Not applicable	Not applicable	Not applicable
Switzerland	None	None	Not applicable	Not applicable	Not applicable
Syria	None	None	Not applicable	Not applicable	Not applicable
Taiwan	None	None	Not applicable	Not applicable	Not applicable
Tajikistan	All areas below 2000 m (6561 ft)	Very low	Chloroquine	*P. vivax* 90%; *P. falciparum* 10%	Atovaquone-proguanil, doxycycline, mefloquine, or primaquine
Tanzania	All areas at altitudes below 1800 m (5906 ft)	Moderate	Chloroquine	*P. falciparum* >85%; *P. ovale* >10%; *P. malariae* and *P. vivax* rare	Atovaquone-proguanil, doxycycline, or mefloquine
Thailand	Rural, forested areas that border Burma (Myanmar), Cambodia, and Laos; rural, forested areas in districts of Phang Nga and Phuket; none in the cities of Bangkok, Chiang Mai, Chiang Rai, Koh Phangan, Koh Samui, Pattaya, Phang Nga, and Phuket	Low	Chloroquine and mefloquine	*P. falciparum* 50% (up to 75% some areas); *P. vivax* 50% (up to 60% some areas); remainder *P. ovale*	Atovaquone-proguanil or doxycycline
Timor-Leste (East Timor)	All	No data	Chloroquine	*P. falciparum* 50%; *P. vivax* 50%; *P. ovale* <1%; *P. malariae* <1%	Atovaquone-proguanil, doxycycline, or mefloquine

(*Continued*)

Table 2.2 (Continued) Country-Specific Antimalarial Prevention Recommendations

Country	Areas with Malaria	Estimated Relative Risk of Malaria for U.S. Travelers	Drug Resistance	Malaria Species	Recommended Chemoprophylaxis
Togo	All	High	Chloroquine	*P. falciparum* 85%; *P. ovale* 5%–10%; remainder *P. vivax*	Atovaquone-proguanil, doxycycline, or mefloquine
Tokelau (New Zealand)	None	None	Not applicable	Not applicable	Not applicable
Tonga	None	None	Not applicable	Not applicable	Not applicable
Trinidad and Tobago	None	None	Not applicable	Not applicable	Not applicable
Tunisia	None	None	Not applicable	Not applicable	Not applicable
Turkey	Present in southeastern part of the country; none on the Incerlik U.S. Air Force Base and on typical cruise itineraries	Very low	None	*P. vivax* predominantly; *P. falciparum* sporadically	Atovaquone-proguanil, chloroquine, doxycycline, mefloquine, or primaquine
Turkmenistan	None	None	Not applicable	Not applicable	Not applicable
Turks and Caicos Islands (U.K.)	None	None	Not applicable	Not applicable	Not applicable
Tuvalu	None	None	Not applicable	Not applicable	Not applicable
Uganda	All	High	Chloroquine	*P. falciparum* >85%; remainder *P. malariae*, *P. ovale*, and *P. vivax*	Atovaquone-proguanil, doxycycline, or mefloquine
Ukraine	None	None	Not applicable	Not applicable	Not applicable
United Arab Emirates	None	None	Not applicable	Not applicable	Not applicable

(*Continued*)

Table 2.2 (Continued) Country-Specific Antimalarial Prevention Recommendations

Country	Areas with Malaria	Estimated Relative Risk of Malaria for U.S. Travelers	Drug Resistance	Malaria Species	Recommended Chemoprophylaxis
United Kingdom (with Channel Islands and Isle of Man)	None	None	Not applicable	Not applicable	Not applicable
United States	None	None	Not applicable	Not applicable	Not applicable
Uruguay	None	None	Not applicable	Not applicable	Not applicable
Uzbekistan	None	None	Not applicable	Not applicable	Not applicable
Vanuatu	All	Moderate	Chloroquine	*P. falciparum* 60%; *P. vivax* 35%–40%; *P. ovale* <1%	Atovaquone-proguanil, doxycycline, or mefloquine
Venezuela	Rural areas of the following states: Amazonas, Anzoategui, Apure, Bolivar, Delta Amacuro, Monagas, Sucre and Zulia; present in Angel Falls; none in the city of Caracas and Margarita Island	Low	Chloroquine	*P. vivax* 83%; *P. falciparum* 17%	Atovaquone-proguanil, doxycycline, or mefloquine

(*Continued*)

Table 2.2 (Continued) Country-Specific Antimalarial Prevention Recommendations

Country	Areas with Malaria	Estimated Relative Risk of Malaria for U.S. Travelers	Drug Resistance	Malaria Species	Recommended Chemoprophylaxis
Vietnam	Rural areas only, except none in the Red River Delta; rare cases in the Mekong Delta; none in Da Nang, Haiphong, Hanoi, Ho Chi Minh City (Saigon), Nha Trang, and Qui Nhon	Low	Chloroquine, mefloquine	*P. falciparum* 50%–90%; remainder *P. vivax*	Southern part of the country in the provinces of Dac Lac, Gia Lai, Khanh Hoa, Kon Tum, Lam Dong, Ninh Thuan, Song Be, Tay Ninh: atovaquone-proguanil or doxycycline; other areas with malaria except Mekong Delta: atovaquone-proguanil, doxycycline or mefloquine; Mekong Delta: mosquito avoidance
Virgin Islands, British	None	None	Not applicable	Not applicable	Not applicable
Virgin Islands, U.S.	None	None	Not applicable	Not applicable	Not applicable
Wake Island, U.S.	None	None	Not applicable	Not applicable	Not applicable
Wallis and Futuna Islands (France)	None	None	Not applicable	Not applicable	Not applicable

(*Continued*)

Table 2.2 (Continued) Country-Specific Antimalarial Prevention Recommendations

Country	Areas with Malaria	Estimated Relative Risk of Malaria for U.S. Travelers	Drug Resistance	Malaria Species	Recommended Chemoprophylaxis
West Bank (includes Palestinian Territories)	None	None	Not applicable	Not applicable	Not applicable
Western Sahara	Rare cases	No data	Chloroquine	Unknown	Mosquito avoidance only
Yemen	All areas at altitudes below 2000 m (6562 ft); none in Sana'a	Low	Chloroquine	*P. falciparum* 95%; *P. malariae*, *P. ovale*, and *P. vivax* 5% combined	Atovaquone-proguanil, doxycycline, or mefloquine
Zambia	All	Moderate	Chloroquine	*P. falciparum* >90%; *P. vivax* up to 5%; *P. ovale* up to 5%	Atovaquone-proguanil, doxycycline, or mefloquine
Zimbabwe	All	Moderate	Chloroquine	*P. falciparum* >90%; *P. vivax* up to 5%; *P. ovale* up to 5%	Atovaquone-proguanil, doxycycline, or mefloquine

Source: Centers for Disease Control and Prevention (CDC), 2014, "Malaria information and prophylaxis, by country," accessed June 4, 2015, http://www.cdc.gov/malaria/travelers/country_table.

Table 2.3 Information on Antimalarial Drugs for Prevention

Medication	Dosing Regimen (Adults)	Dosing Frequency and Duration	Contraindications	Side Effects	Use in Special Populations	Drug Interactions
			Areas without Drug Resistance			
Chloroquine	300 mg chloroquine base once weekly	Commence 1 week before departure, continue weekly during trip, and for 4 weeks after return	Hypersensitivity to chloroquine, history of epilepsy; may exacerbate psoriasis and myasthenia gravis	Gastrointestinal disturbances (e.g., nausea, vomiting, abdominal pain), headache	May be taken in pregnancy and in breastfeeding women	Amiodarone, cyclosporin, digoxin, mefloquine, moxifloxacin, intradermal human diploid cell rabies vaccine
			Areas with Chloroquine-Resistant Malaria			
Mefloquine	250 mg once weekly	Commence 2 weeks before departure, continue weekly during trip, and for 4 weeks after return	History of neuropsychiatric disorders, epilepsy, hypersensitivity to mefloquine or related compounds (e.g., quinine); concomitant use of halofantrine use with caution in patient with cardiac conduction abnormalities	Gastrointestinal disturbances, vivid dreams, dizziness, neuropsychiatric side effects	May be taken in pregnancy and in breastfeeding women; avoid pregnancy while on mefloquine, and for 3 months after last dose	Antiarrhythmics, chloroquine, quinine/quinidine (avoid mefloquine within 12 hours of quinine treatment), antiepileptics, antipsychotics, beta-blockers, calcium channel blockers, cardiac glycosides

(Continued)

Table 2.3 (Continued) Information on Antimalarial Drugs for Prevention

Medication	Dosing Regimen (Adults)	Dosing Frequency and Duration	Contraindications	Side Effects	Use in Special Populations	Drug Interactions
Doxycycline	100 mg tablet once daily	Commence 1 day before departure, continue daily during trip, and for 4 weeks after return	Hypersensitivity to tetracyclines; children under 8 years old, pregnancy	Skin photosensitivity, vaginal infections, esophagitis	Women of childbearing age should avoid pregnancy while on doxycycline and for 1 week after last dose; low theoretical risk of adverse events to infant if taken during breastfeeding	Carbamazepine, phenytoin, warfarin, cyclosporine; vaccination with oral typhoid vaccine should be delayed for at least 24 hours after taking doxycycline
Atovaquone-proguanil	1 tablet (atovaquone 250 mg/ proguanil 100 mg) once daily	Commence 1 or 2 days before departure, continue daily during trip, and 1 week after return	Hypersensitivity to atovaquone or proguanil; severe renal impairment (creatinine clearance <30 ml/min), children <5 kg	Gastrointestinal disturbances, headache, diarrhea	Not recommended in pregnancy, breastfeeding, or children <5 kg; pregnancy should be avoided while on atovaquone-proguanil and for 3 weeks after last dose	Rifampicin, antiretrovirals
Primaquine (primary prevention in areas with primarily *P. vivax*)	30 mg base daily	Commence 1–2 days before travel, continue taking daily during trip, and for 1 week after leaving malarious area	G6PD deficiency, pregnancy, lactation, concurrent administration with quinacrine	Gastrointestinal disturbances	Avoid use in pregnancy and breastfeeding	Concurrent administration with quinacrine is contraindicated

(*Continued*)

Table 2.3 (Continued) Information on Antimalarial Drugs for Prevention

Medication	Dosing Regimen (Adults)	Dosing Frequency and Duration	Contraindications	Side Effects	Use in Special Populations	Drug Interactions
			Areas with Mefloquine-Resistant Malaria			
Atovaquone-proguanil	As per above					
Doxycycline	As per above					
			Terminal Prophylaxis (to Reduce Risk of Relapse from *P. vivax* or *P. ovale*)			
Primaquine	30 mg base daily	14 days after departure from malarious area	G6PD deficiency, pregnancy, lactation, concurrent administration with quinacrine	Gastrointestinal disturbances	Avoid use in pregnancy and breastfeeding	Concurrent administration with quinacrine is contraindicated

Source: Chiodini, P.L. et al., 2015, "Guidelines for malaria prevention in travellers from the United Kingdom, 2015," Public Health England, accessed October 6, 2015, https://www.gov.uk/government/uploads/system/uploads/attachment_data/file/461295/2015.09.16_ACMP_guidelines_FINAL.pdf; World Health Organization (WHO), 2014, "Chapter 7. Malaria," accessed May 17, 2014, http://www.who.int/ith/ITH_chapter_7.pdf?ua=1; Arguin, P.M., and K.R. Tan, 2013, "Chapter 3: Infectious diseases related to travel—Malaria," accessed June 8, 2015, http://wwwnc.cdc.gov/travel/yellowbook/2014/chapter-3-infectious-diseases-related-to-travel/malaria#3939. With permission.

Note: G6PD deficiency, glucose-6-phosphate dehydrogenase deficiency.

malaria, and the adult dosing regimens and respective precautions to consider. Dosings for children are usually based on body weight.

It is to be emphasized that these regimens do not provide 100% protection and should be accompanied by use of protective measures such as using insect repellents, wearing long sleeves and/or pants, or using insecticide-treated nets. For areas with limited malaria transmission, personal protective measures are recommended with no need for preventive medications (Arguin and Tan 2013). Medical consult should be sought immediately if fever occurs a week or more after entering an area with malaria risk and up to 3 months after leaving a malarious area.

2.3 EMERGENCY STANDBY TREATMENT

Emergency standby treatment (Table 2.4) should be recommended for travelers taking antimalarial preventive medications and traveling to remote areas where they may not have access to medical facilities within 24 hours. It is not intended to replace antimalarial preventive medications. The traveler should be well informed for appropriate use of this treatment, and a traveler information leaflet should be provided (see the following page) (Chiodini et al. 2015). Emergency standby treatment should be started if it is not possible to access medical consult within 24 hours, completed, and antimalarial prevention medications should be resumed a week after the first emergency standby treatment dose (except when mefloquine is used for prevention and quinine was used for standby treatment; mefloquine should

Table 2.4 Emergency Standby Treatment

Situation	Medication	Adult Dosing Regimen	Remarks
Chloroquine or multidrug resistant falciparum malaria	20 mg artemether + 120 mg lumefantrine (available as a combination preparation)	4 tablets initially, followed by 5 further doses of 4 tablets each given at 8, 24, 36, 48, and 60 hours (total of 24 tablets over 60 hours)	Take with food to enhance drug absorption
	250 mg atovaquone + 100 mg proguanil (available as a combination preparation)	4 tablets as a single dose daily for 3 consecutive days	NA
	300 mg quinine + 100 mg doxycycline	Quinine, 2 tablets 3 times a day for 3 days; doxycycline, 1 tablet twice a day for 7 days	NA
Pregnancy	300 mg quinine + 150 mg clindamycin	Quinine, 2 tablets 3 times a day for 5–7 days; clindamycin, 3 tablets 3 times a day for 5 days	Pregnant travelers should avoid malarious areas

Source: Adapted from Chiodini, P.L. et al., 2015, "Guidelines for malaria prevention in travellers from the United Kingdom, 2015." © Crown copyright. Reproduced with permission of Public Health England.

Note: NA, not applicable.

be taken at least 12 hours after last quinine dose). Medications used for standby treatment should not be the same as that used for prevention to reduce toxicity and minimize development of resistance. Medical advice should be sought as soon as possible for accurate diagnosis and management, as fever may arise from other illnesses besides malaria.

Emergency Standby Medication: Traveler Information Leaflet

You have been advised to carry emergency standby antimalarial medication with you on your forthcoming trip. This leaflet provides you with advice on when and how to use it. Please keep it safely with your medication. If you are traveling with a friend, please ask them to read this leaflet, as they may be able to assist you in following its advice in the event of your becoming ill.

INCUBATION PERIOD OF MALARIA

The minimum period between being bitten by an infected mosquito and developing symptoms of malaria is 8 days, so a febrile illness starting within the first week of your arrival in a malarious area is not likely to be due to malaria.

SYMPTOMS AND SIGNS OF MALARIA

Malaria usually begins with a fever. You may then feel cold, shivery, shaky, and very sweaty. Headache, feeling sick, and vomiting are common with malaria, and you are also likely to experience aching muscles. Some people develop jaundice (yellowness of the whites of the eyes and the skin). It is not necessary for all these symptoms to be present before suspecting malaria, as fever alone may be present at first.

WHEN TO TAKE YOUR EMERGENCY STANDBY MEDICATION

If you develop a fever of 38°C or higher more than one week after being in a malarious area, please seek medical attention straight away. If you are unable to get medical attention within 24 hours of your fever starting, start your standby medication and set off to find and consult a doctor.

HOW TO TAKE YOUR EMERGENCY STANDBY MEDICATION

1. Take medication (usually paracetamol) to lower your fever. If your fever is controlled, it makes it less likely that you will vomit your antimalarial drugs.
2. Take the first dose of your emergency standby antimalarial medication right after the fever-lowering medication.

   If you do vomit and it is **within 30 minutes** of taking the antimalarial drugs:

   - Repeat the first dose (but do not repeat the paracetamol).

   If you vomit **30–60 minutes after taking the first dose** of the antimalarial drugs:

   - Repeat the treatment, but take only half the first dose.

3. Continue the treatment as instructed for the particular drugs prescribed for you.
   - Please remember that this emergency standby medication has been prescribed based on your particular medical history and should be taken only by you, as it may not be suitable for others.
   - Once you have completed your emergency standby medication you should restart your malaria prevention drug(s) one week after your first treatment dose of emergency standby medication.
   - If your preventive medication consists of mefloquine and your standby treatment included quinine, you should wait at least 12 hours after completing the course of quinine before you restart mefloquine.

Adapted from Chiodini, P.L. et al., 2015, "Guidelines for malaria prevention in travellers from the United Kingdom, 2015." © Crown copyright. Reproduced with permission of Public Health England.

In summary, the various antimalarial medications available are presented in detail. Adherence to these medication regimens must be ensured after the medications have been well tolerated. Weekly dose regimens are more likely to be adhered to than daily dose regimens (Steffen et al. 1990). Moreover, the longer the duration of travel, the higher the likelihood for reduced adherence (Held et al. 1994). The existence of counterfeit antimalarial medications, which is another cause of morbidity, mortality, and development of resistance (Ambroise-Thomas 2012), also prompts the need to purchase antimalarial medications from reliable sources. Judicious use of these drugs, together with personal protective measures, is therefore of great importance. Failing to do so may result in treatment or prevention failure, as well as development of resistance. Healthcare professionals play a vital role, from the careful prescribing of the appropriate regimen tailored to the individual by medical practitioners, to the education of travelers on the proper use of antimalarial medications and importance of nonpharmacological measures of personal protection by pharmacists.

REFERENCES

Ambroise-Thomas, P. 2012. "The tragedy caused by fake antimalarial drugs." *Mediterranean Journal of Hematology and Infectious Diseases* 4 (1):e2012027.

Arguin, P.M., and K.R. Tan. 2013. "Chapter 3: Infectious diseases related to travel—Malaria." Accessed June 8, 2015. http://wwwnc.cdc.gov/travel/yellowbook/2014/chapter-3-infectious-diseases-related-to-travel/malaria#3939.

Centers for Disease Control and Prevention (CDC). 2014. "Malaria information and prophylaxis, by country." Accessed June 4, 2015. http://www.cdc.gov/malaria/travelers/country_table.

Chiodini, P.L., D. Patel, C.J.M. Whitty, and D.G. Lalloo. 2015. "Guidelines for malaria prevention in travellers from the United Kingdom, 2015." Public Health England. Accessed October 6, 2015. http://www.gov.uk/government/uploads/system/uploads/attachment_data/file/461295/2015.09.16_ACMP_guidelines_FINAL.pdf.

Curd, F.H.S., D.C. Davey, and F.L. Rose. 1945. "Study on synthetic antimalarial drugs X. Some biguanide derivatives as new types of antimalarial substances with both therapeutic and causal prophylactic activity." *Annals of Tropical Medicine and Parasitology* 39:208–216.

Eckstein-Ludwig, U., R.J. Webb, I.D. Van Goethem, J.M. East, A.G. Lee, M. Kimura, P.M. O'Neill, P.G. Bray, S.A. Ward, and S. Krishna. 2003. "Artemisinins target the SERCA of *Plasmodium falciparum*." *Nature* 424 (6951):957–961.

Falco, E.A., L.G. Goodwin, G.H. Hitchings, I.M. Rollo, and P.B. Russell. 1951. "2:4-Diaminopyrimidines—A new series of antimalarials." *British Journal of Pharmacology* 6 (2):185–200.

Fidock, D.A., P.J. Rosenthal, S.L. Croft, R. Brun, and S. Nwaka. 2004. "Antimalarial drug discovery: Efficacy models for compound screening." *Nature Reviews Drug Discovery* 3:509–520.

Fry, M., and M. Pudney. 1992. "Site of action of the antimalarial hydroxynaphthoquinone, 2-[trans-4-(4′-chlorophenyl) cyclohexyl]-3-hydroxy-1,4-naphthoquinone (566C80)." *Biochemical Pharmacology* 43 (7):1545–1553.

Held, T.K., T. Weinke, U. Mansmann, M. Trautmann, and H.D. Pohle. 1994. "Malaria prophylaxis: Identifying risk groups for non-compliance." *QJM: An International Journal of Medicine* 87 (1):17–22.

Jambou, R., E. Legrand, M. Niang, N. Khim, P. Lim, B. Volney, M.T. Ekala et al. 2005. "Resistance of *Plasmodium falciparum* field isolates to *in-vitro* artemether and point mutations of the SERCA-type PfATPase6." *Lancet* 366 (9501):1960–1963.

Meshnick, S.R. 2002. "Artemisinin: Mechanisms of action, resistance and toxicity." *International Journal for Parasitology* 32 (13):1655–1660.

Müller, I.B., and J.E. Hyde. 2010. "Antimalarial drugs: Modes of action and mechanisms of parasite resistance." *Future Microbiology* 5:1857–1873.

O'Neill, P.M., V.E. Barton, and S.A. Ward. 2010. "The molecular mechanism of action of artemisinin—The debate continues." *Molecules (Basel, Switzerland)* 15:1705–1721.

Steffen, R., R. Heusser, R. Mächler, U. Naef, D. Chen, A. Hofmann, and B. Somaini. 1990. "Malaria chemoprophylaxis among European tourists in tropical Africa: Use, adverse reactions, and efficacy." *Bulletin of the World Health Organization* 68 (3):313–322.

Travassos, M., and M.K. Laufer. 2012. "Antimalarial drugs: An overview." Accessed August 18, 2014. http://www.uptodate.com/contents/antimalarial-drugs-an-overview?source=search_result&search=malaria&selectedTitle=15~150#H17.

World Health Organization (WHO). 2010. "Guidelines for the treatment of malaria. Second edition." Accessed March 4, 2015. http://whqlibdoc.who.int/publications/2010/9789241547925_eng.pdf?ua=1.

World Health Organization (WHO). 2014. "Chapter 7. Malaria." Accessed May 17, 2014. http://www.who.int/ith/ITH_chapter_7.pdf?ua=1.

CHAPTER 3

Research on Medicinal Plants for Malaria

3.1 THE NEED FOR MEDICINAL PLANTS

Medicinal plants are often used in the treatment of diseases, including malaria. Most endemic areas have limited access and financial capabilities to support the use of commercialized medicines, and hence depend heavily on natural sources from traditional healers (Muthaura et al. 2011). Eighty percent of the population in some parts of Asia and Africa use traditional medicine as their primary source of healthcare, with herbal medicines being the most popular form of traditional medicine (World Health Organization [WHO] 2008a). The global market for traditional medicine is approximately $60 billion per year (WHO 2008b). Traditional plants are first-line therapy for malaria in 25% to 75% of people in endemic areas. Willcox and Bodeker (2004) documented more than 1200 species of plants from 160 families used for malaria and fever. Reports of medicinal plants, their indications, and usage have been extensive, most of which are in the form of ethnobotanical surveys and databases such as NAPRAlert (University of Illinois 2012), and Dr. Duke's Phytochemical and Ethnobotanical Database (Duke 1996). Clinical research in this area, however, has been far from extensive. To date, only a handful of traditional plants used in malaria have been evaluated in controlled clinical trials (Research Initiative on Traditional Antimalarial Methods [RITAM] 2013), for example, *Argemone mexicana* (Graz et al. 2010; Willcox et al. 2007), *Nauclea pobeguinii* (Mesia, Tona, Mampunza, Ntamabyaliro, Muanda, Muyembe, Cimanga et al. 2012; Mesia, Tona, Mampunza, Ntamabyaliro, Muanda, Muyembe, Musuamba et al. 2012), *Vernonia amygdalina* (Challand and Willcox 2009), *Artemisia annua* (Mueller et al. 2004), *Cochlospermum planchonii* (Benoit-Vical et al. 2003), and *Cryptolepis sanguinolenta* (Bugyei, Boye, and Addy 2010).

Isolation of compounds from crude extracts, with the well-known examples of artemisinin and quinine, is further evidence that medicinal plants are a potential source of new drugs and lead compounds in drug discovery. Additionally, atovaquone, another antimalarial drug used as a prophylactic drug in travel medicine, is a synthetic derivative of the natural product lapachol, a naphthoquinone (Wells 2011). Given that two of the most effective antimalarials are derived from plant sources, it is highly possible that there are other potent compounds yet to be discovered from plants.

The sections that follow present the drug discovery process from medicinal plants, ethnobotanical research, and bioassay-guided fractionation to the eventual isolation of active compounds.

3.2 ETHNOBOTANICAL RESEARCH OF PLANTS USED TO TREAT MALARIA

Ethnobotany is the scientific study of the relationship between people and plants. One of the primary approaches in ethnobotanical research is to carry out ethnobotanical surveys, in which information regarding the use of plants for malaria in a specified location or community is collected using either questionnaires or interviews. Many such surveys and reviews in which a specific plant species, genus, or the use of plants in the treatment of specific disease(s) is evaluated have been carried out and published. With regard to malaria, Willcox and Bodeker (2004) did a systematic review of plants reported to be used in malaria, and evaluated the clinical safety and efficacy of antimalarial herbal preparations. However, the results were published a decade ago and included plant species used for fever and splenomegaly. Milliken (1997) published a bibliographic survey on medicinal plant species used for malaria and for fever. However, his work focused on the plants in Latin America, and it also included plants used for fever, and it was published more than 10 years ago. Fever may arise due to multiple pathophysiologies, and it is not specific to malaria. There have been no recent studies reviewing the results of ethnobotanical data and databases, nor is there an extensive database of antimalarial medicinal plants used globally to date, with detailed information on their usage. To the best of our knowledge, no review of plants used for malaria that are also found in Singapore has been carried out. Therefore, in this section, a literature search was done, first, to assess the global use of all medicinal plants documented to be used for malaria to date, second, to determine which medicinal plants are most commonly used for malaria, and third, to create a comprehensive database documenting how these medicinal plants are prepared and used specifically for malaria.

Through Scopus, ScienceDirect, and PubMed, all ethnobotanical surveys up to March 2015 that were specifically done on medicinal plants used in malaria were searched for using the keywords "ethnobotanical," "survey," "ethnopharmacological," and "malaria." In addition, Dr. Duke's Phytochemical and Ethnobotanical Database, as well as books on medicinal plants from the Library of Botany and Horticulture at the Singapore Botanic Gardens, and the Medical and Science Libraries at the National University of Singapore were used in the search. Plants that were reported to be used solely for fever (other than malarial fever), external application, or insect repelling property were excluded. Data collected were entered on an Excel sheet and analyzed according to family, genus, location of use, method of preparation, part used, and indication (treatment and/or prevention).

3.2.1 Global Use of Medicinal Plants for Malaria

A total of 1854 plants from 196 families and 1012 genera were reported to be used globally for malaria (Appendix, Table A.1). This is an evident increase from the numbers reported in 2004 (1277 species from 160 families) by Willcox and Bodeker, indicating continued interest in the search for new antimalarials in the past decade. However, this does not correlate with the progress of research and development of ethnobotanicals used in malaria, as only a handful of these go on to preclinical or clinical development, as seen in the cases of *Dichroa febrifuga*, *Argemone mexicana*, and PR 259 CT1. The most common families and genera of the reported antimalarial plants are shown in Table 3.1. It is not surprising to note that the plants from which artemisinin and quinine were isolated belong to the second and third most commonly reported families (i.e., Compositae and Rubiaceae). For the purpose of this work, the classifications of family and genus are as per the published reports.

3.2.2 Prioritizing Plants for Further Research

In the quest for new antimalarials, medicinal plants may be prioritized for research by several ways, such as by the geographical extent of usage, frequency of citation in ethnobotanical surveys, taxonomy, chemotaxonomy (where classification is based on chemical compounds specific to a particular family, genus, or species), *in vitro* or *in vivo* pharmacology, or clinical trials. In this study, taxonomy and geographical extent of usage are used to prioritize plants. Taxonomically, medicinal plants could be prioritized for further research either by family (e.g., Fabaceae) or genus (e.g., *Ficus*, followed by *Vernonia*), although the former may be less targeted, especially if large families (e.g., Fabaceae and Compositae) are investigated. A chemotaxonomic investigation in the *Ficus* or *Vernonia* genus would be a logical next step if one were to prioritize plants according to genus. Classes of bioactive phytochemicals in the genus *Ficus* seem to be nonspecific (Lansky and Paavilainen 2010), while the two major classes of compounds in *Vernonia* are sesquiterpenes (sesquiterpene lactones being the most studied in this genus) and flavonoids, with many isolated compounds from different *Vernonia* species possessing antiplasmodial activity (Toyang and Verpoorte 2013).

The rationale for using geographical extent of usage is based on the postulation that the significance of use of a particular plant is strengthened if it is used in more than one community for a particular indication (Wells 2011). This may indicate possible efficacy.

Willcox and Bodeker (2004) used the IV_{mal} (importance value for the treatment of malaria) system in their review on herbal antimalarials and found that 11 species were used in all three tropical continents to treat malaria or fever: *Annona muricata* (Annonaceae), *Mangifera indica* (Anacardiaceae), *Kalanchoe pinnata* (Crassulaceae), *Momordica charantia* (Curcubitaceae), *Jatropha curcas* (Euphorbiaceae), *Ricinus communis* (Euphorbiaceae), *Senna occidentalis* (Fabaceae), *Senna tora* (Fabaceae), *Sida rhombifolia* (Malvaceae), *Cissampelos pareira* (Menispermaceae), and *Zingiber*

Table 3.1 Antimalarial Plants Used Globally, Classified by Family and Genus

Classification	Number of Plant Species
Family	
Fabaceae	181
Compositae	156
Rubiaceae	129
Euphorbiaceae	73
Apocynaceae	60
Labiatae	52
Rutaceae	52
Annonaceae	39
Verbenaceae	36
Menispermaceae	34
Moraceae	33
Solanaceae	32
Malvaceae	32
Combretaceae	31
Meliaceae	29
Anacardiaceae	25
Loganiaceae	23
Acanthaceae	22
Bignoniaceae	22
Zingiberaceae	21
Gentianaceae	20
Myrtaceae	20
Simaroubaceae	20
Poaceae	19
Lauraceae	18
Sapindaceae	18
Genus	
Ficus	21
Vernonia	18
Croton	17
Acacia	16
Combretum	16
Solanum	15
Zanthoxylum	15
Senna	13
Aristolochia	12
Indigofera	12

(*Continued*)

Table 3.1 (Continued) Antimalarial Plants Used Globally, Classified by Family and Genus

Classification	Number of Plant Species
Genus	
Piper	12
Remijia	12
Strychnos	11
Terminalia	11
Artemisia	10
Ocimum	10

Note: Plants in families constituting less than 1% and plants in genera constituting less than 0.5% are not shown.

officinale (Zingiberaceae). In the light of the current work, 66 plants are reportedly used for malaria in three or more continents (refer to following list).

Plants reported to be used for malaria in three or more continents:

Abrus precatorius L.
Allium sativum L.
Andrographis paniculata Wall. ex Nees
Annona muricata L.
Argemone mexicana L.
Artemisia vulgaris L.
Azadirachta indica (A. Juss) L.
Bidens pilosa L.
Bixa orellana L.
Buxus sempervirens L.
Caesalpinia bonduc (L.) Roxb.
Caesalpinia pulcherrima (L.) Sw.
Cajanus cajan (L.) Millsp.
Cananga odorata (Lam.) Hook. f. and Thomson
Capsicum frutescens L.
Carica papaya L.
Cassia fistula L.
Catharanthus roseus (L.) G. Don
Cedrela odorata L.
Ceiba pentandra (L.) Gaertn.
Cinchona officinalis Linn. F.
Cinchona pubescens M. Vahl
Cissampelos pareira L.
Citrus aurantiifolia L.
Citrus medica L.
Citrus sinensis (L.) Osbeck

Cocos nucifera L.
Coffea arabica L.
Curcuma longa L.
Cymbopogon citratus (DC.) Stapf.
Eucalyptus globulus Labill.
Gynandropsis gynandra (L.) Briq.
Hamelia patens Jacq.
Helianthus annus L.
Hydrangea macrophylla (Thunb.) Ser.
Hyptis suaveolens (L.) Poit.
Ipomoea pes-caprae (L.) Sweet
Jatropha curcas L.
Lantana camara L.
Lantana trifolia L.
Leonotis nepetifolia (L.) R. Br.
Mangifera indica L.
Momordica charantia L.
Myristica fragrans Houtt.
Neurolaena lobata (L.) R. Br.
Nicotiana tabacum L.
Ocimum americanum L.
Ocimum gratissimum L.
Ocimum sanctum L.
Parthenium hysterophorus L.
Persea americana Mill.
Phyllanthus amarus Schum & Thonn.
Phyllanthus niruri L.
Piper nigrum L.
Polygala paniculata L.
Quassia amara L
Scoparia dulcis L.
Senna alata (L.) Roxb.
Senna occidentalis (L.) Link
Sida acuta Burm. F.
Sida rhombifolia L.
Swietenia mahagoni (L.) Jacq.
Tithonia diversifolia (Hemsl.) A. Gray
Trichilia havanensis Jacq.
Xanthium spinosum L.
Zingiber officinale Roscoe

In view of this, priority for further research could be also given to the 14 plants from the current study found to be used in four continents for malaria (excluding the two *Cinchona* species): *Allium sativum*, *Caesalpinia bonduc*, *Capsicum frutescens*, *Carica papaya*, *Cissampelos pareira*, *Coffea arabica*, *Eucalyptus globulus*, *Leonotis nepetifolia*, *Momordica charantia*, *Mangifera indica*, *Senna occidentalis*, and *Sida acuta*, four of which are mentioned in the list by Willcox and Bodeker. *Annona muricata*, *Sida rhombifolia*, *Jatropha curcas*, and *Zingiber officinale*, which are mentioned to be used in three tropical continents in 2004, are also reported in the current study to be used in three continents. *Ricinus communis* and *Senna tora* are reported to be used in two continents and one continent, respectively, in the current study; whereas *Kalanchoe pinnata* is not reported in the current list, probably owing to different search strategies and sources used. It is of note that the IV_{mal} system included plants used to treat fever. In the current study, plants solely used for fever other than "malarial fever" were excluded, for the purpose of having a more conservative definition of medicinal plants used for malaria.

From the global list, 271 plants (14.6%) can be found in Singapore (Table 3.2). The names of these plants are highlighted in bold in Table A.1 in the Appendix. The most commonly reported families and genera of antimalarial plants found locally are shown in Table 3.3. Eight of the 271 plants are reportedly used in four continents. They are *Allium sativum*, *Caesalpinia bonduc*, *Capsicum frutescens*, *Carica papaya*, *Mangifera indica*, *Momordica charantia*, *Senna occidentalis*, and *Sida acuta* (Figure 3.1). Although putative antiplasmodial compounds have been reported from *Allium sativum* (allicin and ajoene) (Coppi et al. 2006; Perez, De la Rosa, and Apitz 1994), *Caesalpinia bonduc* (cassane furanoditerpenoids) (Pudhom et al. 2007), *Carica papaya* (carpaine and derivatives) (Julianti et al. 2013, 2014), and *Sida acuta* (cryptolepine) (Banzouzi et al. 2004), further investigations are still warranted in lead optimization and drug development. For *Mangifera indica*, *Momordica charantia*, and *Senna occidentalis*, antiplasmodial studies have been done on the crude extracts and/or fractions (Awe et al. 1998; Gbeassor et al. 1990; Malann, Matur, and Akinnagbe 2014; Malann, Matur, and Mailafia 2013; Muñoz et al. 2000; Olasehinde et al. 2014; Ramalhete et al. 2008; Tona et al. 2001; Yousif 2014) with positive findings to substantiate their use in malaria, but there are no reports to date of isolated antimalarial compounds. There are yet no published reports of *Capsicum frutescens* and antiplasmodial activity, despite its documented use for malaria through ethnobotanical records (see Table A.1 of Appendix). Singapore is Asia's leading biomedical research hub, and being a tropical country located near malaria-endemic regions such as Cambodia and Thailand, it is strategically excellent for malaria research. Local research should, therefore, consider using these eight plants as starting points in antimalarial drug discovery via ethnobotanical approaches.

Table 3.2 Number of Medicinal Plant Species Used for Malaria Globally and Number of Such Plants Found in Singapore

Number of Continents in Which Plant Has Been Reported for Use in Malaria	Number of Antimalarial Plants Reported Globally	Number of Such Medicinal Plants Found in Singapore
4	14	8
3	52	37
2	152	55
1	1584	166
Insufficient data	52	5
Total	1854	271

Table 3.3 Antimalarial Plants Found in Singapore, Classified by Family and Genus

Classification	Number of Plant Species
Family	
Fabaceae	31
Compositae	16
Euphorbiaceae	15
Malvaceae	14
Meliaceae	10
Apocynaceae	8
Verbenaceae	8
Labiatae	7
Rutaceae	7
Lauraceae	6
Rubiaceae	6
Bignoniaceae	5
Myrtaceae	5
Poaceae	5
Solanaceae	5
Acanthaceae	4
Anacardiaceae	4
Annonaceae	4
Moraceae	4
Palmae	4
Piperaceae	4
Zingiberaceae	4
Convolvulaceae	3
Gramineae	3
Lecythidaceae	3
Lythraceae	3
Sapindaceae	3
Simaroubaceae	3
Genus	
Citrus	5
Hibiscus	5
Senna	5
Caesalpinia	4
Cinnamomum	4
Piper	4
Annona	3

(Continued)

Table 3.3 (Continued) Antimalarial Plants Found in Singapore, Classified by Family and Genus

Classification	Number of Plant Species
Genus	
Clerodendrum	3
Ipomoea	3
Khaya	3
Phyllanthus	3
Acacia	2
Allium	2
Alstonia	2
Aristolochia	2
Artemisia	2
Capsicum	2
Cassia	2
Eucalyptus	2
Euphorbia	2
Euterpe	2
Hedyotis	2
Jatropha	2
Mangifera	2
Manihot	2
Ocimum	2
Passiflora	2
Pterocarpus	2
Sida	2
Solanum	2
Swietenia	2
Tabebuia	2
Triplaris	2
Vitex	2
Ziziphus	2

Note: Plants in families constituting less than 1% and plants in genera constituting less than 0.5% are not shown.

3.2.3 Methods of Preparation, Plant Parts Used, and Dosing Regimens

Most of the preparations reported are decoctions, followed by infusions, and both are heat-based methods (Figure 3.2). A small proportion of these are prepared by alcoholic infusion or maceration (0.3%, data not shown). Analysis by plant part showed that the leaf is most frequently used, followed by the root/rhizome, and bark

Figure 3.1 Eight plants reported to be used for malaria that are found in Singapore: (a) *Allium sativum*, (b) *Caesalpinia bonduc* seed in pod, (c) *C. bonduc* leaves. (*Continued*)

(d)

(e)

Figure 3.1 (Continued) Eight plants reported to be used for malaria that are found in Singapore: (d) *Capsicum frutescens*, (e) *Carica papaya*. (*Continued*)

(f)

(g)

(h)

Figure 3.1 (Continued) Eight plants reported to be used for malaria that are found in Singapore: (f) *Mangifera indica*, (g) *Momordica charantia* flower and leaves, (h) *M. charantia* fruit. (*Continued*)

(i)

(j)

(k)

Figure 3.1 (Continued) Eight plants reported to be used for malaria that are found in Singapore: (i) *Senna occidentalis* leaves, (j) *S. occidentalis* flower, and (k) *Sida acuta*. (Photos [a] and [e–g] are from the authors; photos [b] and [c] courtesy of Ria Tan, http://wildsingapore.com; photos [d] and [i–k] courtesy of Kwan-Han Ong, http://www.natureloveyou.sg.)

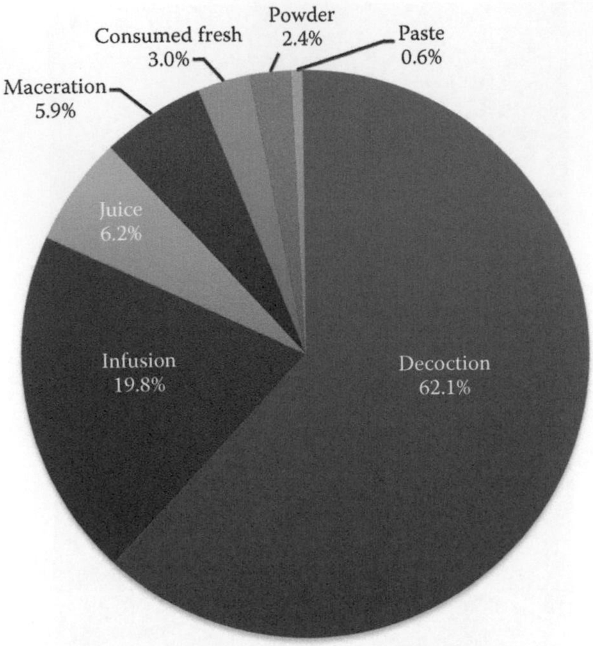

Figure 3.2 Methods of preparations used in malaria treatment.

(Figure 3.3). Different parts of the same plant may be used. The same plant can also be used differently for malaria (i.e., usage of more than one method of preparation, or used both internally and externally). From a scientific viewpoint in general, these may be explained by the presence of the bioactive compound(s) in more than one part of the same plant and also by the possibility of extracting the same bioactive compound(s) using different extraction methods. This can be seen in the case of artemisinin, as discussed in Chapter 4, Section 4.1.1.

There is a great diversity in methods of preparation and dosing regimens reported, as seen in Table A.2 (see Appendix). Certain plants may be used as a standalone therapy or as part of a combination of other herbs (e.g., *Annona senegalensis*). This may suggest that depending on how it is processed or prepared, the extract of a particular plant may contain active principles in a sufficiently potent quantity to be used as monotherapy, whereas the same plant that is prepared in a different way may require the inclusion of other herbs for their synergistic effects, to improve bioavailability of the active constituent(s), or to attenuate toxicity (Willcox, Bodeker, and Rasoanaivo 2005). Some plants have additional insect-repelling properties from the smoke produced as they are burnt on charcoal (Namsa, Mandal, and Tangjang 2011). Apart from being potential sources of novel antimalarial agents, new insecticidal agents may also be developed from these plants.

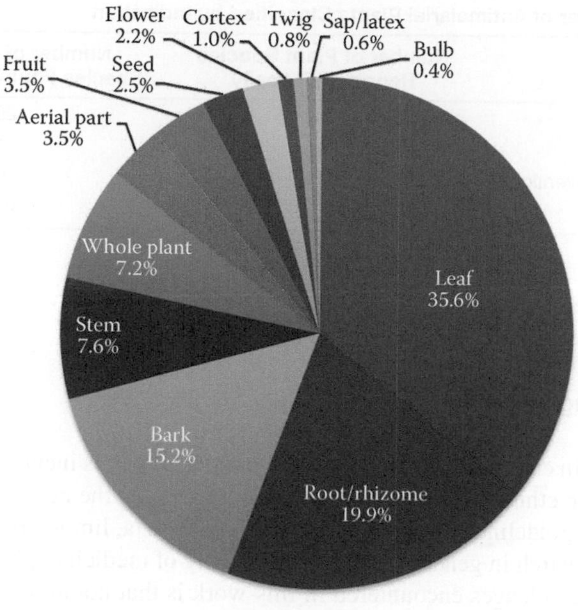

Figure 3.3 Plant parts used in malaria treatment.

3.2.4 Indications for Use

From Table 3.4, a majority (97.7%) of the plants are used solely for curative purposes. About 1.9% of the plants are both curative and prophylactic. Only eight plants (0.4%) are solely prophylactic. Eight (0.4%) are also used as adjuvants to chloroquine or quinine. These plants are reportedly taken with a subtherapeutic dose of chloroquine to prevent resistance and possibly potentiate effects of chloroquine (Rasoanaivo et al. 1992). Thirty-five plants were used as substitutes to *Cinchona* sp. to treat malarial fever. It is interesting to note that there were also four plants used for cerebral malaria intranasally (Akendengue 1992) or orally (Nanyingi et al. 2008), because the current standard of therapy involves only parenteral quinine, quinidine, or artesunate (UpToDate 2014). If these plants are found to be truly efficacious for cerebral malaria via these noninvasive routes of administration, drugs derived from these plants may revolutionize cerebral malaria therapy.

The effectiveness of medicinal plants may be attributed to synergistic and potentiating interactions between the constituents in a crude extract instead of a single entity, as observed in *Artemisia annua*. This may be one of the reasons why combinations of constituents have been found to be more active than individual compounds present in an extract, some of which are even inactive when used on its own. Some of these positive interactions may result in favorable outcomes including immunomodulation, pharmacodynamic or pharmacokinetic synergy (and hence increased

Table 3.4 Number of Antimalarial Plants Classified by Indication

Indication	Number of Plant Species Reported Globally	Number of Such Plant Species Found in Singapore
Treatment only	1811	256
Prevention only	8	1
Treatment and prevention	35	14
Total	1854	271

efficacy), reduced dosages, attenuation of adverse effects, or delay in drug-resistance development (Rasoanaivo et al. 2011).

3.2.5 Challenges

Challenges in ethnobotanical research on medicinal plants include language barriers, incomplete ethnobotanical data in published reports, the need for harmonized methodological guidelines, limited clinical and safety data, limited resources to support malaria research in general, and the complexity of medicinal plants.

One of the challenges encountered in this work is that not all articles are available in English. Articles that are not written in English are not included. Medicinal plants used in traditional Chinese or Ayurvedic medicine that were only documented in Mandarin or Hindi, respectively, would therefore not be included. In addition, the surveys done often lacked complete data. Ethnobotanical studies that provided more complete information (e.g., details on preparation and/or dose, frequency, duration) are included in the Appendix in Table A.2. Most of the studies did not provide details on how the medicinal plant was prepared for use. In one study, the authors mentioned that the traditional healers were unwilling to disclose the information (Karunamoorthi and Tsehaye 2012).

If the medicinal plants are only reported to be used for external application, fever, or as insect repellents (exclusion criteria for the current study), then they are not included. Reported plants were excluded only if the excluded indications were specified. A few of the surveys did not distinguish whether the plants were used solely for fever in malaria or for malaria itself. There were other surveys that identified plants that were used for "symptoms of malaria." Therefore plants from these surveys cannot be excluded. Harmonized methodological guidelines that address basic required parameters in ethnobotanical studies will allow researchers to have a basis for comparison, and therefore improve the quality of ethnobotanical data. For instance, the identity of the plant species used should be confirmed with a botanical expert. The signs and symptoms of malaria in various communities should be described in as much detail as possible in the event that resources are unavailable to make an objective diagnosis of malaria. A unified dosing measurement may also help (for example, the number of leaves per dose could be reported by weight). Having more well-designed ethnobotanical studies, with specific indications, parts used, methods of processing, and preparation and dosing regimens will certainly help researchers in identifying plants that should be screened for antiplasmodial activity *in vitro* as well as in the clinical setting, allowing for the concurrent validation of traditional usage of these plants.

Clinical data on safety and efficacy of medicinal plants used, however, are sparse (Willcox and Bodeker 2004). One of the reasons for the lack of follow-up studies after ethnobotanical surveys is that plant species can have variable amounts of active principles, which are affected by many factors, including climate, soil, and geographical location, making it difficult to ensure reproducibility. This can be minimized by having a designated collection area for the plant of interest and subjecting it to quality control measures such as good agricultural practices; good supply practices; good manufacturing practices; and the use of high-performance liquid chromatography (HPLC), gas chromatography-mass spectrometry (GC-MS), and other spectrophotometric methods.

Despite the widespread ethnopharmacological studies done, not many are further developed in the pipeline due to lack of harmonization of ethnobotanical approaches, quality control, and standardization of usage of medicinal plants for malaria. Furthermore, data is only available for communities in which ethnobotanical studies have been done. There is an age-old traditional saying that the cure for a disease is often found within six feet of the patient (meaning, in the local community). There are probably many more plants with antimalarial potential yet undocumented, revealing another gap in knowledge that needs to be bridged.

As discussed in Chapter 1, although international funding for control of malaria has increased, the resources are still inadequate. Millions of people are still lacking the necessary interventions to curb malaria. The Bill and Melinda Gates Foundation is a strong advocate for malaria research. More continuous and sustainable funding to develop affordable, efficacious, and safe intervention for prevention and treatment of malaria is critical to save millions at risk of malaria infection. Medicinal plants are complex mixtures of biological origin. The active ingredients are often unknown and elusive. The activity of the plant may be due to a single component or may be due to synergy between multiple components acting on multiple receptors. Hence it is a challenge, albeit a fulfilling one, to discover new drugs from medicinal plants.

3.3 BIOASSAY-GUIDED FRACTIONATION OF PLANTS FOR ANTIMALARIAL DRUG DISCOVERY

In natural product drug discovery, plant extracts are screened for a specific biological activity using bioassays. The most active plant extracts are fractionated using techniques such as column chromatography or preparative high-performance liquid chromatography, and subsequent fractions obtained are subjected to further bioassay screening. The most active fractions are subjected to another round of fractionation. This separation process, known as bioassay-guided fractionation, is repetitive, with the ultimate goal of isolating one or a group of active compounds from the extract. The most promising compounds eventually obtained are further improved upon (lead optimization) and developed into useful drugs.

Various antimalarial screening methods have been reviewed (Aguiar et al. 2012; Fidock et al. 2004; Flannery, Chatterjee, and Winzeler 2013). This section gives a general overview on how bioassay-guided fractionation to identify antimalarial compounds is carried out, starting with extraction of plant material, followed by some of the more common biological screening methods for antimalarial activity.

3.3.1 Extraction of Plant Materials

First, to obtain the plant extracts, the part of the plant that is investigated is harvested or collected, cleaned by washing off potential microorganisms such as bacteria or fungi that would interfere with the results, and then dried. The dried plant part is then blended and subjected to various extraction techniques. Blending reduces the size of plant material so that extraction is carried out more efficiently. In extraction, solvents of different polarity are used to separate compounds in the plant material based on the principle of "like dissolves like." An example of some solvents commonly used for plant extraction would be, in increasing order of polarity, hexane or petroleum ether, dichloromethane, ethyl acetate, ethanol and methanol.

Some commonly used extraction techniques include maceration (soaking the plant material in an organic solvent for a period of time), Soxhlet extraction (a specialized form of reflux using the Soxhlet apparatus), ultrasonic-assisted extraction, microwave-assisted extraction, and pressurized liquid extraction. Besides extraction techniques, extraction time, volume of solvent, cost, and temperature can be varied. For Soxhlet and microwave-assisted extractions, high temperatures are used in the extraction process, and this may degrade compounds that are susceptible to heat. The extraction solvent and method of choice are usually decided both empirically and with scientific rationale and it is most helpful if there is existing literature on the plant of interest to give the researcher an idea of the compounds present in the plant. It is also possible that the antiplasmodial constituents may be novel compounds. Plant extracts or isolated compounds are usually first screened *in vitro*, and the most promising ones are then subjected to *in vivo* screening. Screening assays should also complement cytotoxicity assays to ensure active extracts or compounds do not harm host cells.

Further purification can be carried out using column chromatography, or semi-preparative or preparative HPLC. Structural elucidation is carried out using HPLC, GC-MS, nuclear magnetic resonance (NMR), liquid chromatography-mass spectrometry (LC-MS), and elemental analysis.

Extracts, semipurified compounds, and isolated compounds obtained are then evaporated to dryness using a rotary evaporator and subjected to bioassay screening.

3.3.2 Antimalarial Bioassay Screening

Most antimalarial screening assays in laboratory studies require the culturing of *Plasmodium* parasites. *P. falciparum* is the only *Plasmodium* species that can be continuously grown or cultured in human red blood cells *in vitro* (Trager and Jensen 1976).

Short-term *in vitro* cultures of *P. vivax* and *P. ovale* have been developed (Basco and Le Bras 1994).

An example of a *P. falciparum* culture protocol adapted from Cranmer et al. (1997) is described as follows: Parasites are grown in red blood cells free of white blood cells at 2% hematocrit at 37°C at controlled conditions (3%–5% oxygen, 5% carbon dioxide, and 90%–92% nitrogen) at the Roswell Park Memorial Institute (RPMI). The 1640 tissue culture medium is supplemented with 50 μg/ml hypoxanthine, 25 mM 4-(2-hydroxyethyl)piperazine-1-ethanesulfonic acid (HEPES), 25 mM sodium bicarbonate ($NaHCO_3$), 2.5 μg/ml gentamicin, and 0.5% w/v Albumax II (lipid-rich bovine serum albumin). Human serum may be used in place of Albumax. The antimalarial screening can be targeted at the blood stage or liver stage.

3.3.2.1 Blood-Stage Screening

3.3.2.1.1 In Vitro *Assays*

Most *in vitro* screening assays target the blood stage of the parasite and involve the exposure of *Plasmodium* parasites to the compound of interest, assessing the extent of growth inhibition and correlating this to the potency of the compound. Drug-sensitive *P. falciparum* strains (e.g., 3D7, FCR3, D6) and drug-resistant strains (e.g., K1, W2, Dd2) may be used (Table 3.5).

Drug plates were prepared as previously described by Russell et al. (2003). Briefly, a stock solution of each drug was prepared to 1 mg/ml in 70% ethanol and subsequently diluted to obtain the drug concentration desired for testing. Twenty-five microliters of the final drug solution were added to each well of a 96-well plate (Nunc). The drug plates were dried in a laminar flow hood overnight and stored at 4°C.

The protocol for assessing compound efficacy against *P. falciparum in vitro* was modified from the WHO microtest (Russell et al. 2003), as previously described, and is based on microscopic detection on Giemsa-stained slides. Briefly, 200 μL of parasite suspension was added to each well of predosed drug plate at 37°C, typically for 46 to 48 hours, and the assay ends when the parasites in the negative control have

Table 3.5 Types of *P. falciparum* Strains

Name	Clone	Origin	Resistant to
3D7	Yes (from NF54)	Africa	NA
D6	Yes	Sierra Leone	NA
D10	Yes	Papua New Guinea	NA
K1	No	Thailand	Chloroquine, pyrimethamine
Dd2	Yes	Indochina	Chloroquine, quinine, pyrimethamine, sulfadoxine
W2	Yes	Indochina	Chloroquine, quinine, pyrimethamine, sulfadoxine

Note: NA, not applicable.

matured to schizonts. The number of schizonts (excluding rings and gametocytes) is then determined by microscopy and normalized to the control well. By carrying out using twofold dilutions of the compounds in a 96-well plate format, the concentration at which there is 50% inhibition of parasitic growth (IC_{50}) can be determined. The lower the IC_{50}, the more active the compound or extract. Chloroquine and/or artesunate may be used as positive controls.

Microscopy is tedious, time consuming, requires trained microscopists, and is prone to interreader variability (Malleret et al. 2011). Alternative methods are being developed, such as the use of flow cytometry. Other common bioassays used to measure parasite viability include the [^{3}H]-hypoxanthine incorporation assay, the use of fluorescent DNA-binding dyes (e.g., Picogreen), and the parasite lactate dehydrogenase assay (Kuypers et al. 2006; Noedl, Wongsrichanalai, and Wernsdorfer 2003). However, each of these methods has its drawbacks. Use of radiolabelled hypoxanthine is expensive and not environmentally friendly. The various DNA-binding dyes have differing advantages and disadvantages individually in terms of cost, sensitivity, toxicity, and reproducibility, and the spectrophotometric detection of lactate dehydrogenase is far from standardized (Aguiar et al. 2012; Fidock et al. 2004).

3.3.2.1.2 In Vivo *Animal Model Studies*

The 4-day suppression test originally developed by Peters (1975) is commonly used for *in vivo* assay for drug susceptibility. The rodent is infected with a small amount of infected red blood cells, and test compounds are administered via various routes (for example, by mouth) and injected into the peritoneum, vein, or fat tissue. Four daily doses of test compounds are administered to infected mice, and the parasitemia in blood on day 4 postinfection and mean survival time in test and control mice are measured and correlated with efficacy of the compounds. Promising compounds can then be subjected to further tests to determine the optimal treatment dose, the length of time it takes for recrudescence to occur, or if the compound is able to prevent malaria. *P. berghei*, *P. yoelii*, *P. chabaudi*, and *P. vinckei*, which are rodent malaria parasites, can be used as surrogates of *P. falciparum* (Nogueira and Rosário 2010). It should be noted that there are significant differences between these species and human parasites, such as synchronicity, virulence, and drug sensitivity (Fidock et al. 2004).

3.3.2.2 Liver-Stage Screening

The hepatic stage is an attractive target for antimalarial drug discovery due to fewer parasites involved and thus a reduced likelihood of drug resistance development. Furthermore, drugs that act on hypnozoites (dormant parasites) would serve to prevent relapses caused by *P. vivax* and *P. ovale*. However, few antimalarial drugs act on this stage, and the discovery progress is hindered by the lack of reliable and sensitive high-throughput screening methods (Gego et al. 2006).

Plasmodium sporozoites are used to infect either hepatocytes (liver cells) or hepatoma (liver cancer) cell lines *in vitro*. *P. berghei* and *P. yoelii* are used for both *in vitro* and *in vivo* assays, although the use of *P. falciparum* has been described. Availability of primary hepatocytes, however, is limited, and it is difficult to maintain their functional phenotype over a prolonged duration *in vitro* (March et al. 2013). Hepatoma cell lines exhibit intrinsic differences in physiology and metabolic activity and, therefore, may not accurately represent liver function. Moreover, large amounts of sporozoites have to be produced in and isolated from *Anopheles* mosquitoes, and purification of sporozoites is required, as mosquito debris and microbial contaminants would affect *in vitro* assays and influence immune responses when immunocompromised rodents are used for *in vivo* investigations (Kennedy et al. 2012).

3.4 CHEMICAL CLASSES OF COMPOUNDS ISOLATED

There have been several studies reviewing antimalarial compounds isolated from natural products. Active principles elucidated, some of which are secondary metabolites, include alkaloids, terpenes, quassinoids, flavonoids, limonoids, chalcones, peptides, xanthones, quinones, and coumarins, as discussed in a comprehensive review on antimalarial compounds isolated from natural products (Kaur et al. 2009). Not only does this scientifically explain the traditional use of these plants for malaria, it also means that the vast chemical diversity offered by natural products provides novel pharmacophores that may potentially act on undiscovered targets in the *Plasmodium* parasite. To the best of our knowledge, a review on the antimalarial principles isolated from plants (from 2009 to early 2015) has not been published. Therefore, a literature search was performed using PubMed with the keywords "anti malarial plasmodium compounds plants NOT marine" of articles from 2009 to 2015. Compounds with $IC_{50} > 25$ μM are excluded (Cos et al. 2006). Results of the current search are presented in Table 3.6. In addition to currently known structures, there are novel structures discovered over the past 6 years to have antimalarial activities, such as phloroglucinols, depsides, and anacardic acids.

As can be seen from Table 3.6, the collated chemical components that have been found to possess antimalarial activities are very diverse. More research needs to be carried out to further develop these constituents into useful antimalarial drugs. Figure 3.4 shows the chemical structures of several compounds with IC_{50} values of less than 1 μM. Most of these are comprised of complex structures, although there are also some comparatively simpler molecules such as ellagic acid (a polyphenol isolated from *Anogeissus leiocarpus*) and psychorubin (a quinone isolated from *Pentas longiflora*). Ellagic acid has been further studied for *in vivo* antimalarial activity and its plausible mechanism of action as a prooxidant (Soh et al. 2012), whereas simalikalactone D has demonstrated stage-specific activity on midtrophozoites of *P. falciparum* and synergistic action with atovaquone (Bertani et al. 2012). The different chemical structure classes of these potent compounds possibly reflect a diverse range of pharmacophores and action on different targets in the parasite. More can and should be done on these most promising structures, such as lead optimization approaches to improve their physicochemical properties where necessary.

Table 3.6 Compounds Isolated from Plants That Are Reported to Have Antimalarial Activity

Subdivision of Class Structure	Reported Active Constituent	Plant Species	Remarks (*P. falciparum* Strains, Other *Plasmodium* Species, Targets)	References
		Alkaloids		
Carbazole alkaloids	O-Methylmukonal	*Clausena excavata*	MIC = 30 μM (K1)	Sripisut and Laphookhieo 2010
	Heptaphylline, mukonal, 7-methoxymukonal	*Clausena harmandiana*	Heptaphylline: IC_{50} = 11.5–22.9 μM (K1) Mukonal: IC_{50} = 15.5 μM (K1) 7-Methoxymukonal: IC_{50} = 12.2 μM (K1)	Thongthoom et al. 2010; Yenjai et al. 2000
	Clausenawalline A	*Clausena wallichii*	IC_{50} = 4.4 μM (strain not specified)	Maneerat et al. 2011
Benzylisoquinolines	1-(4-Hydroxybenzyl)-6,7-methylenedioxy-2-methylisoquinolinium trifluoroacetate	*Doryphora sassafras*	IC_{50} = 3.0 and 4.4 μM (3D7 and Dd2, respectively)	Buchanan et al. 2009
	(–)-Reticuline	*Dehaasia longipedicellata*	IC_{50} < 30.4 μM (K1)	Zahari et al. 2014
Bisbenzyl-isoquinolines	(–)-O-O-dimethylgrisabine	*Dehaasia longipedicellata*	IC_{50} = 0.031 μM (K1)	Zahari et al. 2014
	Scoulerine, cheilanthifoline, protopine	*Corydalis dubia*	IC_{50} = 5.44, 2.77, 4.10 μM, respectively (TM4/8.2, a wild-type chloroquine and antifolate sensitive strain); IC_{50} = 3.18, 3.75, 3.90 μM, respectively (K1CB1, multidrug resistant strain)	Wangchuk et al. 2012
	2-Norcepharanthine, cepharanoline, and fangchinoline	*Stephania rotunda*	IC_{50} = 0.3, 0.2, and 0.3 μM, respectively (W2)	Chea et al. 2010
Protoberberines	Pseudopalmatine	*Stephania rotunda*	IC_{50} = 2.8 μM (W2)	Baghdikian et al. 2013

(*Continued*)

Table 3.6 (Continued) Compounds Isolated from Plants That Are Reported to Have Antimalarial Activity

Subdivision of Class Structure	Reported Active Constituent	Plant Species	Remarks (*P. falciparum* Strains, Other *Plasmodium* Species, Targets)	References
Aporphines	(–)-Anonaine	*Goniothalamus australis*	IC_{50} = 7 μM (3D7)	Levrier et al. 2013
	(–)-Boldine and (–)-norboldine	*Dehaasia longipedicellata*	IC_{50} = 2.60 and 9.28 μM (K1)	Zahari et al. 2014
Indoles	Geissolosimine, geissopermine, geissoschizoline, geissochizone	*Geissospermum vellosii*	IC_{50} = 0.96, 5.02, 13.96, and 10.29 μM (D10), respectively	Mbeunkui et al. 2012
	α-Dihydrocadambine	*Neonauclea purpurea*	IC_{50} = 6.6 μM (K1)	Karaket et al. 2012
	3-Hydroxylongicaudatine Y, bisnordihydrotoxiferine, divarine, longicaudatine, longicaudatine Y, and longicaudatine F	*Strychnos malacoclados*	IC_{50} range from 0.39 to 3.65 μM and 0.30 to 12.80 μM for 3D7 and W2 strains, respectively	Tchinda et al. 2012
	20-*epi*-Dasycarpidone	*Aspidosperma ulei*	IC_{50} = 16.7 μM (K1)	dos Santos Torres et al. 2013
	Strychnochrysine	*Strychnos nux-vomica*	IC_{50} = 8.5 and 10.3 μM against 3D7 and W2, respectively	Jonville et al. 2013
	Strychnobaillonine	*Strychnos icaja*	IC_{50} = 1.1 μM (3D7)	Tchinda et al. 2014
	3′-oxo-Tabernaelegantine A	*Muntafara sessilifolia*	IC_{50} = 4.4 μM (FcB1)	Girardot et al. 2012
	Aspidoscarpine, uleine, apparicine, and N-methyl-tetrahydrolivacine	*Aspidosperma olivaceum*	IC_{50} = 14.6, 21.6, 11.4, and 21.6 μM (W2)	Chierrito et al. 2014
Piperidine alkaloid	Carpaine	*Carica papaya*	IC_{50} = 0.2 μM (K1), 11.9% reduction of *P. berghei* parasitaemia in mice	Julianti et al. 2014

(*Continued*)

Table 3.6 (Continued) Compounds Isolated from Plants That Are Reported to Have Antimalarial Activity

Subdivision of Class Structure	Reported Active Constituent	Plant Species	Remarks (*P. falciparum* Strains, Other *Plasmodium* Species, Targets)	References
Miscellaneous alkaloids	5-Hydroxy-6-methoxyonychine	*Mitrephora diversifolia*	IC_{50} = 9.9 and 11.4 µM (3D7 and Dd2, respectively)	Mueller et al. 2009
	Sauristolactam	*Goniothalamus australis*	IC_{50} = 9 µM (3D7)	Levrier et al. 2013
	Arborinine, xanthoxoline	*Zanthoxylum leprieurii*	IC_{50} = 15.8 and 17.0 µM, respectively (3D7)	Tchinda et al. 2009
	(2R)-2,3-Epoxy-N-methylatanine	*Drummondita calida*	74% inhibition at 80 µM (Dd2)	Yang et al. 2011
	(+)-*N*-Methylisococlaurine, atherosperminine, 2-hydroxyathersperminine	*Cryptocarya nigra*	IC_{50} = 5.40, 5.80, and 0.75 µM, respectively (K1)	Nasrullah et al. 2013
	(+)-5,6-Dehydrolycorine	*Lycoris radiata*	IC_{50} = 2.3 and 1.9 µM against D6 and W2, respectively	Hao, Shen, and Zhao 2013
	Cassiarins J and K	*Cassia siamea*	IC_{50} = 0.3 and 1.4 µM, respectively (3D7)	Deguchi et al. 2012
	Cassiarins C-E	*Cassia siamea*	IC_{50} = 24.2, 3.6, and 7.3 µM, respectively (3D7)	Oshimi et al. 2009
	Hydroxynoracronycine, 1,5-dihydroxy-2,3-dimethoxy-10-methyl-9-acridone	*Citropsis articulata*	IC_{50} = 2.8 and 9.96 µM, respectively (FcB1/Colombia)	Lacroix et al. 2011
	Chaconine	*Solanum tuberosum*	Parasitaemia suppression of 42.66% at 3.75 mg/kg (*P. yoelii*)	Chen et al. 2010
	3α-(1-Methylitaconyl)-6β-senecioyloxytropane, 3α-(1-methylmesaconyl)-6β-angeloyloxytropane	*Schizanthus tricolor*	IC_{50} = 22.8 and 24.8 µM (K1)	Cretton et al. 2010

(*Continued*)

Table 3.6 (Continued) Compounds Isolated from Plants That Are Reported to Have Antimalarial Activity

Subdivision of Class Structure	Reported Active Constituent	Plant Species	Remarks (*P. falciparum* Strains, Other *Plasmodium* Species, Targets)	References
	Prosopilosidine, isoprosopilosidine	*Prosopis glandulosa* var. *glandulosa*	IC_{50} = 0.062 and 0.067 μM, respectively, against D6; 0.15 and 0.19 μM, respectively, against W2; Prosopilosidine ED_{50} = 2 mg/kg/day (*P. berghei*)	Samoylenko et al. 2009
	Caesalminines A and B	*Caesalpinia minax*	IC_{50} = 0.42 and 0.79 μM (K1)	Ma et al. 2014
	(+)-Sebiferine and (−)-milonine	*Dehaasia longipedicellata*	IC_{50} = 22.46 and 0.097 μM (K1)	Zahari et al. 2014
	Nitidine	*Zanthoxylum chalybeum*	IC_{50} = 0.20 and 0.07 μM against 3D7 and F32 respectively	Muganga et al. 2014
		Quinones		
	2,6-Dimethoxy-1,4-benzoquinone	*Neonauclea purpurea*	IC_{50} = 11.3 μM (K1)	Karaket et al. 2012
	Pentalongin, psychorubrin	*Pentas longiflora*	Pentalongin and psychorubrin: IC_{50} < 4.7 μM (W2, D6) (Selectivity indices against W2 and D6 respectively: pentalongin, 2.96 and 3.48; Psychorubrin, 0.98 and 1.09)	Endale et al. 2012
	Rubiadin	*Pentas lanceolata*	IC_{50} = 21.5 μM (D6)	
	Vismiaquinone A	*Vismia laurentii*	IC_{50} = 1.42 μM (W2)	Noungoue et al. 2009
	Scutianthraquinones A–D	*Scutia myrtina*	IC_{50} = 1.23, 1.14, 3.14, and 3.68 μM, respectively (Dd2); 1.2, 5.4, 15.4, and 5.6 μM, respectively (FCM29)	Hou et al. 2009
	4-Methoxy-1-methyl-quinolin-2-one	*Peltostigma guatemalense*	65% inhibition at 35 ppm (strains of *P. falciparum* not specified)	Cuca Suarez et al. 2009

(Continued)

Table 3.6 (Continued) Compounds Isolated from Plants That Are Reported to Have Antimalarial Activity

Subdivision of Class Structure	Reported Active Constituent	Plant Species	Remarks (*P. falciparum* Strains, Other *Plasmodium* Species, Targets)	References
		Anthraquinones		
	Jozikniphololone A	*Kniphofia foliosa*	IC_{50} = 0.4 and 0.3 μM (D6 and W2, respectively)	Induli et al. 2013
	Jozikniphololone B		IC_{50} = 2.5 and 1.5 μM (D6 and W2, respectively)	
	Chryslandicin		IC_{50} = 4.0 and 2.8 μM (D6 and W2, respectively)	
	10-Hydroxy-10-(chrysophanol-7′-yl) chrysophanol anthrone		IC_{50} = 3.3 and 1.3 μM (D6 and W2, respectively)	
	10-Methoxy-10-(chrysophanol-7′-yl) chrysophanol anthrone		IC_{50} = 8.9 and 2.6 μM (D6 and W2, respectively)	
	Asphodelin		IC_{50} = 16.1 and 12.6 μM (D6 and W2, respectively)	
	Knipholone		IC_{50} = 23.2 and 18.4 μM (D6 and W2, respectively)	
	Isoknipholone		IC_{50} = 19.7 and 18.2 μM (D6 and W2, respectively)	
	Knipholone anthrone		IC_{50} = 9.7 and 8.5 μM (D6 and W2, respectively)	
	10-Acetonylknipholone cyclooxanthrone		IC_{50} = 9.2 and 6.5 μM (D6 and W2, respectively)	
	Knipholone cyclooxanthrone		IC_{50} = 9.5 and 14.5 μM (D6 and W2, respectively)	
		Terpenes		
Sesquiterpenes	Urospermal A-15-O-acetate	*Dicoma tomentosa*	IC_{50} = 2.87 and 2.41 μM (3D7 and W2, respectively); selectivity index = 3.3	Jansen et al. 2010

(Continued)

Table 3.6 (Continued) Compounds Isolated from Plants That Are Reported to Have Antimalarial Activity

Subdivision of Class Structure	Reported Active Constituent	Plant Species	Remarks (*P. falciparum* Strains, Other *Plasmodium* Species, Targets)	References
	Vernangulides A and B	*Distephanus angulifolius*	IC_{50} = 1.9 and 1.55 μM, respectively, against D10; 1.55 and 2.10 μM, respectively, against W2	Pedersen et al. 2009
	Vernodalol, vernodalin		IC_{50} = 3.82 and 1.75 μM, respectively, against D10; 4.94 and 2.69 μM, respectively, against W2	
	Ivaxillin, 11(13)-dehydroivaxillin, C-11 epimer of ivaxillin	*Carpesium cernum*	IC_{50} = 17.1, 2.01, and 9.32 μM (D10)	Chung and Moon 2009
	β-Dictyopterol	*Carpesium divaricatum*	IC_{50} = 13.1 μM (D10)	Chung, Seo et al. 2010
	Vernopicrin, vernomelitensin	*Vernonia guineensis*	IC_{50} = 1.36 and 1.77 μM, respectively, against Hb3; IC_{50} = 2.33 and 1.64 μM, respectively, against Dd2	Toyang et al. 2013
	Dehydrobrachylaenolide	*Dicoma anomala* subsp. *gerrardii*	IC_{50} = 1.865 μM (D10), 4.095 μM (K1)	Becker et al. 2011
	Peruvin, psilostachyin	*Ambrosia tenuifolia*	IC_{50} = 1.1 μM (F32); 2.1 and 6.4 μM (F32 and W2), respectively	Sülsen et al. 2011
	Ishwarane	*Bixa orellana*	EC_{50} < 3 μM (3D7), and <12 μM (K1)	Zhai et al. 2014
	Salaterpenes A–D	*Salacia longipes* var. *camerunensis*	IC_{50} range from 1.71 to 2.637 μM (W2)	Mba'Ning et al. 2013
	1-β-(p-Cumaroyloxyl)-polygodial,1-β-(p-methoxycinnamoyl)-polygodial	*Drimys brasiliensis*	IC_{50} = 1.01 and 4.87 μM, respectively (F32)	Claudino et al. 2013
	Sanandajin, kamolonol acetate and methyl galbanate	*Ferula pseudalliacea*	IC_{50} = 2.6, 16.1, and 7.1 μM, respectively (K1)	Attioua et al. 2012

(*Continued*)

Table 3.6 (Continued) Compounds Isolated from Plants That Are Reported to Have Antimalarial Activity

Subdivision of Class Structure	Reported Active Constituent	Plant Species	Remarks (*P. falciparum* Strains, Other *Plasmodium* Species, Targets)	References
	(15-Acetoxy-8β-[(2-methylbutyryloxy)]-14-oxo-4,5-cis-acanthospermolide), (9α-acetoxy-15-hydroxy-8β-(2-methylbutyryloxy)-14-oxo-4,5-trans-acanthospermolide)	*Acanthospermum hispidum*	IC_{50} = 2.9 and 2.23 μM, respectively (3D7)	Ganfon et al. 2012
	Athrolides C and D	*Athrosima proteiforme*	IC_{50} = 6.6 and 7.2 μM, respectively, against HB3; 5.5 and 4.2 μM, respectively, against Dd2	Pan et al. 2011
	Phomoarcherin B	*Phomopsis archeri*	IC_{50} = 2.06 μM (K1)	Hemtasin et al. 2011
	Muzigadial	*Canella winterana*	IC_{50} = 1.25 μM (D10)	Grace et al. 2010
	4β-Hydroxy-11,12,13-trinor-5-eudesmen-1,7-dione, homalomenol C, oxo-T-cadinol, 1β, 4β, 6β-trihydroxyeudesmane	*Teucrium ramosissimum*	IC_{50} = 15.9, 4.7, 18.6, and 19.5 μM, respectively (FcB1)	Henchiri et al. 2009
	4aαH-3,5α,8aβ-Trimethyl-4,4a,8a-trihydronaphtho-([2,3b]-dihydrofuran-2-one)-8-one	*Siphonochilus aethiopicus*	IC_{50} = 13.9 and 7.17 μM (D10 and K1, respectively)	Lategan et al. 2009
	β-Caryophyllene	*Murraya koenigii*	IC_{50} = 40.4 μM (3D7)	Kamaraj et al. 2014

(*Continued*)

Table 3.6 (Continued) Compounds Isolated from Plants That Are Reported to Have Antimalarial Activity

Subdivision of Class Structure	Reported Active Constituent	Plant Species	Remarks (*P. falciparum* Strains, Other *Plasmodium* Species, Targets)	References
	5-O-Methyl-5-epiisogoyazensolide, 15-O-methylgoyazensolide, 1-oxo-3,10-epoxy-8-(2-methylacr1 0-epoxy-8-(2-methylacryloxy)-15-acetoxygermacra-2,4,11(13)-trien-6(12)-olide and 5-epiisogoyazensolide	*Piptocoma antillana*	IC_{50} = 6.2, 22, 2.2, and 9.0 µM (3D7)	Liu et al. 2014b
	N-Acetyl-8α-polyveolinone and *N*-acetyl-polyveoline	*Polyalthia oliveri*	IC_{50} = 7.6 and 29.1 µM (NF54)	Kouam et al. 2014
Triterpenes	Tormentic acid	*Cecropia pachystachya*	IC_{50} = 23 – 31 µM (W2)	Uchôa et al. 2010
	Betulinic acid	*Hypericum lanceolatum*	IC_{50} = 4.98 µM (W2mef [multidrug resistant strain]), 4.46 µM (SHF4, field isolate)	Zofou, Kowa et al. 2011
	Betulinic acid	*Beilschmiedia zenkeri*	IC_{50} = 5.2 µM (W2)	Lenta et al. 2009
	3β-O-cis-Coumaroyl betulinic acid, 3β-O-trans-coumaroyl betulinic acid	*Cornus florida*	IC_{50} = 10.4 and 15.3 µM, respectively (D10)	Graziose et al. 2012
	2β, 3β, 19α-Trihydroxy-urs-12-en-28-oic acid	*Kigelia africana*	IC_{50} = 1.6 µM (W2)	Zofou, Kengne et al. 2011
	19α-Hydroxy-3-oxo-ursa-1,12-dien-28-oic acid	*Canthium multiflorum*	IC_{50} = 55.6 µM (3D7)	Traoré-Coulibaly et al. 2009
	Ursolic acid	*Keetia leucantha*	IC_{50} = 32.4 µM (3D7)	Bero et al. 2013
	16β-Hydroxylupane-1,20(29)-dien-3-one	*Parinari excelsa*	IC_{50} = 28.3 µM (K1)	Attioua et al. 2012

(*Continued*)

Table 3.6 (Continued) Compounds Isolated from Plants That Are Reported to Have Antimalarial Activity

Subdivision of Class Structure	Reported Active Constituent	Plant Species	Remarks (*P. falciparum* Strains, Other *Plasmodium* Species, Targets)	References
	Gedunin, xyloccensin-I	*Xylocarpus granatum*	MIC = 20.7 and 15.5 μM, respectively (3D7)	Lakshmi et al. 2012
	Alisol A, alisol B 11-monoacetate, alisol B 23-monoacetate and alisol G	*Alisma plantago-aquatica*	IC_{50} range from 5.4 to 13.8 μM (K1)	Adams et al. 2011
	Tirucalla-7,24-dien-3-one	*Vismia laurentii*	IC_{50} = 1.18 μM (W2)	Noungoue et al. 2009
	Balsaminoside A, karavilagenin E	*Momordica balsamina*	IC_{50} = 4.6 and 7.4 μM (3D7); and 4.0 and 8.2 μM (Dd2), respectively	Ramalhete et al. 2010
	Hautriwaic acid lactone	*Baccharis dracunculifolia*	IC_{50} = 0.6 and 7.0 μM (D6 and W2, respectively)	da Silvo Filho et al. 2009
	Neomacrolactone, 22α-acetoxyneomacrolactone, 6-hydroxyneomacrolactone, 22α-acetoxy-6-hydroxyneomacrolactone, 6,7-epoxyneomacrolactone, 22α-acetoxy-6,7-epoxyneomacrolactone, 4-methylen-neomacrolactone, neomacroin, 22-de-O-acetyl-26-deoxyneoboutomellerone	*Neoboutonia macrocalyx*	IC_{50} = 2.10, 2.46, 1.48, 2.74, 9.44, 11.0, 1.91, 3.22, and 2.16 μM, respectively (FcB1/Colombia strain)	Namukobe et al. 2014
	Turranoic acid, turraenine and triptocallic acid B	*Turraea* sp.	IC_{50} = 5.2, 16.6, and 16.4 μM, respectively (FCM29)	Rasamison et al. 2014
	Congoensins A and B, gladoral A	*Entandrophragma congoënse*	IC_{50} = 5.5, 6.1, and 2.4 μM, respectively (NF54)	Happi et al. 2015
Diterpenes	Gomphostenin-A	*Gomphostemma niveum*	92.65% chemosuppression at 200 mg/kg/day in *P. berghei* infected mice	Sathe et al. 2010

(*Continued*)

Table 3.6 (Continued) Compounds Isolated from Plants That Are Reported to Have Antimalarial Activity

Subdivision of Class Structure	Reported Active Constituent	Plant Species	Remarks (*P. falciparum* Strains, Other *Plasmodium* Species, Targets)	References
	Otostegindiol	*Otostegia integrifolia*	Dose dependent (25, 50, and 100 mg/kg/day) suppression of parasitaemia (50.13%–73.16%) in *P. berghei* infected mice	Endale et al. 2013
	Atisinium chloride	*Aconitum orochryseum*	IC_{50} = 4 μM and 3.6 μM, respectively, against TM4 and K1 strains	Wangchuk et al. 2010
	2α-Hydroxyjatropholone, caniojane, jatropholone A	*Jatropha integerrima*	IC_{50} = 13.1, 9.6, and 18.2 μM (K1)	Sutthivaiyakit et al. 2009
	4-*epi*-Triptobenzene L, 12-O-deacetyl-6-O-acetyl-19-acetyloxycoleon Q, 9α-13α-epidioxyabiet-8(14)-en-18-oic acid, 12-O-deacetyl-6-O-acetyl-18-acetyloxycoleon Q	*Anisochilus hamandii*	IC_{50} = 15.6, 6.5, 7.1, and 16.1 μM, respectively (K1)	Lekphrom, Kanokmedhakul, and Kanokmedhakul 2010
	Neophytadiene	*Carpesium divaricatum*	IC_{50} = 23.1 μM (D10)	Chung, Seo et al. 2010
	Mellerin B	*Neoboutonia macrocalyx*	IC_{50} = 19.0 μM (FcB1/Colombia strain)	Namukobe et al. 2014
	Aphadilactones A–D	*Aphanamixis grandifolia*	IC_{50} = 0.19, 1.35, 0.17, and 0.12 μM, respectively (Dd2)	Liu et al. 2014a
		Quassinoids		
	Simalikalactone D	*Quassia amara*	IC_{50} = 0.01 μM (W2)	Houël et al. 2009
		Flavonoids		
	Artemetin	*Artemisia gorgonum*	IC_{50} = 9.01 μM (FcB1)	Ortet et al. 2011

(*Continued*)

Table 3.6 (Continued) Compounds Isolated from Plants That Are Reported to Have Antimalarial Activity

Subdivision of Class Structure	Reported Active Constituent	Plant Species	Remarks (*P. falciparum* Strains, Other *Plasmodium* Species, Targets)	References
	Burttinols-A, -B, -C, -D; eryvarin H, 4′-O-methylsigmoidin B, abyssinone V, abyssinone V methyl ether, calopocarpin	*Erythrina burttii*	IC_{50} range from 5.7 to 19.4 μM (D6) and 6.6 to 21.1 μM (W2)	Yenesew et al. 2012
	Myristicyclins A and B	*Horsfieldia spicata*	IC_{50} = 35, 43, 54 μM for myristicyclin A, and 10, 6.6, 7.9 μM for myristicyclin B against ring, trophozoite, and schizont stages respectively	Lu et al. 2014
	5,4′-Dihydroxy-7-dimethoxyflavanone	*Senecio roseiflorus*	IC_{50} = 3.2 and 4.4 μM against D6 and W2, respectively	Kerubo et al. 2013
	(+)-Chamaejasmin, (−)-diphysin and an inseparable mixture of 7,7″-di-O-methylchamaejasmin and 7,7″-di-O-methylisochamaejasmin	*Ormocarpum trichocarpum*	IC_{50} = 14.01, 16.90, and 4.03 μM, respectively (D10)	Chukwujekwu et al. 2012
	Isochamaejasmin	*Ormocarpum kirkii*	IC_{50} = 13.5 μM (K1)	Dhooghe et al. 2010
	3′,4′,7-Trihydroxyflavone	*Albizia zygia*	IC_{50} = 0.29 μM (3D7)	Abdalla and Laatsch 2011
	Mucusisoflavone A	*Ficus mucuso*	IC_{50} = 7.69 μM (Pf enoyl-ACP reductase)	Bankeu et al. 2011
	Apigenin 7-O-glucoside	*Achillea millefolium*	IC_{50} = 23.4 and 14.1 μM against D10 and W2, respectively	Vitalini et al. 2011

(*Continued*)

Table 3.6 (Continued) Compounds Isolated from Plants That Are Reported to Have Antimalarial Activity

Subdivision of Class Structure	Reported Active Constituent	Plant Species	Remarks (*P. falciparum* Strains, Other *Plasmodium* Species, Targets)	References
	Styracifolins A and B, artoheterophyllins A and B, artonins A, B, and F, and heterophyllin	*Artocarpus styracifolius*	IC_{50} range from 1.1 to 13.7 μM (FcB1)	Bourjot et al. 2010
	Eupalestin, 5,6,7,5′-tetramethoxy-3′,4′-methylenedioxyflavone, 5,6,7,3′,4′,5′-hexamethoxyflavone, ageconyflavone C	*Ageratum conyzoides*	IC_{50} = 10.99, 11.04, 7.43, and 9.25 μM (K1)	Nour et al. 2010
	Dalparvone	*Dalbergia parviflora*	IC_{50} = 24.8 μM (K1)	Songsiang et al. 2009
	Lonchocarpol A	*Erythrina fusca*	IC_{50} = 2.25 μM (K1)	Innok, Rukachaisirikul, and Suksamrarn 2009
	5-Hydroxy-7,8-dimethoxyflavanone	*Beilschmiedia zenkeri*	IC_{50} = 3.7 μM (W2)	Lenta et al. 2009

(*Continued*)

Table 3.6 (Continued) Compounds Isolated from Plants That Are Reported to Have Antimalarial Activity

Subdivision of Class Structure	Reported Active Constituent	Plant Species	Remarks (*P. falciparum* Strains, Other *Plasmodium* Species, Targets)	References
	7-O-α-d-Glucopyranosyl-3,4′-dihydroxy-3′-(4-hydroxy-3-methylbutyl)-5,6-dimethoxyflavone, 7-O-α-d-glucopyranosyl(6‴-p-hydroxcinnamoyl)-3,4′-dihydroxy-3′-(4-hydroxy-3-methylbutyl)-5,6-dimethoxyflavone, 3,7,4′-trihydroxy-3′-(4-hydroxy-3-methylbutyl)-5,6-dimethoxyflavone, 3,7-dihydroxy-3′-(4-hydroxy-3-methylbutyl)-5,6,4′-trimethoxyflavone, 5,7-dihydroxy-3′-(2-hydroxy-3-methyl-3-butenyl)-3,6,4′-trimethoxyflavone, 3,7-dihydroxy-3′-(2-hydroxy-3-methyl-3-buten-yl)-5,6,4′-trimethoxyflavone and 7-O-α-d-glucopyranosyl-3,5-dihydroxy-3′-(4″-acetoxy-3″-methylbutyl)-6,4′-dimethoxyflavone	*Duranta repens*	IC_{50} range from 5.2 to 13.5 μM and 5.9 to 13.1 μM, respectively, against D6 and W2	Ijaz et al. 2010
		Lignans		
	Sesamin	*Artemisia gorgonum*	IC_{50} = 9.51 μM (FcB1)	Ortet et al. 2011
	Nortrachelogenin	*Carissa edulis*	IC_{50} = 5.21 μM (D6)	Kebenei, Ndalut, and Sabah 2011

(*Continued*)

Table 3.6 (Continued) Compounds Isolated from Plants That Are Reported to Have Antimalarial Activity

Subdivision of Class Structure	Reported Active Constituent	Plant Species	Remarks (*P. falciparum* Strains, Other *Plasmodium* Species, Targets)	References
		Chalcones		
	Cajachalcone	*Cajanus cajan*	IC_{50} = 7.4 μM (K1)	Ajaiyeoba et al. 2013
		Xanthones		
	5-Hydroxy-3-methoxyxanthone	*Hypericum lanceolatum*	IC_{50} = 3.26 μM (W2mef [multidrug resistant strain]), 1.43 μM (SHF4, field isolate)	Zofou, Kowa et al. 2011
	1,5,6-Trihydroxy-3-methoxy-7-geranyl-xanthone, 2-(1′-1′-dimethylprop-2′-enyl)-1,4,5-trihydroxyxanthone, pyrojacareubin, isogarcinol, 7-epi-isogarcinol	*Rheedia acuminata*	IC_{50} = 10.5, 15.1, 11.4, 3.5, and 3.2 μM, respectively (FcB1)	Marti, Eparvier, Litaudon et al. 2010
	Butyraxanthones A and B, mangostanin, 1,3,6-trihydroxy-7-methoxy-2,8-diprenylxanthone, rubraxanthone, garcinone E, gartanin	*Pentadesma butyracea*	IC_{50} = 6.27, 5.84, 4.65, 7.07, 8.29, 6.03, and 7.82 μM, respectively (FcB1)	Zelefack et al. 2009
	Pentadexanthone, cratoxylone, α-mangostin, garcinone E	*Pentadesma butryacea*	IC_{50} = 3, 2.9, 2.8, and 0.43 μM, respectively (W2)	Lenta et al. 2011

(*Continued*)

Table 3.6 (Continued) Compounds Isolated from Plants That Are Reported to Have Antimalarial Activity

Subdivision of Class Structure	Reported Active Constituent	Plant Species	Remarks (*P. falciparum* Strains, Other *Plasmodium* Species, Targets)	References
	α-Mangostin, β-mangostin, 3-isomangostin	*Garcinia mangostana*	IC_{50} = 11.40, 7.42, and 7.88 μM, respectively (D6), 10.20, 4.71, and 6.15 μM, respectively (W2)	Lyles et al. 2014
Coumarins				
	Marmesinin, magnolioside	*Angelica gigas*	IC_{50} = 5.3 and 8.2 μM (D10)	Moon et al. 2011
	1,2-seco-dihydromethylumbelliferone methyl ester	*Toddalia asiatica*	IC_{50} = 17.4 μM (K1)	Phatchana and Yenjai 2014
Azaphilones				
	Longirostrerones A–C	*Chaetomium longirostre*	IC_{50} = 0.63, 3.73, and 0.62 μM, respectively (K1)	Panthama et al. 2011
Glycosides				
	Poliothrysoside	*Flacourtia indica*	IC_{50} = 7 ± 1 μM (W2)	Kaou, Mahiou-Leddet, Canlet, Debrauwer, Hutter, Laget et al. 2010
	Homaloside D	*Flacourtia indica*	IC_{50} = 20 ± 3 μM (W2)	
	Specicoside	*Kigelia africana*	IC_{50} = 1.54 μM (W2)	Zofou, Kengne et al. 2011
	Cupacinoside, 6-de-O-acetylcupacinoside	*Molinaea retusa*	IC_{50} = 4.0 and 6.4 μM (Dd2)	Eaton et al. 2013
	Datiscoside, datiscosides I-O, datiscoside B	*Datisca glomerata*	IC_{50} range from 7.7 to 33.3 μM (D10)	Graziose et al. 2013
	Luteoside A	*Tecoma mollis*	45% inhibition against D6	Abdel-Mageed et al. 2012

(Continued)

Table 3.6 (Continued) Compounds Isolated from Plants That Are Reported to Have Antimalarial Activity

Subdivision of Class Structure	Reported Active Constituent	Plant Species	Remarks (*P. falciparum* Strains, Other *Plasmodium* Species, Targets)	References
	4-O-(3′-Methylgalloyl) norbergenin, 4-O-galloylnorbergenin, 11-O-p-hydroxy-benzoyl-norbergenin	*Diospyros sanza-minika*	IC_{50} = 1.25, 8.36, and 11.3 µM, respectively (K1)	Tangmouo et al. 2010
	Jacaglabrosides A–D	*Jacaranda glabra*	IC_{50} = 1.42, 0.94, 0.76, and 0.75 µM, respectively (K1)	Gachet et al. 2010
	Robustasides G and D	*Grevillea* ("Poorinda Queen")	IC_{50} range from 2.1 to 5.5 and 3.0 to 5.5 µM, respectively, against D6, 3D7, Dd2, K1, TM90-C2B, TM93-C1088	Ovenden et al. 2011
	Dianellin	*Kniphofia foliosa*	IC_{50} = 10.4 µM (D6), 6.2 µM (W2)	Induli et al. 2013
		Phloroglucinols		
Phloroglucinol derivatives	Otogirone, erectquione B	*Hypericum erectum*	IC_{50} = 5.6 and 7.2 µM, respectively (D10)	Moon 2010
Dimeric phloroglucinols	Mallotojaponins B and C	*Mallotus oppositifolius*	IC_{50} = 0.75 and 0.14 µM, respectively (Dd2)	Harinantenaina et al. 2013
Polycyclic polyprenylated acylphloroglucinols	7-epi-Isogarcinol, 14-deoxy-7-epi-isogarcinol, symphonones A-I, 7-epi-coccinone B, 7-epi-garcinol	*Symphonia globulifera*	IC_{50} range from 2.1 to 10.1 µM (FcB1)	Marti, Eparvier, Moretti et al. 2010
		Lactones		
Khellactones	(+)-4-decanoyl-cis-khellactone, (+)-3-decanoyl-cis-khellactone	*Angelica purpuraefolia*	IC_{50} = 1.5 and 2.4 µM, respectively (D10)	Chung, Ghimire et al. 2010

(*Continued*)

Table 3.6 (Continued) Compounds Isolated from Plants That Are Reported to Have Antimalarial Activity

Subdivision of Class Structure	Reported Active Constituent	Plant Species	Remarks (*P. falciparum* Strains, Other *Plasmodium* Species, Targets)	References
Pyrones	Lippialactone	*Lippia javanica*	IC_{50} = 24.7 μM (D10)	Ludere, Ree, and Vleggaar 2013
		Sterols		
	Stigmasterol	*Bixa orellana*	EC_{50} < 3 μM (3D7 and K1)	Zhai et al. 2014
	25-(Acetyloxy)-2-(β-d-glucopyranosyloxy)-3,16-dihydroxy-9-methyl-19-norlanosta-5,23-dien-22-one	*Picrorhiza scrophulariiflora*	IC_{50} = 8.3 μM (3D7)	Wang et al. 2013
	3-O-[β-Glucopyranosyl(1→2)-O-β-xylopyranosyl]-stigmasterol	*Caesalpinia volkensii*	IC_{50} = 4.44 and 2.74 μM (D6 and W2, respectively)	Ochieng et al. 2013
		Miscellaneous		
β-triketones	Watsonianones A–C	*Corymbia watsoniana*	IC_{50} = 5.3, 0.29, and 1.07 μM, respectively, against 3D7; 8.8, 0.44, and 1.18 μM, respectively, against Dd2	Carroll et al. 2013
Depside	Atranorin	*Kigelia africana*	IC_{50} = 4.41 μM (W2)	Zofou, Kengne et al. 2011
Spiro heterocycles	Decarboxyportentol acetate, 3,4-dehydrotheaspirone	*Laumoniera bruceadelpha*	IC_{50} = 16 and 0.027 μM, respectively (3D7)	Morita et al. 2012
Anacardic acids	6-(8′Z-Pentadecenyl)-salicylic acid, 6-(8′Z,11′Z, 14′Z-heptadecatrienyl)-salicylic acid	*Viola websteri*	IC_{50} = 10.1 and 13.3 μM, respectively (D10)	Lee, Park, and Moon 2009
Tocopherols	δ-Tocotrienol	*Bixa orellana*	EC_{50} < 3 μM (3D7) and <12 μM (K1)	Zhai et al. 2014
Resorcinols	Malabaricone A	*Knema glauca*	IC_{50} = 8.5 μM (K1)	Rangkaew et al. 2009

(*Continued*)

Table 3.6 (Continued) Compounds Isolated from Plants That Are Reported to Have Antimalarial Activity

Subdivision of Class Structure	Reported Active Constituent	Plant Species	Remarks (*P. falciparum* Strains, Other *Plasmodium* Species, Targets)	References
Trinorcadalenes	Parviflorals B and F	*Decaschistia parviflora*	IC_{50} = 11.45 and 6.85 µM, respectively (K1)	Wongsa et al. 2013
Polyphenols	Geraniin	*Phyllanthus muellerianus*	IC_{50} = 11.9 µM (3D7)	Ndjonka et al. 2012
	Ellagic acid, gallic acid, gentisic acid	*Anogeissus leiocarpus*	IC_{50} = 0.87, 12.13, and 7.67 µM, respectively (3D7)	
Lanostanes	Ganoderic acid TR, ganoderic acid TR 1, ganoderic aldehyde TR, ganoderic acid S, ganoderic acid DM, ganodermanondiol, 23-hydroxyganoderic acid S	*Ganoderma lucidum*	IC_{50} = 20, 18, 6, 11, 13, and 11 µM, respectively (strain not mentioned)	Adams et al. 2010
Chromone	Anhydrobarakol	*Cassia siamea*	IC_{50} = 2.3 µM (3D7)	Oshimi et al. 2009
Benzophenones	Guttiferone E, isoxanthochymol, guttiferone H	*Garcinia xanthochymus*	IC_{50} = 7.90, 6.9, and 5.31 µM, respectively (D6); 7.47, 7.90, and 5.31 µM, respectively (W2)	Lyles et al. 2014
	Bipendensin	*Entandrophragma congoënse*	IC_{50} = 24.5 µM (NF54)	Happi et al. 2015

(*Continued*)

Table 3.6 (Continued) Compounds Isolated from Plants That Are Reported to Have Antimalarial Activity

Subdivision of Class Structure	Reported Active Constituent	Plant Species	Remarks (*P. falciparum* Strains, Other *Plasmodium* Species, Targets)	References
	(2S)-1,2-di-O-[(9Z)-Octadeca-9-enoyl]-3-O-β-d-galactopyranosyl glycerol, (2S)-1,2-di-O-[(9Z,12Z,15Z)-octadeca-9,12,15-trienoyl]-3-O-(6-sulpho-α-D) quinovopyranosyl glycerol, 1-O-β-d-glucopyranosyl-(2S,3R,8E)-2-[(2′R)-2-hydroxy-palmitoylamino]-8-octadecene-1,3-diol, 3-O-β-d-glucopyranosyl-3,4-dihydroxybenzoic acid	*Conyza sumatrensis*	IC_{50} = 44.3, 21.9, 23.3, and 79.1 μM (NF54)	Boniface et al. 2015
	Neonthrene	*Neoboutonia macrocalyx*	IC_{50} = 36.3 μM (FcB1/Colombia strain)	Namukobe et al. 2014
	Myristic acid	*Murraya koenigii*	IC_{50} = 46.0 μM (3D7)	Kamaraj et al. 2014
	Pipyahyine	*Beilschmiedia zenkeri*	IC_{50} = 3.7 μM (W2)	Lenta et al. 2009
	2-Iopropenyl-6-acetyl-8-methoxy-1,3-benzodioxin-4-one	*Carpesium divaricatum*	IC_{50} = 2.3 μM (D10)	Chung, Seo et al. 2010
	Methyl p-hydroxy benzoate	*Peltostigma guatemalense*	80% inhibition at 35 ppm (strain of *P. falciparum* not specified)	Cuca Suarez et al. 2009
	2,3,6-Trihydroxy benzoic acid, 2,3,6-trihydroxy methyl benzoate	*Sorindeia juglandifolia*	IC_{50} = 16.5 and 13.0 μM, respectively (W2)	Kamkumo et al. 2012

(Continued)

Table 3.6 (Continued) Compounds Isolated from Plants That Are Reported to Have Antimalarial Activity

Subdivision of Class Structure	Reported Active Constituent	Plant Species	Remarks (*P. falciparum* Strains, Other *Plasmodium* Species, Targets)	References
	Fimbricalyx B, fimbricalyxanhydride A, and fimbricalyx A	*Strophioblachia fimbricalyx*	IC_{50} = 0.019, 5.7, and 3.9 μM, respectively (K1)	Seephonkai et al. 2013
	Cryptobeilic acids A–D, tsangibeilin B	*Beilschmiedia cryptocaryoides*	IC_{50} = 17.7, 5.35, 14, 10.8, and 8.2 μM, respectively (NF52)	Talontsi et al. 2013
	Physalins B, F, and G	*Physalis angulata*	IC_{50} = 2.8, 2.2, and 6.7 μM (W2)	Sá et al. 2011
	Z-Antiepilepsirine	*Piper capense*	IC_{50} = 27 μM (W2)	Kaou, Mahiou-Leddet, Canlet, Debrauwer, Hutter, Azas et al. 2010
	Phoyunnanin E, densiflorol B, phoyunnanin C, gigantol, batatasin III	*Dendrobium venustum*	IC_{50} = 1.1, 1.3, 5.8, 12.2, and 39.3 μM, respectively (K1)	Sukphan et al. 2014

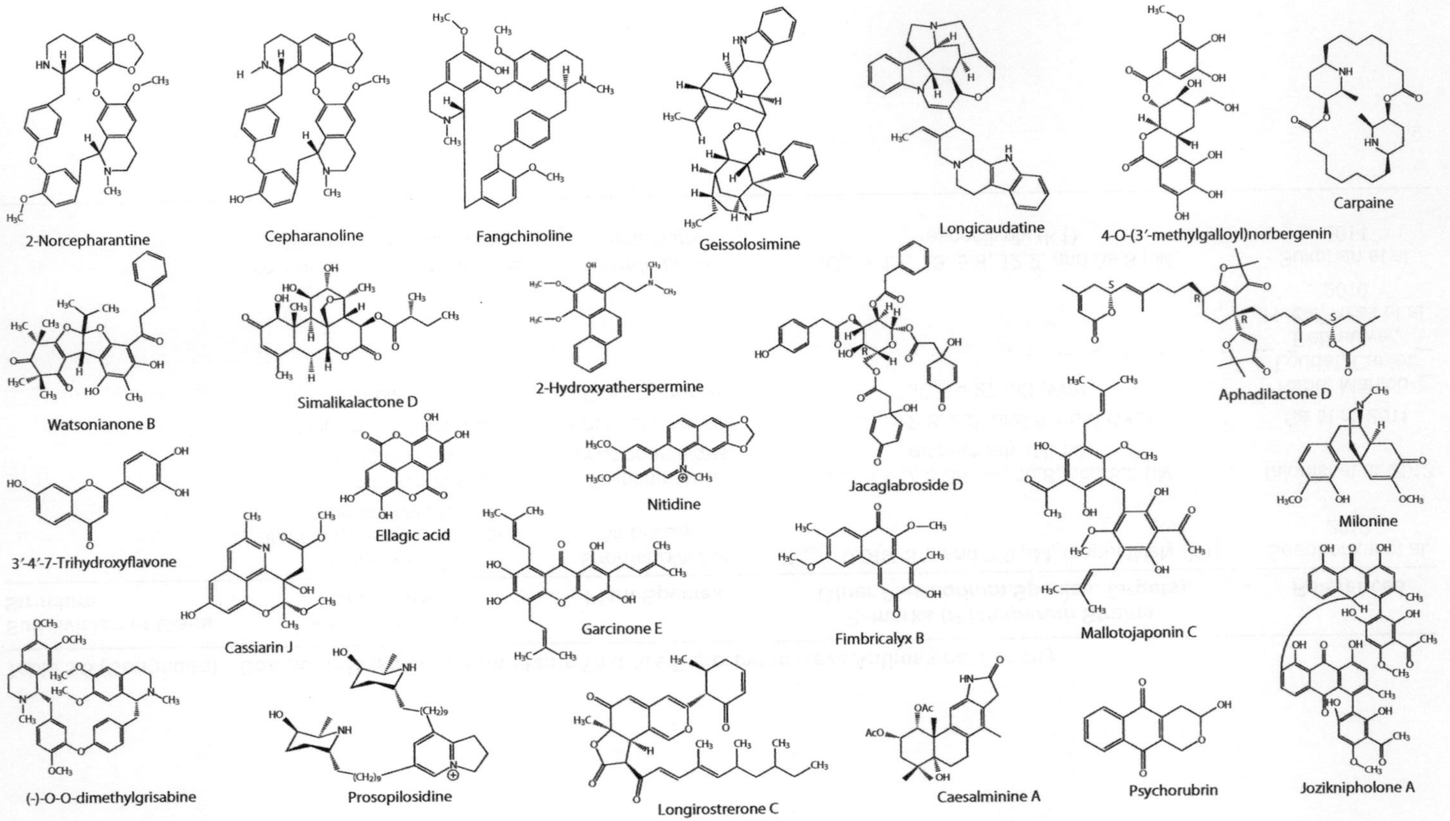

Figure 3.4 Structures of compounds with IC_{50} values of less than 1 μM. Where there is more than one very potent isomer, only the most potent isomer is shown.

3.5 CONSERVATION OF MEDICINAL PLANTS

3.5.1 Introduction

Worldwide, about 77,000 plant species are used as medicinal plants (Hedberg 1993). This represents only 25.8% of the predicted total plant species on earth (Mora et al. 2011). In a survey of 200 medicinal plant users in Singapore (Siew et al. 2014), as many as 102 species of plants were used for general health purposes. Although not all plants may prove beneficial to humans, there remains a significant portion that is under-researched and, hence, not fully harnessed for their potential benefits to humans, be it to improve health or for other purposes. With the growing popularity of traditional medicine and the increasing resistance of diseases such as malaria to allopathic drugs (see Chapter 1), medicinal plant consumption is not likely to decrease in the short to medium term (Smith-Hall, Larsen, and Pouliot 2012). Concerted action for their conservation must be put in place to ensure a continuing supply of these plants for the future, especially in light of the fact that many of such plants are being threatened (Uwe, Leaman, and Cunningham 2002). There is a need to conserve medicinal plants in the long run to ensure sustainability, especially when these plants are collected from the wild. Information gained from many years of research in medicinal plants has significantly contributed to scientific advancements in the field of medicine. As expounded earlier in Chapter 3, many of the current medications used in allopathic medicine have been derived indirectly or directly from plant sources. An estimated 25% to 30% of Western medicines have been derived from higher plants (Rao et al. 2012). In 1980, 25% of dispensed prescription drugs in the United States had at least one ingredient developed from plants (Hedberg 1993). Newman and Cragg found that 64% of new drugs from 1981 to 2010 are either derived directly from or inspired by natural products (Cragg and Newman 2013; Newman and Cragg 2012). In the same study, seven out of nine antimalarial drugs were either natural products or derived from natural products. Besides being important sources of novel phytochemicals, many medicinal plants are also used as raw materials in folk remedies and traditional medicine.

3.5.2 Brief History

Although the history of modern plant conservation long precedes the Chiang Mai Declaration in 1988, this international conference that convened politicians, health experts, and plant conservationists from around the world was an important milestone in the history of medicinal plant conservation. It led to the development of the "Guidelines on the Conservation of Medicinal Plants" published jointly by WHO, the International Union for Conservation of Nature (IUCN), and the World Wide Fund for Nature (WWF) in 1993. This set the stage for further developments in this field as exemplified by the Bangalore Declaration in 1998 and also related guidelines that were subsequently published (Shanker 1998; WHO 1998, 2003). The 1993 guidelines are due for a revision (Kathe 2006). However, to the best of our knowledge, no newer guidelines have been published.

3.5.3 Methods for Plant Conservation

Conservation methods have generally been divided into two categories: *ex situ* and *in situ* conservation. *In situ* conservation involves preserving the plants in environments in which they are "subject to continuing selection pressures" (Frankel, Brown, and Burdon 1995). An example would be the creation of national reserves or parks where the medicinal plants are allowed to grow in their native habitats. It has been highlighted, however, that such methods, albeit with good intentions, might not work if the attitudes and livelihood of the local indigenous population are not taken into consideration. An outright ban and strict laws against harvesting in these areas might lead to conflicts, as was the case when the Bwindi Impenetrable forest in Uganda was designated as a national park in 1991 (Blomkey 2003). Hence, it is suggested that the heart of successful conservation programs lie in having a robust management system that takes into account of the needs of various stakeholders from producers to consumers at the site where these medicinal plants are found (Hamilton 2004).

Acting as a complement to *in situ* conservation, *ex situ* conservation involves conserving the genetic resources of plants at sites distinct from their habitats. For example, medicinal plants of interest may be planted in botanical gardens. *Ex situ* conservation may also involve seed, pollen, or DNA storage, and biotechnological techniques such as cryopreservation and tissue culture (Kasagana and Karumuri 2011). In particular, plant tissue culture is touted to be able to produce genetically identical clones to cater to the needs of consumers. However, the ability of this technique to create plants with similar bioactive compounds as their counterparts in the wild has been called into question (Nair and Ganapathi 1998; Uwe, Leaman, and Cunningham 2002), although recent research (Moyo et al. 2013; White, Davies-Coleman, and Ripley 2008) revealed that it might be possible to produce greenhouse plants with comparable bioactive quality as wild plants. Regardless of the nature of such conservation methods, it is important that the factors that threaten the survival of the plants in the wild be removed or lessened; if not, the contributions of *ex situ* conservation to *in situ* conservation can be limited (Hamilton 2004).

3.5.4 Conservation Status of Plants Used in Malaria Treatment

Although recommendations made for medicinal plant conservation strategies may vary between different organizations and guidelines, there is an overall agreement that there is an increasing loss of medicinal plants due to unsustainable harvesting. In addition, climate change and uncontrolled population growth in forest lands leading to habitat loss might also contribute to a decrease in the supply of these plants (Rao et al. 2012). Although few medicinal plant species are known to have become extinct, it has been estimated that between 4160 and 10,000 medicinal plants are globally threatened (Hamilton 2004; Uwe, Leaman, and Cunningham 2002). It is believed that many medicinal plants are facing severe genetic losses, although detailed information is lacking (IUCN, WHO, and WWF 1993). There is also limited information on the plants used, their genetic diversity and distribution, and, most important, the annual sustained yield of plants that can be harvested without being detrimental to the growing wild population (Uwe, Leaman, and Cunningham 2002).

The first medicinal plant believed to be extinct globally is Silphium, a plant in the genus *Ferula* (its exact identity remains debatable) prized by the ancient Greeks for use in maladies such as cough, sore throat, and fever (Parejko 2003). Of note is the mention of its use in quartan fever (periodic fevers recurring every 72 hours, somewhat characteristic of malaria caused by *P. malariae*), as advised by ancient Greek physicians (Haas 2008). In the years that followed, reports of other medicinal plants being extinct are few and far between (Hamilton 2004). To the best of our knowledge, there has been no mention of the global extinction of a plant species vital in the treatment of malaria. Although this is reassuring for now, it is important that conservation of useful medicinal plants, including antimalarial plants, on the national and even regional level is ensured and not overlooked. Table 3.7 shows the criteria and trade implications for plant and animal species listed in the Convention on International Trade in Endangered Species of Wild Fauna and Flora (CITES).

Information on the conservation status of plants used in malaria treatment is scarce, and little has been explicitly explored. As such, it is of interest to review whether any of the medicinal plants reported to have antimalarial activity are listed in Table A.1 of the Appendix. Comparison of Table A.1 to species of medicinal plants listed in the CITES appendices (CITES 2015) reveals that there are indeed medicinal plants with reported antimalarial properties that are being threatened. Table 3.8 shows the plants with reported use in malaria treatment that have been included in the CITES appendices.

Using *Prunus africana* as an example, it was included in the CITES appendix II in 1995 and is also listed as "vulnerable" in both the IUCN Red List (IUCN 2014) and the Red List of South African Plants (South Africa National Biodiversity Institute 2014). Otherwise known as the African cherry, its bark has been used in the treatment of benign prostate hyperplasia, malaria, and other disorders. At the peak of its international trade, its bark extracts, which are entirely wild collected, were estimated to be worth approximately $4.36 million a year. Worldwide exports of its dried bark had been estimated at 1350 to 1525 metric tons per year, peaking at 3225 tons in 1997 (Stewart 2003). However, due to overexploitation, *Prunus africana* was depleted and this led to the closure of its extraction factory in Cameroon by Plantecam, a French company that was at that time one of the main exporters of the bark (Stewart 2003).

Table 3.7 Criteria and Trade Implications for Plant and Animal Species Listed in CITES Appendices I, II, and III

Appendix	Criteria and Trade Implications
I	Includes species threatened with extinction where their trade is only permitted under exceptional circumstances.
II	Includes species not necessarily threatened with extinction, but in which trade must be controlled in order to avoid utilization incompatible with their survival.
III	Contains species that are protected in at least one country, which has asked CITES for assistance in controlling the trade.

Table 3.8 Plants with Reported Use in Malaria That Have Been Included in the CITES Appendices

Plant Species	Family	CITES Appendix
Aloe species[a]	Xanthorrhoeaceae	I and II
Diospyros species[b]	Ebenaceae	II
Gnetum species[c]	Gnetaceae	III
Luisia teres	Orchidaceae	II
Panax ginseng	Araliaceae	II
Picrorhiza kurrooa	Scrophulariaceae	II
Prunus africana	Rosaceae	II
Rauvolfia serpentine	Apocynaceae	II
Swietenia macrophylla	Meliaceae	II
Swietenia mahagoni	Meliaceae	II

[a] Except *aloe vera*.
[b] Populations of Madagascar.
[c] Only populations of *Gnetum montanu* in Nepal.

In East Africa, several important antimalarial trees and shrubs such as *Ekebergia capensi* and *Harrisonia abyssinica* are reported to be threatened due to overexploitation (Dharani et al. 2010). A common sight once in the wet montane forests of Kenya and Tanzania, the plant *Ocotea usambarensis* has also become increasingly rare (Dharani et al. 2010). Its roots and bark are used to treat malaria. In Uganda, four key antimalarial tree species, namely, *Hallea rubrostipula*, *Warburgia ugandensis*, *Zanthoxylum chalybeum*, and *Syzygium guineense*, have been reported to be threatened by extensive debarking (Galabuzi et al. 2015). In particular, it was found that *H. rubrostipulata* suffers medium to high damage from bark harvest and has a low recruitment and is, hence, particularly vulnerable. Trees with known antimalarial properties that also play an important role in providing raw materials for timber production are also at stake. For instance, the African wild olive (*Olea europaea* ssp. *africana*) had been logged extensively for lumber, and restorative methods were explored for its conservation as the number of trees dwindled in Ethiopia (Negash 2003).

Elsewhere in India, *Acorus calamus* or calamus is another plant with antiplasmodial properties that is under threat (Sharma, Singh, and Chaudhary 2014). It is widely used in the treatment of dysentery, fever, chronic diarrhea, and fever. Although listed in the IUCN Red List as "least concern," this species is reported to be "vulnerable" in Tamil Nadu and "endangered" in Kerala on the National Red Lists database (National Red List 2000). Unsustainable harvesting of this plant from the wild has led to a decrease in the natural population in the wild. Currently, plant tissue culture is being investigated as a conservation strategy for this plant (Sharma, Singh, and Chaudhary 2014).

Clearly, there remains many more plants with antimalarial properties that are being threatened, but research in this area is lacking. It is imperative that important medicinal plants for malaria that are vulnerable or at the brink of extinction be swiftly identified so that suitable conservation strategies can be put in place to safeguard their continuous survival and sustainability, while harnessing their health benefits.

REFERENCES

Abdalla, M.A., and H. Laatsch. 2011. "Flavonoids from Sudanese *Albizia zygia* (Leguminosae, subfamily Mimosoideae), a plant with antimalarial potency." *African Journal of Traditional, Complementary, and Alternative Medicine* 9 (1):56–58.

Abdel-Mageed, W.M., E.Y. Backheet, A.A. Khalifa, Z.Z. Ibraheim, and S.A. Ross. 2012. "Antiparasitic antioxidant phenylpropanoids and iridoid glycosides from *Tecoma mollis*." *Fitoterapia* 83:500–507.

Adams, M., M. Christen, I. Plitzko, S. Zimmermann, R. Brun, M. Kaiser, and M. Hamburger. 2010. "Antiplasmodial lanostanes from the *Ganoderma lucidum* mushroom." *Journal of Natural Products* 73:897–900.

Adams, M., S. Gschwind, S. Zimmermann, M. Kaiser, and M. Hamburger. 2011. "Renaissance remedies: Antiplasmodial protostane triterpenoids from *Alisma plantago-aquatica* L. (Alismataceae)." *Journal of Ethnopharmacology* 135:43–47.

Aguiar, A.C.C., E.M.M.D. Rocha, N.B.D. Souza, T.C.C. França, and A.U. Krettli. 2012. "New approaches in antimalarial drug discovery and development: A review." *Memórias do Instituto Oswaldo Cruz* 107 (7):831–845.

Ajaiyeoba, E.O., O.O. Ogbole, O.O. Abiodun, J.S. Ashidi, P.J. Houghton, and C.W. Wright. 2013. "Cajachalcone: An antimalarial compound from *Cajanus cajan* leaf extract." *Journal of Parasitology Research* 2013: Article ID 703781, 5 pages. doi:10.1155/2013/703781.

Akendengue, B. 1992. "Medicinal plants used by the Fang traditional healers in Equatorial Guinea." *Journal of Ethnopharmacology* 37:165–173.

Attioua, B., D. Yeo, L. Lagnika, R. Harisolo, C. Antheaume, B. Weniger, M. Kaiser, A. Lobstein, and C. Vonthron-Sénécheau. 2012. "In vitro antileishmanial, antiplasmodial and cytotoxic activities of a new ventiloquinone and five known triterpenes from Parinari excelsa." *Pharmaceutical Biology* 50:801–806.

Awe, S.O., O.A. Olajide, O.O. Oladiran, and J.M. Makinde. 1998. "Antiplasmodial and antipyretic screening of *Mangifera indica* extract." *Phytotherapy Research* 12 (6):437–438.

Baghdikian, B., V. Mahiou-Leddet, S. Bory, S.S. Bun, A. Dumetre, F. Mabrouki, S. Hutter, N. Azas, and E. Ollivier. 2013. "New antiplasmodial alkaloids from *Stephania rotunda*." *Journal of Ethnopharmacology* 145:381–385.

Bankeu, J.J.K., R. Khayala, B.N. Lenta, D.T. Noungou, S.A. Ngouela, S.A. Mustafa, K. Asaad et al. 2011. "Isoflavone dimers and other bioactive constituents from the figs of *Ficus mucuso*." *Journal of Natural Products* 74:1370–1378.

Banzouzi, J.T., R. Prado, H. Menan, A. Valentin, C. Roumestan, M. Mallié, Y. Pelissier, and Y. Blach. 2004. "Studies on medicinal plants of Ivory Coast: Investigation of *Sida acuta* for in vitro antiplasmodial activities and identification of an active constituent." *Phytomedicine* 11 (4):338–341.

Basco, L.K., and J. Le Bras. 1994. "Short-term in vitro culture of *Plasmodium vivax* and *P. ovale* for drug-susceptibility testing." *Parasitology Research* 80:262–264.

Becker, J.V.W., M.M. van der Merwe, A.C. van Brummelen, P. Pillay, B.G. Crampton, E.M. Mmutlane, C. Parkinson et al. 2011. "In vitro anti-plasmodial activity of *Dicoma anomala* subsp. *gerrardii* (Asteraceae): Identification of its main active constituent, structure-activity relationship studies and gene expression profiling." *Malaria Journal* 10:295.

Benoit-Vical, F., A. Valentin, B. Da, Z. Dakuyo, L. Descamps, and M. Mallié. 2003. "N'Dribala (*Cochlospermum planchonii*) versus chloroquine for treatment of uncomplicated *Plasmodium falciparum* malaria." *Journal of Ethnopharmacology* 89:111–114.

Bero, J., M.-F. Hérent, G. Schmeda-Hirschmann, M. Frédérich, and J. Quetin-Leclercq. 2013. "In vivo antimalarial activity of *Keetia leucantha* twigs extracts and in vitro antiplasmodial effect of their constituents." *Journal of Ethnopharmacology* 149 (1):176–183.

Bertani, S., E. Houel, V. Jullian, G. Bourdy, A. Valentin, D. Stien, and E. Deharo. 2012. "New findings on Simalikalactone D, an antimalarial compound from *Quassia amara* L. (Simaroubaceae)." *Experimental Parasitology* 130 (4):341–347.

Blomkey, T. 2003. "Natural resource conflict management: The case of Bwindi Impenetrable and Mgahinga Gorilla National Parks, southwestern Uganda." *CARE International Uganda* 231–250.

Boniface, P.K., S. Verma, A. Shukla, H.S. Cheema, S.K. Srivastava, F. Khan, M.P. Darokar, and A. Pal. 2015. "Bioactivity-guided isolation of antiplasmodial constituents from *Conyza sumatrensis* (Retz.) E.H. Walker." *Parasitology International* 64 (1):118–123.

Bourjot, M., C. Apel, M.T. Martin, P. Grellier, V.H. Nguyen, F. Guéritte, and M. Litaudon. 2010. "Antiplasmodial, antitrypanosomal, and cytotoxic activities of prenylated flavonoids isolated from the stem bark of *Artocarpus styracifolius*." *Planta Medica* 76:1600–1604.

Buchanan, M.S., R.A. Davis, S. Duffy, V.M. Avery, and R.J. Quinn. 2009. "Antimalarial benzylisoquinoline alkaloid from the rainforest tree *Doryphora sassafras*." *Journal of Natural Products* 72:1541–1543.

Bugyei, K.A., G.L. Boye, and M.E. Addy. 2010. "Clinical efficacy of a tea-bag formulation of *Cryptolepis sanguinolenta* root in the treatment of acute uncomplicated falciparum malaria." *Ghana Medical Journal* 44:3–9.

Carroll, A.R., V.M. Avery, S. Duffy, P.I. Forster, and G.P. Guymer. 2013. "Watsonianone A-C, anti-plasmodial beta-triketones from the Australian tree, *Corymbia watsoniana*." *Organic and Biomolecular Chemistry* 11:453–458.

Challand, S., and M. Willcox. 2009. "A clinical trial of the traditional medicine *Vernonia amygdalina* in the treatment of uncomplicated malaria." *Journal of Alternative and Complementary Medicine* 15:1231–1237.

Chea, A., S.-S. Bun, N. Azas, M. Gasquet, S. Bory, E. Ollivier, and R. Elias. 2010. "Antiplasmodial activity of three bisbenzylisoquinoline alkaloids from the tuber of *Stephania rotunda*." *Natural Product Research* 24:1766–1770.

Chen, Y., S. Li, F. Sun, H. Han, X. Zhang, Y. Fan, G. Tai, and Y. Zhou. 2010. "In vivo antimalarial activities of glycoalkaloids isolated from Solanaceae plants." *Pharmaceutical Biology* 48:1018–1024.

Chierrito, T.P.C, A.C.C. Aguiar, I.M. de Andrade, I.P. Ceravolo, R.A.C. Gonçalves, A.J.B. de Oliveira, and A.U. Krettli. 2014. "Anti-malarial activity of indole alkaloids isolated from *Aspidosperma olivaceum*." *Malaria Journal* 13:142.

Chukwujekwu, J.C., C.A. De Kock, P.J. Smith, F.R. Van Heerden, and J. Van Staden. 2012. "Antiplasmodial and antibacterial activity of compounds isolated from *Ormocarpum trichocarpum*." *Planta Medica* 78:1857–1860.

Chung, I.-M., and H.-I. Moon. 2009. "Antiplasmodial activities of sesquiterpene lactone from *Carpesium cernum*." *Journal of Enzyme Inhibition and Medicinal Chemistry* 24:131–135.

Chung, I.-M., B.K. Ghimire, E.Y. Kang, and H.I. Moon. 2010a. "Antiplasmodial and cytotoxic activity of khellactone derivatives from *Angelica purpuraefolia* Chung." *Phytotherapy Research* 24:469–471.

Chung, I.-M., S.-H. Seo, E.-Y. Kang, W.-H. Park, S.-D. Park, and H.-I. Moon. 2010b. "Antiplasmodial activity of isolated compounds from *Carpesium divaricatum*." *Phytotherapy Research* 24:451–453.

CITES. 2015. "Appendices I, II and III." Accessed April 22, 2015. http://www.cites.org/eng/app/appendices.php.

Claudino, V.D., K.C. da Silva, V. Cechinel Filho, R.A. Yunes, F. Delle Monache, A. Giménez, E. Salamanca, D. Gutierrez-Yapu, and A. Malheiros. 2013. "Drimanes from *Drimys brasiliensis* with leishmanicidal and antimalarial activity." *Memórias do Instituto Oswaldo Cruz* 108:140–144.

Coppi, A., M. Cabinian, D. Mirelman, and P. Sinnis. 2006. "Antimalarial activity of allicin, a biologically active compound from garlic cloves." *Antimicrobial Agents and Chemotherapy* 50:1731–1737.

Cos, P., A.J. Vlietinck, D. Vanden Berghe, and L. Maes. 2006. "Anti-infective potential of natural products: How to develop a stronger in vitro 'proof-of-concept.'" *Journal of Ethnopharmacology* 106:290–302.

Cragg, G.M., and D.J. Newman. 2013. "Natural products: A continuing source of novel drug leads." *Biochimica et Biophysica Acta* 1830 (6):3670–3695.

Cranmer, S.L., C. Magowan, J. Liang, R.L. Coppel, and B.M. Cooke. 1997. "An alternative to serum for cultivation of *Plasmodium falciparum* in vitro." *Transactions of the Royal Society of Tropical Medicine and Hygiene* 91:363–365.

Cretton, S., G. Glauser, M. Humam, D. Jeannerat, O. Muñoz, L. Maes, P. Christen, and K. Hostettmann. 2010. "Isomeric tropane alkaloids from the aerial parts of *Schizanthus tricolor*." *Journal of Natural Products* 73:844–847.

Cuca Suarez, L.E., M.E. Pattarroyo, J.M. Lozano, and F. Delle Monache. 2009. "Biological activity of secondary metabolites from *Peltostigma guatemalense*." *Natural Product Research* 23:370–374.

da Silvo Filho, A.A., D.O. Resende, M.J. Fukui, F.F. Santos, P.M. Pauletti, W.R. Cunha, M.L.A. Silva, L.E. Gregório, J.K. Bastos, and N.P.D. Nanayakkara. 2009. "In vitro antileishmanial, antiplasmodial and cytotoxic activities of phenolics and triterpenoids from *Baccharis dracunculifolia* D. C. (Asteraceae)." *Fitoterapia* 80:478–482.

Deguchi, J., T. Hirahara, Y. Hirasawa, W. Ekasari, A. Widyawaruyanti, O. Shirota, M. Shiro, and H. Morita. 2012. "New tricyclic alkaloids, Cassiarins G, H, J, and K from leaves of *Cassia siamea*." *Chemical & Pharmaceutical Bulletin* 60:219–222.

Dharani, N., G. Rukunga, A. Yenesew, A. Mbora, L. Mwaura, I. Dawson, and R. Jamnadass. 2010. *Common Antimalarial Trees and Shrubs of East Africa: A Description of Species and a Guide to Cultivation and Conservation through Use*. Kenya: The World Agroforestry Centre.

Dhooghe, L., S. Maregesi, I. Mincheva, D. Ferreira, J.P.J. Marais, F. Lemière, A. Matheeussen et al. 2010. "Antiplasmodial activity of (I-3,II-3)-biflavonoids and other constituents from *Ormocarpum kirkii*." *Phytochemistry* 71 (7):785–791.

dos Santos Torres, Z.E., E.R. Silveira, L.F. Rocha e Silva, E.S. Lima, M.C. de Vasconcellos, D.E. de Andrade Uchoa, R.B. Filho, and A.M. Pohlit. 2013. "Chemical composition of *Aspidosperma ulei* Markgr. and antiplasmodial activity of selected indole alkaloids." *Molecules (Basel, Switzerland)* 18:6281–6297.

Duke, J.A. 1996. "Dr. Duke's Phytochemical and Ethnobotanical Database." Accessed September 10, 2014. http://www.ars-grin.gov/duke/.

Eaton, A.L., L. Harinantenaina, P.J. Brodie, M.B. Cassera, J.D. Bowman, M.W. Callmander, R. Randrianaivo et al. 2013. "A new bioactive diterpene glycoside from *Molinaea retusa* from the Madagascar dry forest." *Natural Product Communications* 8 (9):1201–1203.

Endale, M., J.P. Alao, H.M. Akala, N.K. Rono, F.L. Eyase, S. Derese, A. Ndakala et al. 2012. "Antiplasmodial quinones from *Pentas longiflora* and *Pentas lanceolata*." *Planta Medica* 78 (1):31–35.

Endale, A., D. Bisrat, A. Animut, F. Bucar, and K. Asres. 2013. "In vivo antimalarial activity of a labdane diterpenoid from the leaves of *Otostegia integrifolia* Benth." *Phytotherapy Research* 27 (12):1805–1809.

Fidock, D.A., P.J. Rosenthal, S.L. Croft, R. Brun, and S. Nwaka. 2004. "Antimalarial drug discovery: Efficacy models for compound screening." *Nature Reviews Drug Discovery* 3:509–520.

Flannery, E.L., A.K. Chatterjee, and E.A. Winzeler. 2013. "Antimalarial drug discovery—Approaches and progress towards new medicines." *Nature Reviews Microbiology* 11:849–862.

Frankel, O.H., A.H.D. Brown, and J.J. Burdon. 1995. *The Conservation of Plant Biodiversity.* Cambridge: Cambridge University Press.

Gachet, M.S., O. Kunert, M. Kaiser, R. Brun, R.A. Muñoz, R. Bauer, and W. Schühly. 2010. "Jacaranone-derived glucosidic esters from Jacaranda glabra and their activity against *Plasmodium falciparum*." *Journal of Natural Products* 73:553–556.

Galabuzi, C., G.N. Nabanoga, P. Ssegawa, J. Obua, and G. Eilu. 2015. "Double jeopardy bark harvest for malaria treatment and poor regeneration threaten tree population in a tropical forest of Uganda." *African Journal of Ecology* 53 (2):214–222.

Ganfon, H., J. Bero, A.T. Tchinda, F. Gbaguidi, J. Gbenou, M. Moudachirou, M. Frédérich, and J.L. Quetin-Leclercq. 2012. "Antiparasitic activities of two sesquiterpenic lactones isolated from *Acanthospermum hispidum* D.C." *Journal of Ethnopharmacology* 141:411–417.

Gbeassor, M., A.Y. Kedjagni, K. Koumaglo, C.D. Souza, K. Agbo, K. Aklikokou, and K.A. Amegbo. 1990. "In vitro antimalarial activity of six medicinal plants." *Phytotherapy Research* 4 (3):115–117.

Gego, A., O. Silvie, J.F. Franetich, K. Farhati, L. Hannoun, A.J.F. Luty, R.W. Sauerwein, C. Boucheix, E. Rubinstein, and D. Mazier. 2006. "New approach for high-throughput screening of drug activity on *Plasmodium* liver stages." *Antimicrobial Agents and Chemotherapy* 50:1586–1589.

Girardot, M., C. Deregnaucourt, A. Deville, L. Dubost, R. Joyeau, L. Allorge, P. Rasoanaivo, and L. Mambu. 2012. "Indole alkaloids from *Muntafara sessilifolia* with antiplasmodial and cytotoxic activities." *Phytochemistry* 73:65–73.

Grace, M.H., C. Lategan, F. Mbeunkui, R. Graziose, P.J. Smith, I. Raskin, and M.A. Lila. 2010. "Antiplasmodial and cytotoxic activities of drimane sesquiterpenes from *Canella winterana*." *Natural Product Communications* 5:1869–1872.

Graz, B., M.L. Willcox, C. Diakite, J. Falquet, F. Dackuo, O. Sidibe, S. Giani, and D. Diallo. 2010. "*Argemone mexicana* decoction versus artesunate-amodiaquine for the management of malaria in Mali: Policy and public-health implications." *Transactions of the Royal Society of Tropical Medicine and Hygiene* 104 (1):33–41.

Graziose, R., P. Rojas-Silva, T. Rathinasabapathy, C. Dekock, M.H. Grace, A. Poulev, M.A. Lila, P. Smith, and I. Raskin. 2012. "Antiparasitic compounds from *Cornus florida* L. with activities against *Plasmodium falciparum* and *Leishmania tarentolae*." *Journal of Ethnopharmacology* 142 (2):456–461.

Graziose, R., M.H. Grace, T. Rathinasabapathy, P. Rojas-Silva, C. Dekock, A. Poulev, M.A. Lila, P. Smith, and I. Raskin. 2013. "Antiplasmodial activity of cucurbitacin glycosides from *Datisca glomerata* (C. Presl) Baill." *Phytochemistry* 87:78–85.

Haas, C. 2008. "[Silphium from Cyrenaica, an extinct medicinal plant]." *Bulletin de L'Académie Nationale de Médecine* 192 (1):153–160.

Hamilton, A.C. 2004. "Medicinal plants, conservation and livelihoods." *Biodiversity and Conservation* 13 (8):1477–1517.

Hao, B., S.F. Shen, and Q.J. Zhao. 2013. "Cytotoxic and antimalarial amaryllidaceae alkaloids from the bulbs of *Lycoris radiata*." *Molecules* 18:2458–2468.
Happi, G.M., S.F. Kouam, F.M. Talontsi, S. Zuhlke, M. Lamshoft, and M. Spiteller. 2015. "Minor secondary metabolites from the bark of *Entandrophragma congoënse* (Meliaceae)." *Fitoterapia* 102:35–40.
Harinantenaina, L., J.D. Bowman, P.J. Brodie, C. Slebodnick, M.W. Callmander, E. Rakotobe, R. Randrianaivo et al. 2013. "Antiproliferative and antiplasmodial dimeric phloroglucinols from *Mallotus oppositifolius* from the Madagascar dry forest." *Journal of Natural Products* 76:388–393.
Hedberg, I. 1993. "Botanical methods in ethnopharmacology and the need for conservation of medicinal plants." *Journal of Ethnopharmacology* 38 (2–3):121–128.
Hemtasin, C., S. Kanokmedhakul, K. Kanokmedhakul, C. Hahnvajanawong, K. Soytong, S. Prabpai, and P. Kongsaeree. 2011. "Cytotoxic pentacyclic and tetracyclic aromatic sesquiterpenes from *Phomopsis archeri*." *Journal of Natural Products* 74:609–613.
Henchiri, H., B. Bodo, A. Deville, L. Dubost, L. Zourgui, A. Raies, P. Grellier, and L. Mambu. 2009. "Sesquiterpenoids from *Teucrium ramosissimum*." *Phytochemistry* 70:1435–1441.
Hou, Y., S. Cao, P.J. Brodie, M.W. Callmander, F. Ratovoson, E.A. Rakotobe, V.E. Rasamison et al. 2009. "Antiproliferative and antimalarial anthraquinones of *Scutia myrtina* from the Madagascar forest." *Bioorganic and Medicinal Chemistry* 17:2871–2876.
Houël, E., S. Bertani, G. Bourdy, E. Deharo, V. Jullian, A. Valentin, S. Chevalley, and D. Stiena. 2009. "Quassinoid constituents of *Quassia amara* L. leaf herbal tea. Impact on its antimalarial activity and cytotoxicity." *Journal of Ethnopharmacology* 126 (1):114–118.
Ijaz, F., N. Ahmad, I. Ahmad, A. ul Haq, and F. Wang. 2010. "Two new anti-plasmodial flavonoid glycosides from *Duranta repens*." *Journal of Enzyme Inhibition and Medicinal Chemistry* 25:773–778.
Induli, M., M. Gebru, N. Abdissa, H. Akala, I. Wekesa, R. Byamukama, M. Heydenreich, S. Murunga, E. Dagne, and A. Yenesew. 2013. "Antiplasmodial quinones from the rhizomes of *Kniphofia foliosa*." *Natural Product Communications* 8 (9):1261–1264.
Innok, P., T. Rukachaisirikul, and A. Suksamrarn. 2009. "Flavanoids and pterocarpans from the bark of *Erythrina fusca*." *Chemical & Pharmaceutical Bulletin* 57:993–996.
International Union for Conservation of Nature (IUCN). 2014. "IUCN Red List of Threatened Species." Accessed March 8, 2015. http://www.iucnredlist.org/.
International Union for Conservation of Nature (IUCN), World Health Organization (WHO), and World Wide Fund for Nature (WWF). 1993. "Guidelines on the conservation of medicinal plants." Accessed March 8, 2015. http://www.who.int/medicinedocs/documents/s7150e/s7150e.pdf.
Jansen, O., L. Angenot, M. Tits, J.P. Nicolas, P.D. Mol, J. Nikiéma, and M. Frédérich. 2010. "Evaluation of 13 selected medicinal plants from Burkina Faso for their antiplasmodial properties." *Journal of Ethnopharmacology* 130 (1):143–150.
Jonville, M.C., G. Dive, L. Angenot, J. Bero, M. Tits, E. Ollivier, and M. Frédérich. 2013. "Dimeric bisindole alkaloids from the stem bark of *Strychnos nux-vomica* L." *Phytochemistry* 87:157–163.
Julianti, T., M. De Mieri, S. Ebrahimi, M. Neuburger, S. Zimmermann, M. Kaiser, and M. Hamburger. 2013. "Potent antiplasmodial agents in *Carica papaya* L." *Planta Medica* 79-SL6.
Julianti, T., M. De Mieri, S. Zimmermann, S.N. Ebrahimi, M. Kaiser, M. Neuburger, M. Raith, R. Brun, and M. Hamburger. 2014. "HPLC-based activity profiling for antiplasmodial compounds in the traditional Indonesian medicinal plant *Carica papaya* L." *Journal of Ethnopharmacology* 155 (1):426–434.

Kamaraj, C., A.A. Rahuman, S.M. Roopan, A. Bagavan, G. Elango, A.A. Zahir, G. Rajakumar, C. Jayaseelan, T. Santhoshkumar, S. Marimuthu, and A.V. Kirthi. 2014. "Bioassay-guided isolation and characterization of active antiplasmodial compounds from *Murraya koenigii* extracts against *Plasmodium falciparum* and *Plasmodium berghei*." *Parasitology Research* 113 (5):1657–1672.

Kamkumo, R.G., A.M. Ngoutane, L.R.Y. Tchokouaha, P.V.T. Fokou, E.A.K. Madiesse, J. Legac, J.J.B. Kezetas et al. 2012. "Compounds from *Sorindeia juglandifolia* (Anacardiaceae) exhibit potent anti-plasmodial activities in vitro and in vivo." *Malaria Journal* 11:382.

Kaou, A.M., V. Mahiou-Leddet, C. Canlet, L. Debrauwer, S. Hutter, N. Azas, and E. Ollivier. 2010. "New amide alkaloid from the aerial part of *Piper capense* L.f. (Piperaceae)." *Fitoterapia* 81:632–635.

Kaou, A.M., V. Mahiou-Leddet, C. Canlet, L. Debrauwer, S. Hutter, M. Laget, R. Faure, N. Azas, and E. Ollivier. 2010. "Antimalarial compounds from the aerial parts of *Flacourtia indica* (Flacourtiaceae)." *Journal of Ethnopharmacology* 130:272–274.

Karaket, N., K. Supaibulwatana, S. Ounsuk, V. Bultel-Ponce, V.C. Pham, and B. Bodo. 2012. "Chemical and bioactivity evaluation of the bark of *Neonauclea purpurea*." *Natural Product Communications* 7:169–170.

Karunamoorthi, K., and E. Tsehaye. 2012. "Ethnomedicinal knowledge, belief and self-reported practice of local inhabitants on traditional antimalarial plants and phytotherapy." *Journal of Ethnopharmacology* 141:143–150.

Kasagana, V.N., and S.S. Karumuri. 2011. "Conservation of medicinal plants (past, present & future trends)." *Journal of Pharmaceutical Science and Research* 3 (8):1378–1386.

Kathe, W. 2006. "Revision of the 'Guidelines on the Conservation of Medicinal Plants' by WHO, IUCN, WWF and TRAFFIC: Process and Scope." *Frontis* 17:109–120.

Kaur, K., M. Jain, T. Kaur, and R. Jain. 2009. "Antimalarials from nature." *Bioorganic & Medicinal Chemistry* 17 (9):3229–3256.

Kebenei, J.S., P.K. Ndalut, and A.O. Sabah. 2011. "Anti-plasmodial activity of nortrachelogenin from the root bark of *Carissa edulis*." *International Journal of Applied Research in Natural Products* 4 (3):1–5.

Kennedy, M., M.E. Fishbaugher, A.M. Vaughan, R. Patrapuvich, R. Boonhok, N. Yimamnuaychok, N. Rezakhani et al. 2012. "A rapid and scalable density gradient purification method for *Plasmodium* sporozoites." *Malaria Journal* 11:421.

Kerubo, L.O., J.O. Midiwo, S. Derese, M.K. Langat, H.M. Akala, N.C. Waters, M. Peter, and M. Heydenreich. 2013. "Antiplasmodial activity of compounds from the surface exudates of *Senecio roseiflorus*." *Natural Product Communications* 8:175–176.

Kouam, S.F., A.W. Ngouonpe, M. Lamshoft, F.M. Talontsi, J.O. Bauer, C. Strohmann, B.T. Ngadjui, H. Laatsch, and M. Spiteller. 2014. "Indolosesquiterpene alkaloids from the Cameroonian medicinal plant *Polyalthia oliveri* (Annonaceae)." *Phytochemistry* 105:52–59.

Kuypers, K., P. Cos, E. Ortega-Baria, D. Vanden Berghe, and L. Maes. 2006. "Bioassays for some parasitic protozoa screening concepts and standard in vitro and in vivo laboratory model." In *Biological Screening of Plant Constituents*, edited by M.P. Gupta, S.S. Handa and K. Vasisht, 7–18. Trieste: International Centre for Science and High Technology.

Lacroix, D., S. Prado, D. Kamoga, J. Kasenene, and B. Bodo. 2011. "Structure and in vitro antiparasitic activity of constituents of *Citropsis articulata* root bark." *Journal of Natural Products* 74:2286–2289.

Lakshmi, V., S. Srivastava, S.K. Mishra, M.N. Srivastava, K. Srivastava, and S.K. Puri. 2012. "Antimalarial activity in *Xylocarpus granatum* (Koen)." *Natural Product Research* 26:1012–1015.

Lansky, E.P., and H.M. Paavilainen. 2010. *Figs: The Genus Ficus*. Boca Raton, FL: CRC Press.

Lategan, C.A., W.E. Campbell, T. Seaman, and P.J. Smith. 2009. "The bioactivity of novel furanoterpenoids isolated from *Siphonochilus aethiopicus*." *Journal of Ethnopharmacology* 121 (1):92–97.

Lee, S.J., W.H. Park, and H.I. Moon. 2009. "Bioassay-guided isolation of antiplasmodial anacardic acids derivatives from the whole plants of *Viola websteri* Hemsl." *Parasitology Research* 104:463–466.

Lekphrom, R., S. Kanokmedhakul, and K. Kanokmedhakul. 2010. "Bioactive diterpenes from the aerial parts of *Anisochilus harmandii*." *Planta Medica* 76:726–728.

Lenta, B.N., F. Tantangmo, K.P. Devkota, J.D. Wansi, J.R. Chouna, R.C.F. Soh, B. Neumann, H.G. Stammler, E. Tsamo, and N. Sewald. 2009. "Bioactive constituents of the stem bark of *Beilschmiedia zenkeri*." *Journal of Natural Products* 72:2130–2134.

Lenta, B.N., L.M. Kamdem, S.R. Ngouela, F. Tantangmo, K.P. Devkota, F.F. Boyom, P.J. Rosenthal, and E. Tsamo. 2011. "Antiplasmodial constituents from the fruit pericarp of *Pentadesma butyracea*." *Planta Medica* 77:377–379.

Levrier, C., M. Balastrier, K.D. Beattie, A.R. Carroll, F. Martin, V. Choomuenwai, and R.A. Davis. 2013. "Pyridocoumarin, aristolactam and aporphine alkaloids from the Australian rainforest plant *Goniothalamus australis*." *Phytochemistry* 86:121–126.

Liu, J., X.F. He, G.H. Wang, E.F. Merino, S.P. Yang, R.X. Zhu, L.S., Gan, H. Zhang, M.B. Cassera, H.Y. Wang, D.G. Kingston, and J.M. Yue. 2014a. "Aphadilactones A-D, four diterpenoid dimers with DGAT inhibitory and antimalarial activities from a Meliaceae plant." *Journal of Organic Chemistry* 79 (2):599–607.

Liu, Y., L.H. Rakotondraibea, P.J. Brodie, J.D. Wiley, M.B. Cassera, M. Goetzc, and D.G. Kingstona. 2014b. "Antiproliferative and antimalarial sesquiterpene lactones from *Piptocoma antillana* from Puerto Rico." *Natural Product Communications* 9 (10):1403–1406.

Lu, Z., R.M. Van Wagoner, C.D. Pond, A.R. Pole, J.B. Jensen, D. Blankenship, B.T. Grimberg, R. Kiapranis, T.K. Matainaho, L.R. Barrows, and C.M. Ireland. 2014. "Myristicyclins A and B: Antimalarial procyanidins from *Horsfieldia spicata* from Papua New Guinea." *Organic Letters* 16 (2):346–349.

Ludere, M.T., T.V. Ree, and R. Vleggaar. 2013. "Isolation and relative stereochemistry of lippialactone, a new antimalarial compound from *Lippia javanica*." *Fitoterapia* 86:188–192.

Lyles, J.T., A. Negrin, S.I. Khan, K. He, and E.J. Kennelly. 2014. "In vitro antiplasmodial activity of benzophenones and xanthones from edible fruits of *Garcinia* species." *Planta Medica* 80 (8–9):676–681.

Ma, G., Z. Sun, Z. Sun, J. Yuan, H. Wei, J. Yang, H. Wu, and X. Xu. 2014. "Antimalarial diterpene alkaloids from the seeds of *Caesalpinia minax*." *Fitoterapia* 95:234–239.

Malann, Y.D., B.M. Matur, and S. Mailafia. 2013. "Evaluation of the antimalarial activity of aqueous leaf extracts of *Casuarina equistifolia* and *Mangifera indica* against *Plasmodium berghei* in mice." *Journal of Pharmacy & Bioresources* 10 (1).

Malann, Y.D., B.M. Matur, and E.S. Akinnagbe. 2014. "Antiplasmodial activity of extracts and fractions of *Mangifera indica* against *Plasmodium berghei*." *Nigerian Journal of Parasitology* 35 (1–2).

Malleret, B., C. Claser, A.S.M. Ong, R. Suwanarusk, K. Sriprawat, S.W. Howland, B. Russell, F. Nosten, and L. Rénia. 2011. "A rapid and robust tri-color flow cytometry assay for monitoring malaria parasite development." *Scientific Reports* 1:118.

Maneerat, W., T. Ritthiwigrom, S. Cheenpracha, U. Prawat, and S. Laphookhieo. 2011. "Clausenawallines A and B, two new dimeric carbazole alkaloids from the roots of *Clausena wallichii*." *Tetrahedron Letters* 52:3303–3305.

March, S., S. Ng, S. Velmurugan, A. Galstian, J. Shan, D.J. Logan, A.E. Carpenter et al. 2013. "A microscale human liver platform that supports the hepatic stages of *Plasmodium falciparum* and *vivax*." *Cell Host and Microbe* 14:104–115.

Marti, G., V. Eparvier, M. Litaudon, P. Grellier, and F. Guéritte. 2010. "A new xanthone from the bark extract of *Rheedia acuminata* and antiplasmodial activity of its major compounds." *Molecules* 15 (10):7106–7114.

Marti, G., V. Eparvier, C. Moretti, S. Prado, P. Grellier, N. Hue, O. Thoison, B. Delpech, F. Guéritte, and M. Litaudon. 2010. "Antiplasmodial benzophenone derivatives from the root barks of *Symphonia globulifera* (Clusiaceae)." *Phytochemistry* 71:964–974.

Mba'Ning, B.M., B.N. Lenta, D.T. Noungoué, C. Antheaume, Y.F. Fongang, S.A. Ngouela, F.F. Boyom et al. 2013. "Antiplasmodial sesquiterpenes from the seeds of *Salacia longipes* var. *camerunensis*." *Phytochemistry* 96:347–352.

Mbeunkui, F., M.H. Grace, C. Lategan, P.J. Smith, I. Raskin, and M.A. Lila. 2012. "In vitro antiplasmodial activity of indole alkaloids from the stem bark of *Geissospermum vellosii*." *Journal of Ethnopharmacology* 139 (2):471–477.

Mesia, K., L. Tona, M.M. Mampunza, N. Ntamabyaliro, T. Muanda, T. Muyembe, K. Cimanga et al. 2012. "Antimalarial efficacy of a quantified extract of *Nauclea pobeguinii* stem bark in human adult volunteers with diagnosed uncomplicated falciparum malaria. Part 1: A clinical Phase IIA trial." *Planta Medica* 78 (3):211–218.

Mesia, K., L. Tona, M. Mampunza, N. Ntamabyaliro, T. Muanda, T. Muyembe, T. Musuamba et al. 2012. "Antimalarial efficacy of a quantified extract of *Nauclea pobeguinii* stem bark in human adult volunteers with diagnosed uncomplicated falciparum malaria. Part 2: A clinical Phase IIB trial." *Planta Medica* 78 (9):853–860.

Milliken, W. 1997. *Plants for Malaria, Plants for Fever: Medicinal Species in Latin America—A Bibliographic Survey*. United Kingdom: Whitstable Litho.

Moon, H.I. 2010. "Antiplasmodial and cytotoxic activity of phloroglucinol derivatives from *Hypericum erectum* Thunb." *Phytotherapy Research* 24:941–944.

Moon, H.-I., J.-H. Lee, Y.-C. Lee, and K.-S. Kim. 2011. "Antiplasmodial and cytotoxic activity of coumarin derivatives from dried roots of *Angelica gigas* Nakai in vitro." *Immunopharmacology and Immunotoxicology* 33 (4):663–666.

Mora, C., D.P. Tittensor, S. Adl, A.G. Simpson, and B. Worm. 2011. "How many species are there on Earth and in the ocean?" *PLoS Biology* 9 (8):e1001127.

Morita, H., R. Mori, J. Deguchi, S. Oshimi, Y. Hirasawa, W. Ekasari, A. Widyawaruyanti, and A.H.A. Hadi. 2012. "Antiplasmodial decarboxyportentol acetate and 3,4-dehydrotheaspirone from *Laumoniera bruceadelpha*." *Journal of Natural Medicines* 66:571–575.

Moyo, M., A.O. Aremu, J. Gruz, M. Subrtova, L. Szucova, K. Dolezal, and J. Van Staden. 2013. "Conservation strategy for *Pelargonium sidoides* DC: Phenolic profile and pharmacological activity of acclimatized plants derived from tissue culture." *Journal of Ethnopharmacology* 149 (2):557–561.

Mueller, M.S., N. Runyambo, I. Wagner, S. Borrmann, K. Dietz, and L. Heide. 2004. "Randomized controlled trial of a traditional preparation of *Artemisia annua* L. (annual wormwood) in the treatment of malaria." *Transactions of the Royal Society of Tropical Medicine and Hygiene* 98:318–321.

Mueller, D., R.A. Davis, S. Duffy, V.M. Avery, D. Camp, and R.J. Quinn. 2009. "Antimalarial activity of azafluorenone alkaloids from the Australian tree *Mitrephora diversifolia*." *Journal of Natural Products* 72 (8):1538–1540.

Muganga, R., L. Angenot, M. Tits, and M. Frederich. 2014. "In vitro and in vivo antiplasmodial activity of three Rwandan medicinal plants and identification of their active compounds." *Planta Med* 80 (6):482–489.

Muñoz, V., M. Sauvain, G. Bourdy, J. Callapa, I. Rojas, L. Vargas, A. Tae, and E. Deharo. 2000. "The search for natural bioactive compounds through a multidisciplinary approach in Bolivia. Part II. Antimalarial activity of some plants used by Mosetene indians." *Journal of Ethnopharmacology* 69 (2):139–155.

Muthaura, C.N., J.M. Keriko, S. Derese, A. Yenesew, and G.M. Rukunga. 2011. "Investigation of some medicinal plants traditionally used for treatment of malaria in Kenya as potential sources of antimalarial drugs." *Experimental Parasitology* 127 (3):609–626.

Nair, M.N.B., and N. Ganapathi. 1998. *Medicinal Plants: CURE for the 21st Century*. Malaysia: Faculty of Forestry, University Putra Malaysia.

Namsa, N.D., M. Mandal, and S. Tangjang. 2011. "Anti-malarial herbal remedies of northeast India, Assam: An ethnobotanical survey." *Journal of Ethnopharmacology* 133:565–572.

Namukobe, J., B.T. Kiremire, R. Byamukama, J.M. Kasenene, V. Dumontet, F. Gueritte, S. Krief, I. Florent, and J.D. Kabasa. 2014. "Cycloartane triterpenes from the leaves of *Neoboutonia macrocalyx* L." *Phytochemistry* 102:189–196.

Nanyingi, M.O., J.M. Mbaria, A.L. Lanyasunya, C.G. Wagate, K.B. Koros, H.F. Kaburia, R.W. Munenge, and W.O. Ogara. 2008. "Ethnopharmacological survey of Samburu district, Kenya." *Journal of Ethnobiology and Ethnomedicine* 4:14.

Nasrullah, A.A., A. Zahari, J. Mohamad, and K. Awang. 2013. "Antiplasmodial alkaloids from the bark of *Cryptocarya nigra* (Lauraceae)." *Molecules (Basel, Switzerland)* 18:8009–8017.

National Red List. 2000. "*Acorus calamus*." Accessed April 24, 2015. http://www.nationalredlist.org/species-information/?speciesID=111652.

Ndjonka, D., B. Bergmann, C. Agyare, F.M. Zimbres, K. Lüersen, A. Hensel, C. Wrenger, and E. Liebau. 2012. "In vitro activity of extracts and isolated polyphenols from West African medicinal plants against *Plasmodium falciparum*." *Parasitology Research* 111 (2):827–834.

Negash, L. 2003. "Vegetative propagation of the threatened African wild olive [*Olea europaea* L. subsp. *cuspidata* (Wall. ex DC.) Ciffieri]." *New Forests* 26 (2):137–146.

Newman, D.J., and G.M. Cragg. 2012. "Natural products as sources of new drugs over the 30 years from 1981 to 2010." *Journal of Natural Products* 75 (3):311–335.

Noedl, H., C. Wongsrichanalai, and W.H. Wernsdorfer. 2003. "Malaria drug-sensitivity testing: New assays, new perspectives." *Trends in Parasitology* 19 (4):175–181.

Nogueira, F., and V. Rosário. 2010. "Methods for assessment of antimalarial activity in the different phases of the *Plasmodium* life cycle." *Rev Pan-Amaz Saude* 1 (3):109–124.

Noungoue, D.T., M. Chaabi, S. Ngouela, C. Antheaume, F.F. Boyom, J. Gut, P.J. Rosenthal, A. Lobstein, and E. Tsamo. 2009. "Antimalarial compounds from the stem bark of *Vismia laurentii*." *Zeitschrift für Naturforschung C. (Journal for Nature Research)* 64c:210–214.

Nour, A.M.M., S.A. Khalid, M. Kaiser, R. Brun, W.L.E. Abdalla, and T.J. Schmidt. 2010. "The antiprotozoal activity of methylated flavonoids from *Ageratum conyzoides* L." *Journal of Ethnopharmacology* 129 (1):127–130.

Ochieng, C.O., L.A.O. Manguro, P.O. Owuor, and H. Akala. 2013. "Voulkensin C-E, new 11-oxocassane-type diterpenoids and a steroid glycoside from *Caesalpinia volkensii* stem bark and their antiplasmodial activities." *Bioorganic & Medicinal Chemistry Letters* 23 (10):3088–3095.

Olasehinde, G.I., O. Ojurongbe, A.O. Adeyeba, O.E. Fagade, N. Valecha, I.O. Ayanda, A.A. Ajayi, and L.O. Egwari. 2014. "In vitro studies on the sensitivity pattern of *Plasmodium falciparum* to anti-malarial drugs and local herbal extracts." *Malaria Journal* 13:63.

Ortet, R., S. Prado, E.L. Regalado, F.A. Valeriote, J. Media, J. Mendiola, and O.P. Thomas. 2011. "Furfuran lignans and a flavone from *Artemisia gorgonum* Webb and their in vitro activity against *Plasmodium falciparum*." *Journal of Ethnopharmacology* 138 (2):637–640.

Oshimi, S., J. Deguchi, Y. Hirasawa, W. Ekasari, A. Widyawaruyanti, T.S. Wahyuni, N.C. Zaini, O. Shirota, and H. Morita. 2009. "Cassiarins C–E, antiplasmodial alkaloids from the flowers of *Cassia siamea*." *Journal of Natural Products* 72 (10):1899–1901.

Ovenden, S.P.B., M. Cobbe, R. Kissell, G.W. Birrell, M. Chavchich, and M.D. Edstein. 2011. "Phenolic glycosides with antimalarial activity from *Grewillea* 'Poorinda Queen.'" *Journal of Natural Products* 74:74–78.

Pan, E., A.P. Gorka, J.N. Alumasa, C. Slebodnick, L. Harinantenaina, P.J. Brodie, P.D. Roepe, R. Randrianaivo, C. Birkinshaw, and D.G.I. Kingston. 2011. "Antiplasmodial and antiproliferative pseudoguaianolides of *Athroisma proteiforme* from the Madagascar dry forest." *Journal of Natural Products* 74:2174–2180.

Panthama, N., S. Kanokmedhakul, K. Kanokmedhakul, and K. Soytong. 2011. "Cytotoxic and antimalarial azaphilones from *Chaetomium longirostre*." *Journal of Natural Products* 74 (11):2395–2399.

Parejko, K. 2003. "Pliny the Elder's Silphium: First recorded species extinction." *Conservation Biology* 17 (3):925–927.

Pedersen, M.M., J.C. Chukwujekwu, C.A. Lategan, J.V. Staden, P.J. Smith, and D. Staerk. 2009. "Antimalarial sesquiterpene lactones from *Distephanus angulifolius*." *Phytochemistry* 70:601–607.

Perez, H.A., M. De la Rosa, and R. Apitz. 1994. "In vivo activity of ajoene against rodent malaria." *Antimicrobial Agents and Chemotherapy* 38:337–339.

Peters, W. 1975. "The chemotherapy of rodent malaria: XXII. The value of drug-resistant strains of *P. berghei* in screening for blood schizontocidal activity." *Annals of Tropical Medicine and Parasitology* 69 (2):155–171.

Phatchana, R., and C. Yenjai. 2014. "Cytotoxic coumarins from *Toddalia asiatica*." *Planta Medica* 80 (8–9):719–722.

Pudhom, K., D. Sommit, N. Suwankitti, and A. Petsom. 2007. "Cassane furanoditerpenoids from the seed kernels of *Caesalpinia bonduc* from Thailand." *Journal of Natural Products* 70:1542–1544.

Ramalhete, C., D. Lopes, S. Mulhovo, V.E. Rosário, and M.J.U. Ferreira. 2008. Antimalarial activity of some plants traditionally used in Mozambique. *Workshop Plantas Medicinais e Fitoterapêuticas nos Trópicos*. Accessed May 19, 2015. http://www2.iict.pt/archive/doc/C_Ramalhete_wrkshp_plts_medic.pdf.

Ramalhete, C., D. Lopes, S. Mulhovo, J. Molnár, V.E. Rosário, and M.-J.U. Ferreira. 2010. "New antimalarials with a triterpenic scaffold from *Momordica balsamina*." *Bioorganic and Medicinal Chemistry* 18:5254–5260.

Rangkaew, N., R. Suttisri, M. Moriyasu, and K. Kawanishi. 2009. "A new acyclic diterpene acid and bioactive compounds from *Knema glauca*." *Archives of Pharmacal Research* 32 (5):685–692.

Rao, B.R.R., K.V. Syamasundar, D.K. Rajput, G. Nagaraju, and G. Adinarayana. 2012. "Biodiversity, conservation and cultivation of medicinal plants." *Journal of Pharmacognosy* 3 (2):59–62.

Rasamison, V.E., L.H. Rakotondraibe, C. Slebodnick, P.J. Brodie, M. Ratsimbason, K. TenDyke, Y. Shen, L.M. Randrianjanaka, and D.G. Kingston. 2014. "Nitrogen-containing dimeric nor-multiflorane triterpene from a *Turraea* sp." *Organic Letters* 16 (10):2626–2629.

Rasoanaivo, P., A. Petitjean, S. Ratsimamanga-Urverg, and R. Rakoto. 1992. "Medicinal plants used to treat malaria in Madagascar." *Journal of Ethnopharmacology* 37:117–127.

Rasoanaivo, P., C.W. Wright, M.L. Willcox, and B. Gilbert. 2011. "Whole plant extracts versus single compounds for the treatment of malaria: Synergy and positive interactions." *Malaria Journal* 10 (Suppl. 1):S4.

Research Initiative on Traditional Antimalarial Methods (RITAM). 2013. "Home." Accessed October 20, 2014. http://giftsofhealth.org/ritam/.

Russell, B.M., R. Udomsangpetch, K.H. Rieckmann, B.M. Kotecka, R.E. Coleman, and J. Sattabongkot. 2003. "Simple in vitro assay for determining the sensitivity of Plasmodium vivax isolates from fresh human blood to antimalarials in areas where *P. vivax* is endemic." *Antimicrobial Agents and Chemotherapy* 47:170–173.

Sá, M.S., M.N. de Menezes, A.U. Krettli, I.M. Ribeiro, T.C.B. Tomassini, R.R. dos Santos, W.F. de Azevedo Jr., and M.B.P. Soares. 2011. "Antimalarial activity of physalins B, D, F, and G." *Journal of Natural Products* 74 (10):2269–2272.

Samoylenko, V., M.K. Ashfaq, M.R. Jacob, B.L. Tekwani, S.I. Khan, S.P. Manly, V.C. Joshi, L.A. Walker, and I. Muhammad. 2009. "Indolizidine, antiinfective and antiparasitic compounds from *Prosopis glandulosa* var. glandulosa." *Journal of Natural Products* 72:92–98.

Sathe, M., R. Ghorpade, A.K. Srivastava, and M.P. Kaushik. 2010. "In vivo antimalarial evaluation of gomphostenins." *Journal of Ethnopharmacology* 130:171–174.

Seephonkai, P., S.G. Pyne, A.C. Willis, and W. Lie. 2013. "Bioactive compounds from the roots of *Strophioblachia fimbricalyx*." *Journal of Natural Products* 76:1358–1364.

Shanker, D. 1998. "Declaration of the International Conference on Medicinal Plants Held at Bangalore, India." *Botanic Gardens Conservation International* 3 (1):46–47.

Sharma, V., I. Singh, and P. Chaudhary. 2014. "*Acorus calamus* (the healing plant): A review on its medicinal potential, micropropagation and conservation." *Natural Product Research* 28 (18):1454–1466.

Siew, Y.Y., S. Zareisedehizadeh, W.G. Seetoh, S.Y. Neo, C.H. Tan, and H.L. Koh. 2014. "Ethnobotanical survey of usage of fresh medicinal plants in Singapore." *Journal of Ethnopharmacology* 155 (3):1450–1466.

Smith-Hall, C., H.O. Larsen, and M. Pouliot. 2012. "People, plants and health: A conceptual framework for assessing changes in medicinal plant consumption." *Journal of Ethnobiology and Ethnomedicine* 8 (1):43.

Soh, P.N., B. Witkowski, A. Gales, E. Huyghe, A. Berry, B. Pipy, and F. Benoit-Vical. 2012. "Implication of glutathione in the in vitro antiplasmodial mechanism of action of ellagic acid." *Plos One* 7 (9):e45906.

Songsiang, U., S. Wanich, S. Pitchuanchom, S. Netsopa, K. Uanporn, and C. Yenjai. 2009. "Bioactive constituents from the stems of *Dalbergia parviflora*." *Fitoterapia* 80:427–431.

South Africa National Biodiversity Institute. 2014. "Red List of South African Plants." Accessed April 24, 2015. http://redlist.sanbi.org/.

Sripisut, T., and S. Laphookhieo. 2010. "Carbazole alkaloids from the stems of *Clausena excavata*." *Journal of Asian Natural Products Research* 12:614–617.

Stewart, K.M. 2003. "The African cherry (*Prunus africana*): Can lessons be learned from an over-exploited medicinal tree?" *Journal of Ethnopharmacology* 89 (1):3–13.

Sukphan, P., B. Sritularak, W. Mekboonsonglarp, V. Lipipun, and K. Likhitwitayawuid. 2014. "Chemical constituents of *Dendrobium venustum* and their antimalarial and anti-herpetic properties." *Natural Product Communications* 9 (6):825–827.

Sülsen, V., D.G. Yappu, L. Laurella, C. Anesini, A.G. Turba, V. Martino, and L. Muschietti. 2011. "In vitro antiplasmodial activity of sesquiterpene lactones from *Ambrosia tenuifolia*." *Evidence-Based Complementary and Alternative Medicine* 2011: Article ID 352938.

Sutthivaiyakit, S., W. Mongkolvisut, S. Prabpai, and P. Kongsaeree. 2009. "Diterpenes, sesquiterpenes, and a sesquiterpene-coumarin conjugate from *Jatropha integerrima*." *Journal of Natural Products* 72:2024–2027.

Talontsi, F.M., M. Lamshöft, J.O. Bauer, A.A. Razakarivony, B. Andriamihaja, C. Strohmann, and M. Spiteller. 2013. "Antibacterial and antiplasmodial constituents of *Beilschmiedia cryptocaryoides*." *Journal of Natural Products* 76:97–102.

Tangmouo, J.G., R. Ho, A. Matheeussen, A.M. Lannang, J. Komguem, B.B. Messi, L. Maes, and K. Hostettmann. 2010. "Antimalarial activity of extract and norbergenin derivatives from the stem bark of *Diospyros sanza-minika* A. Chevalier (Ebenaceae)." *Phytotherapy Research* 24 (11):1676–1679.

Tchinda, A.T., V. Fuendjiep, A. Sajjad, C. Matchawe, P. Wafo, S. Khan, P. Tane, and M.I. Choudhary. 2009. "Bioactive compounds from the fruits of *Zanthoxylum leprieurii*." *Pharmacologyonline* 1:406–415.

Tchinda, A.T., A.R.N. Ngono, V. Tamze, M.C. Jonville, M. Cao, L. Angenot, and M. Frédérich. 2012. "Antiplasmodial alkaloids from the stem bark of *Strychnos malacoclados*." *Planta Medica* 78:377–382.

Tchinda, A.T., O. Jansen, J.N. Nyemb, M. Tits, G. Dive, L. Angenot, and M. Frederich. 2014. "Strychnobaillonine, an unsymmetrical bisindole alkaloid with an unprecedented skeleton from *Strychnos icaja* roots." *Journal of Natural Products* 77 (4):1078–1082.

Thongthoom, T., U. Songsiang, C. Phaosiri, and C. Yenjai. 2010. "Biological activity of chemical constituents from *Clausena harmandiana*." *Archives of Pharmacal Research* 33:675–680.

Tona, L., K. Mesia, N.P. Ngimbi, B. Chrimwami, Okond'ahoka, K. Cimanga, T. de Bruyne et al. 2001. "In-vivo antimalarial activity of *Cassia occidentalis, Morinda morindoides* and *Phyllanthus niruri*." *Annals of Tropical Medicine and Parasitology* 95 (1):47–57.

Toyang, N.J., and R. Verpoorte. 2013. "A review of the medicinal potentials of plants of the genus *Vernonia* (Asteraceae)." *Journal of Ethnopharmacology* 146:681–723.

Toyang, N.J., M.A. Krause, R.M. Fairhurst, P. Tane, J. Bryant, and R. Verpoorte. 2013. "Antiplasmodial activity of sesquiterpene lactones and a sucrose ester from *Vernonia guineensis* Benth. (Asteraceae)." *Journal of Ethnopharmacology* 147 (3):618–621.

Trager, W., and J.B. Jensen. 1976. "Human malaria parasites in continuous culture." *Science* 193 (4254):673–675.

Traoré-Coulibaly, M., H.L. Ziegler, C.E. Olsen, M.-K. Hassanata, G.I. Pierre, O.G. Nacoulma, T.R. Guiguemdé, and S.B. Christensen. 2009. "19alpha-Hydroxy-3-oxo-ursa-1,12-dien-28-oic acid, an antiplasmodial triterpenoid isolated from *Canthium multiflorum*." *Natural Product Research* 23:1108–1111.

Uchôa, V.T., R.C. De Paula, L.G. Krettli, A.E.G. Santana, and A.U. Krettli. 2010. "Antimalarial activity of compounds and mixed fractions of *Cecropia pachystachya*." *Drug Development Research* 71 (1):82–91.

University of Illinois. 2012. "Natural Products Alert." Accessed March 20, 2014. http://www.napralert.org/.

UpToDate. 2014. "Treatment of severe falciparum malaria." Accessed April 10, 2014. http://www.uptodate.com/contents/treatment-of-severe-falciparum-malaria.

Uwe, S., D.J. Leaman, and A.B. Cunningham. 2002. "Impact of cultivation and gathering of medicinal plants on biodiversity: Global trends and issues." FAO. Accessed April 27, 2015. http://www.fao.org/docrep/005/aa010e/aa010e00.htm.

Vitalini, S., G. Beretta, M. Iriti, S. Orsenigo, N. Basilico, S. Dall'Acqua, M. Iorizzi, and G. Fico. 2011. "Phenolic compounds from *Achillea millefolium* L. and their bioactivity." *Acta Biochimica Polonica* 58:203–209.

Wang, H., W. Zhao, V. Choomuenwai, K.T. Andrews, R.J. Quinn, and Y. Feng. 2013. "Chemical investigation of an antimalarial Chinese medicinal herb *Picrorhiza scrophulariiflora*." *Bioorganic and Medicinal Chemistry Letters* 23:5915–5918.

Wangchuk, P., J.B. Bremner, Samten, B.W. Skelton, A.H. White, R. Rattanajak, and S. Kamchonwongpaisan. 2010. "Antiplasmodial activity of atisinium chloride from the Bhutanese medicinal plant, *Aconitum orochryseum*." *Journal of Ethnopharmacology* 130:559–562.

Wangchuk, P., P.A. Keller, S.G. Pyne, A.C. Willis, and S. Kamchonwongpaisan. 2012. "Antimalarial alkaloids from a Bhutanese traditional medicinal plant *Corydalis dubia*." *Journal of Ethnopharmacology* 143 (1):310–313.

Wells, N.C.T. 2011. "Natural products as starting points for future anti-malarial therapies: Going back to our roots?" *Malaria Journal* 10 (Suppl 1):S3.

White, A.G., M.T. Davies-Coleman, and B.S. Ripley. 2008. "Measuring and optimising umckalin concentration in wild-harvested and cultivated *Pelargonium sidoides* (Geraniaceae)." *South African Journal of Botany* 74 (2):260–267.

Willcox, M.L., and G. Bodeker. 2004. "Traditional herbal medicines for malaria." *British Medical Journal* 329:1156–1159.

Willcox, M., G. Bodeker, and P. Rasoanaivo. 2005. *Traditional Medicinal Plants and Malaria*. Vol. 59. Washington, DC: CRC Press.

Willcox, M.L., B. Graz, J. Falquet, O. Sidibé, M. Forster, and D. Diallo. 2007. "*Argemone mexicana* decoction for the treatment of uncomplicated falciparum malaria." *Transactions of the Royal Society of Tropical Medicine and Hygiene* 101:1190–1198.

Wongsa, N., S. Kanokmedhakul, K. Kanokmedhakul, P. Kongsaeree, S. Prabpai, and S.G. Pyne. 2013. "Parviflorals A–F, trinorcadalenes and bis-trinorcadalenes from the roots of *Decaschistia parviflora*." *Phytochemistry* 95:368–374.

World Health Organization (WHO). 1998. "Quality control methods for medicinal plant materials." Accessed April 26, 2015. http://whqlibdoc.who.int/publications/1998/9241545100.pdf.

World Health Organization (WHO). 2003. "WHO Guidelines on Good Agricultural and Collection Practice (GACP) for medicinal plants." Accessed April 25, 2015. http://whqlibdoc.who.int/publications/2003/9241546271.pdf.

World Health Organization (WHO). 2008a. "Fact sheet: Traditional medicine." Accessed June 9, 2014. http://www.who.int/mediacentre/factsheets/fs134/en/.

World Health Organization (WHO). 2008b. "Herbal medicine research and global health: An ethical analysis." Accessed April 25, 2015. http://www.who.int/bulletin/volumes/86/8/07-042820/en/.

Yang, X., Y. Feng, S. Duffy, V.M. Avery, D. Camp, R.J. Quinn, and R.A. Davis. 2011. "A new quinoline epoxide from the Australian plant *Drummondita calida*." *Planta Medica* 77:1644–1647.

Yenesew, A., H.M. Akala, H. Twinomuhwezi, C. Chepkirui, B.N. Irungu, F.L. Eyase, M. Kamatenesi-Mugisha, B.T. Kiremire, J.D. Johnson, and N.C. Waters. 2012. "The antiplasmodial and radical scavenging activities of flavonoids of *Erythrina burttii*." *Acta Tropica* 123:123–127.

Yenjai, C., S. Sripontan, P. Sriprajun, P. Kittakoop, A. Jintasirikul, M. Tanticharoen, and Y. Thebtaranonth. 2000. "Coumarins and carbazoles with antiplasmodial activity from *Clausena harmandiana*." *Planta Medica* 66 (3):277–279.

Yousif, S.A. 2014. "In vitro screening of antiplasmodium activity of *Momordica charantia*." *Journal of Natural Resources and Environmental Studies* 2 (3):29–33.

Zahari, A., F.K. Cheah, J. Mohamad, S.N. Sulaiman, M. Litaudon, K.H. Leong, and K. Awang. 2014. "Antiplasmodial and antioxidant isoquinoline alkaloids from *Dehaasia longipedicellata*." *Planta Medica* 80 (7):599–603.

Zelefack, F., D. Guilet, N. Fabre, C. Bayet, S. Chevalley, S. Ngouela, B.N. Lenta, A. Valentin, E. Tsamo, and M.G. Dijoux-Franca. 2009. "Cytotoxic and antiplasmodial xanthones from *Pentadesma butyracea*." *Journal of Natural Products* 72:954–957.

Zhai, B., J. Clark, T. Ling, M. Connelly, F. Medina-Bolivar, and F. Rivas. 2014. "Antimalarial evaluation of the chemical constituents of hairy root culture of *Bixa orellana* L." *Molecules* 19 (1):756–766.

Zofou, D., A.B.O. Kengne, M. Tene, M.N. Ngemenya, P. Tane, and V.P.K. Titanji. 2011. "In vitro antiplasmodial activity and cytotoxicity of crude extracts and compounds from the stem bark of *Kigelia africana* (Lam.) Benth (Bignoniaceae)." *Parasitology Research* 108 (6):1381–1390.

Zofou, D., T.K. Kowa, H.K. Wabo, M.N. Ngemenya, P. Tane, and V.P.K. Titanji. 2011. "*Hypericum lanceolatum* (Hypericaceae) as a potential source of new anti-malarial agents: A bioassay-guided fractionation of the stem bark." *Malaria Journal* 10:167.

CHAPTER 4

Selected Antimalarial Plants

4.1 ANTIMALARIAL PLANTS AS A SOURCE OF NOVEL THERAPEUTIC LEADS

This section discusses four antimalarial plants that have been well studied, and from which compounds have been isolated and found to be active against *Plasmodium*. Of note, artemisinin and quinine have been isolated from two of these herbs and undergone further development into clinically used antimalarial agents, for example, artesunate, artemether, arteether, chloroquine, and primaquine.

4.1.1 *Artemisia annua* L. (Compositae)

Also known as qinghao or sweet wormwood, *Artemisia annua* L. (Compositae) is indicated for fever, malaria, and jaundice in the Chinese Pharmacopoeia (Chinese Pharmacopoeia Commission 2010). Nobel Prize winner in Medicine 2015, Youyou Tu, led a team to search for novel malaria treatment and evaluated more than 380 extracts from about 200 Chinese herbs against a rodent malaria model. They showed that *A. annua* significantly inhibited parasite growth but the results were inconsistent (Tu 2011). The original reference they consulted that reported its use for malaria was in Ge Hong's *A Handbook of Prescriptions for Emergencies*, which states: "A handful of qinghao immersed with two liters of water, wring out the juice and drink it all." Extraction at lower temperature with ether resulted in much better activity. In 1971, the Chinese managed to isolate a nontoxic extract that was 100% effective against parasitemia in *P. berghei*–infected mice and *P. cynomolgi*–infected monkeys. Artemisinin, an endoperoxide sesquiterpene lactone, was subsequently identified.

4.1.1.1 *Mode of Action*

The exact mechanism of artemisinin remains to be resolved, although it has been postulated to be involved in interference of heme detoxification (Meshnick 2002). The endoperoxide moiety of artemisinin is cleaved when it interacts with the Fe(II) center of free heme released as the parasite digests hemoglobin in its food vacuole, resulting in formation of free radicals that go on to alkylate macromolecular targets in the parasite. Artemisinin has also been found to inhibit the *P. falciparum* SERCA ortholog PfATP6, leading to impaired folding of parasitic proteins and disruption of calcium homeostasis, causing parasitic death (Eckstein-Ludwig et al. 2003; Jambou et al. 2005; O'Neill, Barton, and Ward 2010). Other mechanisms of action elucidated thus far include immunostimulation and interference of mitochondrial electron transport. The latter causes depolarization of the mitochondrial membrane, leading to disruption of synthesis of pyrimidine, which is required for production of nucleic acid, eventually resulting in death of the parasite (Golenser et al. 2006).

4.1.1.2 Phytochemistry and Comparison between Preparations

A. annua contains nearly 600 secondary metabolites from a diverse range of chemical classes, predominantly terpenoids (sesquiterpene lactones, monoterpenoids, sesquiterpenoids, diterpenes, triterpenes, and sterols), phenylpropanoids, flavonoids, peptides, aliphatic hydrocarbons, aliphatic and aromatic alcohols, aldehydes, ketones, and acids (Brown 2010). The most active principle is the endoperoxide sesquiterpene artemisinin. Artemisinin content in this species ranges from 0.0006% to 1.4% w/w, based on the variety and cultivar. A study by Mannan et al. (2010) showed that *A. annua* leaves and flowers contained the greatest amount of artemisinin (about 0.4% w/w) compared to 13 other *Artemisia* species in Pakistan (Mannan et al. 2010). It is also reported to be present in shoots, stems, and roots of the plant.

The traditional method mentioned in the aforementioned Chinese text showed greater artemisinin concentrations but lower extraction efficiency than the infusion (Willcox 2011). The extract has been shown to be more potent than an equivalent dose of artemisinin. An aqueous infusion contains more artemisinin as compared to a decoction, as artemisinin is heat labile. This fact is also reflected in traditional Chinese medicine prescriptions for malaria treatment where in one source *A. annua* is to be added when the decoction is nearly done (Yan, Fischer, and Fratkin 1997).

Several methoxylated flavonoids such as chrysospenol-D from the plant have been found to enhance the activity of artemisinin *in vitro*, and some of these possess antimalarial activity on their own although to a smaller extent than artemisinin (Liu et al. 1992; Willcox 2009). Other constituents such as casticin and artemetin have similar synergistic effects (Weathers and Towler 2012). Artemisinin is relatively nonpolar and thus poorly water soluble, which explains its low oral bioavailability (approximately 30%) (Titulaer et al. 1990). However, bioavailability of artemisinin is greater in the crude extract as compared to the pure drug (Weathers et al. 2011). A recent study showed that oral administration of the dried whole plant in a rodent malaria model reduced parasitemia 12 to 72 hours after treatment as compared to pure artemisinin, possibly due to synergistic effects between artemisinin and the other constituents (e.g., flavonoids) in the plant, individual antimalarial activity of other constituents, and increased bioavailability attributed to inhibition of intestinal and hepatic cytochrome P450 (CYP450) enzymes by flavonoids or other compounds present (Elfawal et al. 2012).

4.1.1.3 Clinical Trials

Herbal infusions were most commonly used in six trials in Africa, where rates of parasite clearance were about 70% to 100%, with up to a 39% recrudescence rate (Willcox 2011; Wright 2005). In clinical trials in China, ethanol extracts were tested against *P. vivax* and were very effective in clearance of parasitemia and alleviation of malaria symptoms, including some chloroquine-resistant and cerebral malaria. However, recrudescence occurred in some of the patients 3 to 4 weeks after artemisinin treatment. This was due to the short half-life (ca. 2 hours) of the compound. The high recrudescence rate was also observed in a German trial, despite bringing about a prompt resolution of symptoms and parasitemia (Mueller et al. 2004).

4.1.1.4 Derivatives

Dihydroartemisinin, the active metabolite of artemisinins, was found to be more stable and effective than artemisinin, with lower recrudescence rates (Tu 2011). From dihydroartemisinin, ether (artemether, arteether) and ester (artesunate) derivatives have been synthesized to improve formulation and reduce recrudescence of the parent compound (Figure 4.1).

4.1.1.5 Place in Therapy

To reduce development of resistance as well as taking into account the short half-life of artemisinin, artemisinin and its derivatives should not be used as monotherapy for malaria but instead with another antimalarial agent. Today, artemisinin-based combination therapies (ACTs) form the first-line treatment for uncomplicated falciparum malaria (including pediatrics and lactating women), and artesunate is the first-line recommendation for severe malaria in adults and children and in the second and third trimesters of pregnancy (World Health Organization [WHO] 2010).

4.1.1.6 Mechanisms of Resistance

To date, the molecular mechanism of resistance to artemisinins is still not established. A portion of ring-stage *P. falciparum* appears to go into a dormant stage upon exposure to artemisinin *in vitro* (Fairhurst et al. 2012). These then survive and replicate to detectable levels. Thus it has been extrapolated that lines that are less susceptible to artemisinin may consist of more slow-clearing parasites than fast-clearing ones that revive from dormancy. Reduced susceptibility to artemether has been linked to genetic mutations in PfATP6 (Petersen, Eastman, and Lanzer 2011).

4.1.1.7 Adverse Effects, Toxicity, and Use in Pregnancy

Adverse effects reported include mild gastrointestinal disturbances, dizziness, tinnitus, reticulocytopenia, neutropenia, elevated liver enzymes, electrocardiographic

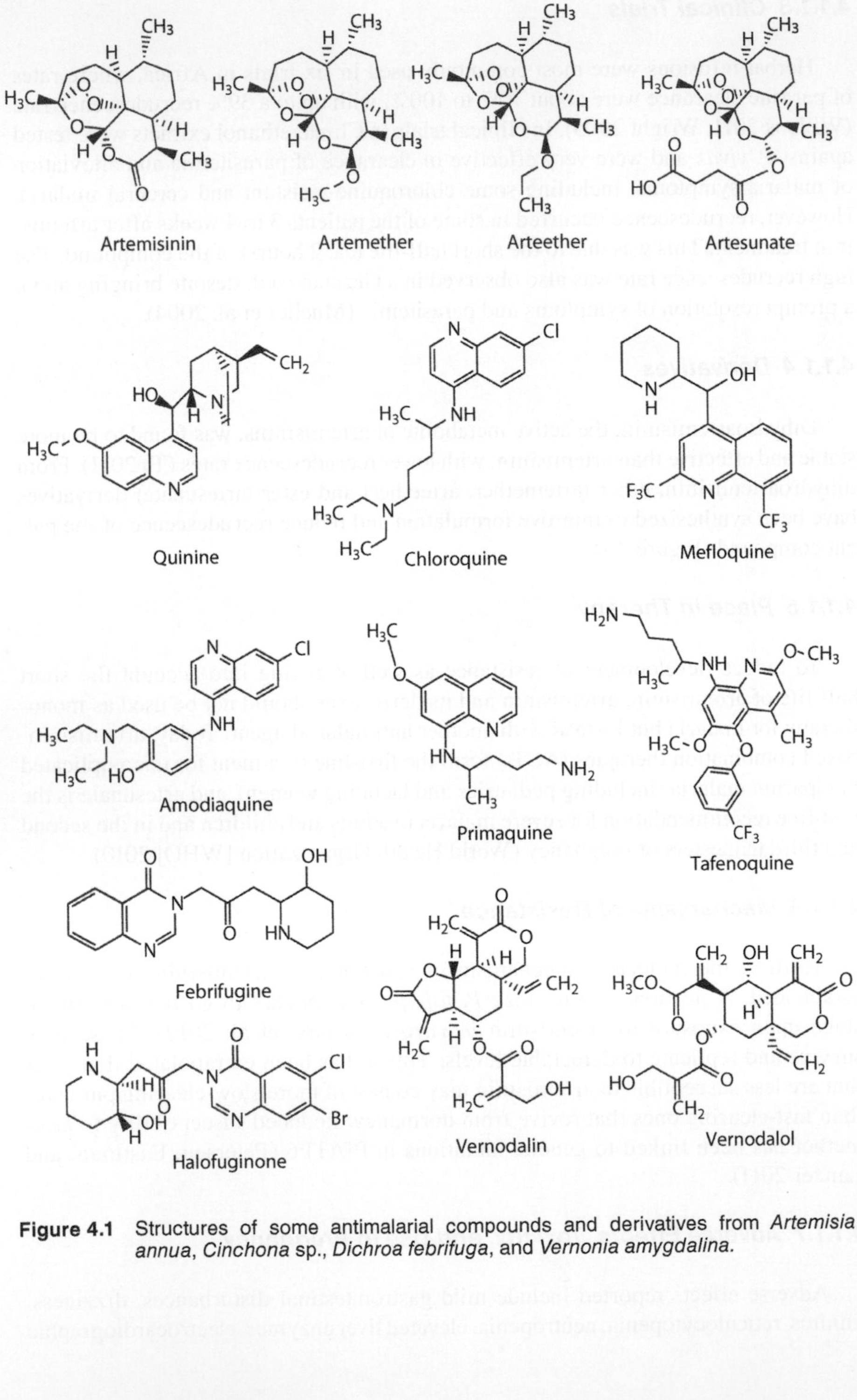

Figure 4.1 Structures of some antimalarial compounds and derivatives from *Artemisia annua*, *Cinchona* sp., *Dichroa febrifuga*, and *Vernonia amygdalina*.

abnormalities (including bradycardia and QT prolongation), and a case report of neurotoxicity from artesunate (WHO 2010).

Use in the first trimester of pregnancy is not recommended. Studies showed fetal resorption in rodents with low-dose artemisinin (13 to 25 mg/kg or 1/200 to 1/400 of LD_{50}) and one incident of miscarriage during the first trimester.

4.1.1.8 Herb–Drug Interactions

Hepatic metabolism is postulated to be the main elimination pathway for artemisinin (Ashton et al. 1998), which has a high hepatic extraction ratio of 0.93 (Gordi et al. 2005). Artemisinin derivatives are metabolized by CYP450 enzymes to the more potent dihydroartemisinin (Kiang, Wilby, and Ensom 2014). Incubations of artemisinin with human liver microsomes showed that artemisinin metabolism appears to be mainly mediated by CYP2B6, followed by CYP3A4 (Svensson and Ashton 1999). The inductive effects of artemisinin (as well as autoinduction) on CYP2B6 and 3A4 have been shown in various studies (Ashton et al. 1998; Simonsson et al. 2003; Xing et al. 2012). It is interesting that artemisinin, artemeter, artesunate, and dihydroartemisinin showed inhibitory effects on microsomal or recombinant CYP1A2, 2B6, 2C19, and 3A4 (Ericsson et al. 2014). It was predicted that artemisinin could lead to a net induction rather than inhibition (Xing et al. 2012). Taking it altogether, the supposed inductive effect of artemisinin(s) may result in increased metabolism, and thus lower plasma drug levels, of concomitantly administered drugs such as anticonvulsants and even other antimalarials such as mefloquine, which may lead to potential treatment failure (Ashton et al. 1998; Na-Bangchang et al. 1995).

4.1.2 Bark of the *Cinchona* Species (Rubiaceae)

The earliest mention of *Cinchona* in literature was in the 17th century in a book by an Augustine monk, who wrote about the antipyretic properties of the "tree of fevers … with cinnamon-coloured bark of which the Lojan cast powders which are drunk in the weight of two small coins, and [thereby] cure fevers and tertians" (Duran-Reynals 1946). The long history leading up to its isolation is described in detail in the literature. Two centuries later, quinine and cinchonine were isolated from the bark by Pelletier and Caventou (Lee 2002).

Several crude extracts of the bark were used in the 20th century. Totaquina type I was prepared by dissolving soluble constituents of the powdered bark of *C. succirubra* Pav. ex Klotzsch (Rubacieae) with hydrochloric acid, and then adding sodium hydroxide to precipitate the alkaloids, which were then dried to obtain the extract. Totaquina type II was the residual alkaloid content from the bark of *C. ledgeriana* (Howard) Bern Moens ex Trimen (Rubiaceae) after quinine extraction. Preparations from the total alkaloids of *C. ledgeriana* containing quinine were also used. From the database compiled in Chapter 3, several preparations of *Cinchona* bark of different species are used worldwide for malaria treatment, ranging from decoctions, infusions, and alcoholic extracts to powdered forms.

4.1.2.1 Mode of Action

Quinine is a blood schizonticide that acts on the mature trophozoite stage. Its exact mechanism of action is not well elucidated, but it is thought to inhibit the detoxification of heme. Being a monoprotic base, it accumulates in the parasite's acidic food vacuole via ion trapping. The protonated compound then binds to hematin or hemozoin in the food vacuole to inhibit the spontaneous process of hemozoin polymerization. It is also gametocytocidal in *P. vivax, P. ovale*, and *P. malariae* (Lee 2002; WHO 2010). Chloroquine, the 4-aminoquinoline derivative, also accumulates in the food vacuole and complexes with hematin. This complex is involved in blocking further sequestration of toxic heme by incorporating into the growing polymer to terminate chain extension (Müller and Hyde 2010).

4.1.2.2 Phytochemistry

Cinchona bark generally contains about 7% to 12% w/w of total alkaloid, of which quinine (Figure 4.1) forms 70% to 90%, cinchonidine (the L-isomer of cinchonine) 1% to 3%, and quinidine (the D-isomer of quinine) up to 1% (McCalley 2002). Content varies with the species of bark used. *C. pubescens* contains almost equal amounts of cinchonine, cinchonidine, and quinine; *C. succirubra* and *C. calisaya* contain all four alkaloids; *C. ledgeriana* contains only quinine (though much greater amounts) and cinchonine; *C. officinalis* contains mainly cinchonine, cinchonidine, and a small amount of quinine; and *C. micrantha* contains cinchonine and cinchonidine, but never quinine (Kacprzak 2013). Cinchonine, cinchonidine, and quinidine are as effective as quinine (Borris and Schaeffer 1992). When combined, these four alkaloids work more effectively than in isolation, possibly by additivity (Wesche and Black 1990). Seven cinchona alkaloids were recently isolated from *C. succirubra* and *C. ledgeriana* for the first time, namely, cinchonanines A–G, although they were evaluated for cytotoxicity (Cheng et al. 2014).

Other nonalkaloid components of the bark include tannins (cinchotannic acid, red cinchonic acid, and quinic acid), chinovin, cinchona red, flavonoids, anthraquinones, and anthocyanosides (Kacprzak 2013; Rajani and Kanaki 2008; Willcox, Bodeker, and Rasoanaivo 2005).

4.1.2.3 Clinical Trials

The earliest clinical trials were done in the 19th century on four alkaloids—cinchonine, cinchonidine, quinine, and quinidine—in more than 3000 patients. All four alkaloids had comparable cure rates of more than 98%, with the main outcome measure of "cessation of febrile paroxysms" (Achan et al. 2011).

Trials on individual alkaloids confirm their efficacy as antimalarials. Quinidine is more effective than quinine in resistant infections. Both quinidine and quinine were effective in Thai children with uncomplicated falciparum malaria, but quinine was recommended for pediatrics in view of cardiac effects due to quinidine (Sabchareon et al. 1988). In a dose-ranging clinical trial, where patients with acute symptomatic uncomplicated falciparum malaria received treatment with either 400 mg or 500 mg of a fixed dose combination of quinine, quinidine, and cinchonine, there was prompt resolution of parasitemia in all patients, with no recrudescence during the period of observation, comparable to use of quinine alone (Sowunmi et al. 1990). The combination drug was also well tolerated. This shows a possible advantage of using a mixture of alkaloids in lower doses over a larger dose of a single alkaloid.

4.1.2.4 Derivatives

Synthetic derivatives of quinine include chloroquine, which is the safest and least expensive drug used for malaria, amodiaquine, bulaquine, mefloquine, piperaquine, primaquine, and tafenoquine (Aguiar et al. 2012). Most of these are currently in use for malaria treatment and prevention, with the exception of bulaquine and tafenoquine, which are in the pipeline.

4.1.2.5 Place in Therapy

Widespread resistance in most parts of the world has rendered chloroquine ineffective against *P. falciparum*, although it still is used against *P. vivax, P. ovale*, and *P. malariae* (WHO 2010). Similar to other 4-aminoquinolines, it does not result in a radical cure. It is to be taken in combination with primaquine in chloroquine-sensitive vivax and ovale malaria and as monotherapy in *P. malariae* infections.

Quinine, when used in combination with clindamycin, is used as the first-line treatment of falciparum malaria in the first trimester of pregnancy and as an alternative to artemisinins in severe falciparum malaria in adults and children.

Quinidine is more cardiotoxic than quinine, and hence should only be used if no other effective parenteral drugs are available.

Amodiaquine is effective against some chloroquine-resistant strains of *P. falciparum*, although there is cross-resistance.

Mefloquine is effective against all forms of malaria.

Primaquine is used in *P. vivax* and *P. ovale* malaria, together with a blood schizonticide for blood-stage parasites. It is effective against the intrahepatic stage of all forms of malaria and thus prevents late relapses.

Piperaquine is recently approved as a combination drug with dihydroartemisinin, as an alternative treatment in uncomplicated falciparum malaria.

Tafenoquine and bulaquine target the liver forms (including hypnozoites) (Aguiar et al. 2012). Tafenoquine is undergoing a Phase IIB trial as a radical cure for *P. vivax* malaria, whereas bulaquine is approved for clinical use in India (GlaxoSmithKline 2015).

4.1.2.6 Mechanisms of Resistance

Mechanisms of resistance of *P. falciparum* to chloroquine have been hypothesized to be associated with increased propensity of chloroquine efflux such that there are insufficient levels to inhibit heme polymerization. It is not clear if this occurs with the other quinoline derivatives (Bloland 2001). Polymorphisms of genes that encode transporter proteins—PfNHE1 (specifically to quinine), PfCRT, PfMDR1, and PfMRP—have been implicated in reduced susceptibility to this class of antimalarials. For *P. vivax*, PvMDR1 has been associated with altered susceptibility to chloroquine and mefloquine (Petersen, Eastman, and Lanzer 2011).

4.1.2.7 Adverse Effects, Toxicity, and Use in Pregnancy

Regular administration of quinine causes cinchonism (characterized by tinnitus, headache, impaired high tone hearing, nausea, dizziness, dysphoria, disturbed vision in its mild form and vomiting, abdominal pain, diarrhea, and severe vertigo in more severe cases) (WHO 2010). Other side effects include hypersensitivity and QT prolongation. Intravenous quinine should be given only by infusion in view of hypotension and cardiac arrest from rapid intravenous injection.

Quinine overdosage may cause oculotoxicity, cardiotoxicity, hypotension, and cardiovascular collapse.

Quinine and quinidine may cause QT prolongation and hyperinsulinaemic hypoglycemia (common in pregnancy, due to stimulation in pancreatic β-cells).

Cinchonine has been reported to cause cinchonism and diarrhea (at larger doses) and visual disturbances.

Chloroquine may cause unpleasant taste, pruritus, headache, skin eruptions, and gastrointestinal disturbances. Less commonly, neurotoxicity, myopathy, photosensitivity, alopecia, and reduced hearing may occur. Chronic use may lead to retinopathy or keratopathy.

Amodiaquine has a similar side effect profile as chloroquine. The risk of pruritus and cardiotoxicity (in overdose) is lower, but agranulocytosis is relatively greater. Reported adverse effects include syncope, spasticity, seizures, and involuntary movements.

The common side effects of mefloquine are minor, such as nausea, vomiting, anorexia, diarrhea, headache, dizziness, insomnia, abnormal dreams, and somnolence. Less common side effects include urticaria, myopathy, alopecia, blood dyscrasias, neuropsychiatric disturbances, and cardiovascular effects.

Primaquine is contraindicated in G6PD-deficient patients due to risk of hemolytic anemia.

For tafenoquine, in one clinical trial vortex keratopathy was observed, which resolved in all affected subjects within a year. In another trial, no such adverse effects were observed (Schrader et al. 2012). Tafenoquine is otherwise well tolerated, although it carries a greater risk of hemolysis compared to primaquine in view of a longer half-life (Rajapakse, Rodrigo, and Fernando 2015). Its safety profile in children, pregnant women, and G6PD-deficient individuals has yet to be investigated.

4.1.2.8 Herb–Drug Interactions

Cinchonine is a P-glycoprotein (P-gp) inhibitor. This has implications in its use to circumvent multidrug resistance development, as well as for drugs that are P-gp substrates (WHO 2010). For instance, it reversed the multidrug resistance activity in refractory or relapsed lymphoid malignancies (Solary et al. 2000).

There is an increased risk of arrhythmia when co-administering chloroquine with drugs that prolong the QT interval.

With quinine there is a risk of increased plasma digoxin levels. Cimetidine increases quinine levels while rifampicin reduces plasma quinine concentrations in view of inhibition and induction of quinine metabolism, respectively.

With chloroquine there is reduced absorption with antacids, possible increased seizure risk with mefloquine, reduced metabolism and clearance with cimetidine, increased risk of acute dystonic reactions with metronidazole, reduced bioavailability of ampicillin and praziquantal, reduced effect of thyroxine, increased plasma cyclosporine concentrations, and possible antagonistic effect of carbamazepine and sodium valproate on seizures.

There is insufficient data concerning amodiaquine.

Mefloquine should not be given with halofantrine due to QT prolongation. There is an increased risk of arrhythmias with beta-blockers, calcium-channel blockers, amiodarone, pimozide, digoxin, or antidepressants; and increased risk of seizures with chloroquine and quinine. Concentrations are increased with ampicillin, tetracycline, and metoclopramide. Caution with alcohol is advised.

With primaquine, drugs that may increase risk of hemolysis or bone marrow suppression should be avoided.

4.1.3 *Dichroa febrifuga* Lour. (Saxifragaceae)

The earliest record of *Dichroa febrifuga* Lour., known in China as Changshan, comes from *Shen Nong Ben Cao* where it mentions the use of Changshan for treating fevers (Unschuld 1992). The roots and leaves of this plant have been used for malaria and fever since ancient China. Its monograph in the Chinese Pharmacopoeia states that it is indicated for malaria, and it has "cold tendency, a bitter and pungent taste" (Chinese Pharmacopoeia Commission 2010). According to *A Materia Medica for Chinese Medicine Plants, Minerals and Animal Products*, it acts to dispel heat and, hence, is used in fever, and in tertian, quartan malaria, and malaria relapses (Hempen and Fischer 2009). *D. febrifuga* is also used as an emetic agent, for "chronic phlegm accumulation in the chest" and "phlegm-induced vomiting." A general dose for this herb is 3 to 6 g boiled in water for 20 minutes. *D. febrifuga* is used in combination with other herbs such as *Zingiber officinale* Roscoe (ginger), possibly to offset the former's strong emetic effects (Poonam and Singh 2009; van Valkenburg 2001; Willcox, Bodeker, and Rasoanaivo 2005).

4.1.3.1 Mode of Action

The exact mechanism of febrifugine has yet to be elucidated. It may play a role in interference of hemozoin formation. Nitrate $\left(NO_3^-\right)$ levels increased in mice treated with febrifugine, although mortality increased 4 days after the last dose, indicating the possible role of febrifugine in host defense against malaria by promoting the production of nitric oxide in activated macrophages (Murata et al. 1998; Zhu et al. 2012).

4.1.3.2 Phytochemistry

Changshan contains the alkaloids isofebrifugine (α-dichroine), febrifugine (β-dichroine), γ-dichroine, dichroidine, 4-quinazolone, and a coumarin, umbelliferone (Jang et al. 1948). The reported alkaloid content of the root and leaf are ca. 0.1% and 0.5% w/w, respectively. The major active constituents are the quinazolinone alkaloids febrifugine and isofebrifugine.

4.1.3.3 Pharmacological Studies

It was found that an aqueous extract of Changshan was highly active against *P. gallinaceum* in chicks. The antimalarial activity was 5 times more in the leaf as compared to the root (Jang et al. 1948).

Febrifugine was about 100 times more active than quinine against *P. lophurae* in ducks, and 50 times more active than quinine in *P. gallinaceum*–infected chicks and in *P. cynomolgi*–infected rhesus monkeys (Tang and Eisenbrand 1992). Oral administration of febrifugine reduced parasitemia and prolonged survival time in *P. berghei*–infected mice (Jiang et al. 2005). In another study, infected mice treated with febrifugine experienced recrudescence and deteriorated 10 days posttreatment, but were inexplicably and completely cured a week after. It was postulated that febrifugine metabolites might persist in the body longer than expected. Another possibility was immunomodulatory effects (Jiang et al. 2005).

4.1.3.4 Clinical Trials

In 1942, the root extract was found to be effective in 13 clinical cases of tertian malaria, although it took one day more as compared to quinine in negating positive smears (Jang et al. 1946).

4.1.3.5 Derivatives

The 3′-keto derivative of febrifugine showed high selectivity against *P. falciparum in vitro* (IC_{50} = 20 nM); and the 4-quinazolinone ring, the nitrogen atom in piperidine, and a propyl group were found to be crucial for activity (Kikuchi et al. 2002; Sen et al. 2010). Contrary to artemisinin and quinine, total synthesis of febrifugine is simpler. However, although it is very potent, febrifugine causes nausea, vomiting, and liver toxicity, limiting its usefulness as a drug (De Smet 1997). Therefore, analogs with better safety profiles and comparable efficacy have been explored (Kikuchi et al. 2014; Mai et al. 2014).

Halofuginone, a synthetic halogenated derivative of febrifugine (Figure 4.1), was found to reduce *P. berghei* sporozoite load in HepG2 cells (human liver carcinoma cells) with high potency (IC_{50} = 17 nM), which indicates effects against early and late liver-stage parasites (Derbyshire, Mazitschek, and Clardy 2012). Against the blood-stage D2 strain of *P. falciparum*, it had a lower IC_{50} (0.7 nM). In another study, halofuginone was equally active on *P. falciparum* ring stages, trophozoites, and schizonts *in vitro* (Geary,

Divo, and Jensen 1989). It inhibits *P. falciparum* prolyl-tRNA synthetase in an ATP-dependent manner (Zhou et al. 2013), which results in growth inhibition via disruption of protein translation (Jain et al. 2014). Direct binding of halofuginone to *P. falciparum* prolyl-tRNA synthetase has been recently demonstrated (Jain et al. 2015).

Besides malaria, febrifugine derivatives are used in cancer, fibrosis, and inflammatory diseases. Halofuginone is in Phase II clinical trials for cancer and fibrosis. The inhibition of prolyl-tRNA synthetase and consequent activation of the amino acid response pathway by halofuginone may explain its role in immune regulation (Keller et al. 2012). It disrupted proline incorporation or uptake, leading to apoptosis from amino acid starvation response in T cells (Chu et al. 2013). Oral administration of halofuginone reduced concanavalin A–induced liver fibrosis possibly via inhibition of collagen I synthesis and inflammatory processes (Liang et al. 2013). It also showed proapoptotic activity on breast cancer cells and downregulated matrix metalloproteinase-9, which is involved in cancer metastasis (Jin et al. 2014).

4.1.3.6 Place in Therapy

D. febrifuga is used in traditional Chinese medicine for malaria. It is also taken together with *A. annua* (Yan, Fischer, and Fratkin 1997), or with Magnoliae cortex, Citri reticulatae viride pericarpium, and Tsao ko fructus for malaria (Hempen and Fischer 2009).

4.1.3.7 Mechanism of Resistance

To date, development of resistance has not been reported for *D. febrifuga.*

4.1.3.8 Adverse Effects, Toxicity, and Use in Pregnancy

Nausea and vomiting is a dose-limiting side effect of *D. febrifuga* (Chinese Pharmacopoeia Commission 2010), possibly due to stimulation of the vagal and sympathetic nerves in the gastrointestinal tract (Chiang 1961). It may also cause hepatotoxicity (Mojab 2012) and renal injury (Huang 1998). It is contraindicated in pregnancy and in very weak patients (Chinese Pharmacopoeia Commission 2010; Hempen and Fischer 2009).

4.1.3.9 Herb–Drug Interactions

Information is not available.

4.1.4 *Vernonia amygdalina* Delile. (Compositae)

Vernonia amygdalina is used traditionally for a myriad of conditions, such as amoebiasis, cough, dermatitis, diarrhea, hepatitis, infertility, malaria, menstrual problems, venereal diseases, and as an abortifacient and anthelmintic (Toyang and Verpoorte 2013; Yeap et al. 2010). Its traditional uses has also been scientifically demonstrated in its diverse biological activities, including but not limited to antibacterial, antifungal, antileishmanial, wound healing, anticancer, antioxidant, hypoglycemic, oxytocic, nephroprotection, lowering of serum lipids, pain relief, and antiplasmodial effects (Ijeh and Ejike 2011). Its leaves are reported to be used for malaria in Africa (Asase, Akwetey, and Achel 2010; Tor-anyiin, Sha'ato, and Oluma 2003).

4.1.4.1 Mode of Action

There have been no studies done to date investigating the mechanism of action of active compounds found in *V. amygdalina* against *Plasmodium*.

4.1.4.2 Phytochemistry

Antiplasmodial activity of *V. amygdalina* has been attributed to the sesquiterpene lactones such as vernodalin (Figure 4.1), vernodalol, vernolide, and hydroxyvernolide, as well as steroidal compounds such as vernonioside B1 and vernonoid B1 in the leaves (Omoregie, Pal, and Sisodia 2011). These are more potent when used in combination than as individual compounds. The most active extract is more potent than the most active principle isolated (vernodalin) (Figure 4.1), but vernodalin is toxic in mice when administered intramuscularly at 5 mg (Ohigashi et al. 1994).

4.1.4.3 Pharmacological Studies

Leaf extracts (extracted using ethanol, water, and 50:50 ethanol:water solvent systems) of *V. amygdalina* were tested in *P. falciparum* 3D7 (chloroquine-sensitive strain) (Omoregie, Pal, and Sisodia 2011). The ethanol extract had the lowest IC_{50} of 9.82 ± 0.43 μg/ml but was also the most cytotoxic (IC_{50} when tested on Vero cell line: 60.33 ± 0.24 μg/ml). Extracts from the leaves and root bark of *V. amygdalina* were studied in mice infected with *P. berghei* (Abosi and Raseroka 2003). The leaf and root bark extracts demonstrated significant suppression of parasitemia (67% and 53.5%, respectively) in a 4-day suppression test. In another study, the aqueous leaf extract showed 73% inhibition in mice infected with *P. berghei* when tested at a dose of 200 mg/kg intraperitoneally daily for 4 days (Njan et al. 2008). There were no clinical toxicity or adverse effects other than a decrease in red blood cell count and dose-dependent increase in serum bilirubin, although these were within control values.

4.1.4.4 Clinical Trials

In a trial which studied the safety and efficacy of *V. amygdalina* leaf infusion (administered 4 times a day for 7 days) in patients aged 12 years and older with uncomplicated malaria, there was 67% of cases with adequate clinical response at day 14, although only 32% of these showed complete clearance and 71% had recrudescence (Challand and Willcox 2009). There were no significant adverse effects or toxicity from the herb. A decrease in hemoglobin between days 0 and 28 was observed although this was not statistically significant.

4.1.4.5 Derivatives

No derivatives have been developed to date.

4.1.4.6 Place in Therapy

More evidence from well-designed randomized controlled trials is required to establish the safety and efficacy of *V. amygdalina* and its active principles before any therapeutic recommendations can be made.

4.1.4.7 Mechanism of Resistance

To date, development of resistance has not been reported for *V. amygdalina.*

4.1.4.8 Adverse Effects, Toxicity, and Use in Pregnancy

From the results of pharmacological studies and a clinical trial done on *V. amygdalina* (see Sections 4.1.4.3 and 4.1.4.4), there were no significant adverse effects or toxicity reported, besides reductions in red blood cell count and serum bilirubin which were within control values.

V. amygdalina is used in Nigeria to facilitate childbirth (Attah et al. 2012). The aqueous extract of the plant was found to induce sustained contractility of human myometrial smooth muscle cell. Therefore, the use of *V. amygdalina* as an antimalarial herb in pregnancy should be contraindicated.

4.1.4.9 Herb–Drug Interactions

The leaf extract was found to decrease the bioavailability of dihydroartemisinin in albino rats when administered both as a single dose (250 and 500 mg/kg simultaneously with dihydroartemisinin) and as a subacute treatment (when given once daily for a week before given together with dihydroartemisinin on day 7) (Eseyin et al. 2012). This necessitates caution when patients on ACTs are on concomitant herbal medications.

V. amygdalina was found to increase rat intestinal absorption of digoxin *in vivo* by inhibition of P-gp (Oga, Sekine, and Horie 2013). Coadministration of *V. amygdalina* and nifedipine in rabbits increased the antihypertensive effect of the latter, due to reduction of metabolism of nifedipine (Owolabi et al. 2013). Patients on P-gp substrates and anti-hypertensives should be advised not to take *V. amygdalina* concomitantly without medical supervision.

4.2 HERBAL PREPARATIONS IN CLINICAL USE

As discussed in the earlier chapters, there are benefits of using the crude extract of a medicinal plant over a single purified active compound. This section gives a brief overview of some medicinal plant preparations that have demonstrated clinical efficacy and are either government-approved in some countries (e.g. N'Dribala, Phyto-laria, and *Argemone mexicana*) or undergoing development (PR 259 CT1). Table 4.1 shows a summary of these herbal preparations.

4.2.1 Sumafoura Tiemoko Bengaly

Sumafoura Tiemoko Bengaly is the name of an *Argemone mexicana* (Papaveraceae) decotion approved as a recommended herbal preparation for malaria in Mali (Graz et al. 2010; Sanogo 2014). It was birthed from the concept of "reverse pharmacology" from India, which aimed to develop new pharmaceuticals from Ayurvedic medicine. It is based on the rationale that although traditional medicine is potentially a viable source of new leads in drug discovery, the process of drug development is one that takes years and requires huge amounts of capital. In contrast, the development of standardized phytomedicine may be a faster, cheaper, and more sustainable means for undeveloped areas. Clinical evaluation forms the initial stage in the latter, whereas isolation of the active principle(s) becomes one of the last steps in its development since it serves to ensure quality control in the standardization of the products (Willcox et al. 2011).

Several alkaloids in the plant are found to have antimalarial activity (berberine, allocryptopine, and protopine, with IC_{50} [μg/ml] against W2 [chloroquine-resistant *P. falciparum* strain] *in vitro* of 0.32, 0.32, and 1.46, respectively) (Avello 2009). However, in some animal models, berberine was found to have poor oral absorption. Protopine and allocryptopine demonstrated good selectivity for *Plasmodium*, with low cytotoxicities. Other compounds include dehydrocorydalmine, jatrorrhizine, columbamine, and oxyberberine from the whole plant (Singh et al. 2010) and rotomexicine, mexitin, 8-methoxydihydrosanguinarine, 13-oxoprotopine, rutin, and quercetrin from the aerial parts of the methanol extract of the plant (Singh, Pandey, and Singh 2012).

Different extracts of the aerial parts of *A. mexicana* were assayed on a chloroquine-resistant (K1) strain of *P. falciparum in vitro*, with the respective IC_{50} (μg/ml): 1.00 (methanol), 1.22 (dichloromethane), 5.89 (aqueous decoction), and 6.22 (aqueous maceration) (Willcox et al. 2011). The aqueous extract was found to be inactive against *P. berghei*–infected mice.

In a dose-escalation study that was done on 80 patients with uncomplicated falciparum malaria, it was found that the dose of one glass twice a day for a week had better efficacy than one glass a day for 3 days (proportion of patients with adequate clinical response increased from 35% to 73%), without increase in side effects. However, at the third and highest dose studied (one glass 4 times a day for the first 4 days, followed by one glass twice a day up to 7 days), there was no added benefit (65% with adequate clinical response) and two patients developed QTc prolongation (Willcox et al. 2007).

In view of the aforementioned trial, the dose of one glass twice a day for a week was used in a randomized controlled trial in 300 patients (Graz et al. 2010). About 89% of patients treated with the herb showed clinical recovery (as compared to 95%

Table 4.1 Summary of Herbal Preparations in Clinical Use

Plant Species (Family)	Proprietary Name	Status	Active Principles, IC_{50}, and Strain of *Plasmodium* Used	Clinical Studies	References
Argemone mexicana (Papaveraceae)	Sumafoura Tiemoko Bengaly	Government approved in Mali	• Berberine: 0.32 µg/ml [0.9 µM] (W2, *P. falciparum*) • Allocryptopine: 0.32 µg/ml [0.8 µM] (W2, *P. falciparum*) • Protopine: 1.46 µg/ml [4.1 µM] (W2, *P. falciparum*)	• Dose-escalation study (n = 80) Doses: 1 glass BD for 1 week versus 1 glass QDS for 4 days, followed by 1 glass BD for up to 7 days Adequate clinical response: 73% versus 65%, respectively QTc prolongation in two patients at maximum dose • Randomized-controlled trial (n = 300) *A. mexicana* decoction BD for 1 week versus ACT (artesunate–amodiaquine) BD for 3 days • Clinical recovery in 89% of patients treated with *A. mexicana* (versus 95% in ACT arm) Relapse from day 15 to 28 in 12.8% in *A. mexicana* arm Most common adverse effects: cough and diarrhea (*A. mexicana*), vomiting (ACT)	Avello 2009; Graz et al. 2010; Willcox et al. 2007
Cochlospermum planchonii (Bixaceae)	N'Dribala	Government approved in Burkina Faso	Not known; essential oils of leaves: 22–35 µg/ml	• Open-label, single center, randomized, uncontrolled trial (n = 85) N'Dribala versus chloroquine Similar parasitemia clearance in both arms at day 5 (52% versus 57%, respectively) Adverse effects: arthralgia-myalgia, anorexia (N'Dribala arm), abdominal pain, diarrhea (chloroquine arm)	Benoit-Vical et al. 1999, 2003

(*Continued*)

Table 4.1 (Continued) Summary of Herbal Preparations in Clinical Use

Plant Species (Family)	Proprietary Name	Status	Active Principles, IC_{50}, and Strain of *Plasmodium* Used	Clinical Studies	References
Cryptolepis sanguinolenta (Apocynaceae)	Phyto-laria	Government approved in Ghana	• Cryptolepine: 0.23 μM and 0.059 μM (K1 and T996 *P. falciparum*, respectively) • Cryptoheptine: 0.8 μM and 1.2 μM (K1 and T996 *P. falciparum*, respectively)	• Open-label study (n = 44) Phyto-laria dose: 1 teabag TDS for 5 days Overall cure rate: 93.5% Mean parasitic clearance time: 82.3 hours 2 cases of recrudescence on day 28 Elevations in alkaline phosphatase and uric acid levels, with latter level normalized after 28 days	Bugyei, Boye, and Addy 2010; Paulo et al. 2000
Nauclea pobeguinii (Rubiaceae)	PR 259 CT1	Undergoing clinical development in DR Congo	Strictosamide as putative active compound (>64 μg/ml or >128.3 μM)	• Phase IIA (open cohort study) (n = 11) Dose: PO 1 g TDS for 3 days followed by 500 mg TDS for 4 days All patients had complete fever and parasitemia clearance One recrudescence after a week Adverse effects: headache and fatigue (self-resolved) • Phase IIB (single-blind, prospective trial) (n = 65) PR 259 CT1 (as per dose above) versus artesunate-amodiaquine (ACT) OD for 3 days 87.9% reduction in parasitemia versus 96.9% in artesunate-amodiaquine arm PR 259 CT1 was better tolerated than ACT	Mesia et al. 2010; Mesia, Tona, Mampunza, Ntamabyaliro, Muanda, Muyembe, Cimanga et al. 2012; Mesia, Tona, Mampunza, Ntamabyaliro, Muanda, Muyembe, Musuamba et al. 2012

Note: ACT, artemisinin-based combination therapy; BD, twice daily; OD, once daily; PO, orally; QDS, four times daily; TDS, three times daily.

in ACT-treated group), although only 9% of patients had clearance of parasitemia. Worsening of the condition to severe malaria occurred in both groups to a similar extent (1.9% of patients aged 0 to 5 years old). No coma or convulsions developed in either treatment group. The most common adverse effects observed in the group treated with *A. mexicana* were cough and diarrhea. Relapse from day 15 to 28 occurred in 12.8% of patients treated with *A. mexicana* (as compared to the ACT-treated group). A Phase IIb/III trial is ongoing to study the phytoconstituents that are responsible for the antimalarial activity in the preparation (Flannery, Chatterjee, and Winzeler 2013).

4.2.2 N'Dribala

A decoction of the tuberculous roots of *Cochlospermum planchonii* Hook f. ex Planch (Bixaceae) is traditionally used in fever and malaria. "N'Dribala" is the name of sachets of 70 g of the root in Burkina Faso. Because more than 85,000 patients are treated with it annually, there is a need to conserve the wild sources of this herb (Willcox, Bodeker, and Rasoanaivo 2005). Essential oils of *C. planchonii* have been reported to possess antiplasmodial activity. Major compounds in the oil include sesquiterpenes (80%), with a greater proportion of hydrocarbons: β-caryophyllene, (E,E)-α-farnesene and tetradecan-3-one, with IC_{50} of 22 to 35 μg/ml for the leaf essential oil extract and 25 to 75 μg/ml for the aqueous leaf extract (FcB1-Columbia [chloroquine-resistant strain]) *P. falciparum*, with low *in vitro* cytotoxicity in K562 cells displayed by the essential oil of the leaves (Benoit-Vical et al. 1999).

An open-label, single center, randomized, uncontrolled trial was done on 58 patients with uncomplicated falciparum malaria, with clearance of parasitemia in 52% of patients treated with N'Dribala after 5 days, which is comparable with that of chloroquine (57%). Adverse effects included abdominal pain, diarrhea (which occurred in slightly more patients treated with chloroquine), and arthralgia-myalgia and anorexia (which were more prevalent in patients treated with N'Dribala). It was government-approved in Burkina Faso as an antimalarial herbal medicine in 2005 (Benoit-Vical et al. 2003).

4.2.3 Phyto-laria

Cryptolepis sanguinolenta (Lindl.) Schltr. (Apocynaceae) has been developed as an herbal tea preparation under the proprietary name of "Phyto-laria" and is regulatory-approved in Ghana. A decoction of the roots is traditionally used to treat malaria (Adebayo and Krettli 2011).

Roots contain about 0.5% alkaloidal content. Cryptolepine, cryptolepinoic acid, methyl cryptolepinoate, hydroxycryptolepine, and quindoline have been isolated from the roots. Cryptolepine, an indoloquinoline derivative, was shown to be the most active (IC_{50}: 0.23 μM and 0.059 μM against K1 [multidrug-resistant strain] and T996 [chloroquine-sensitive] *P. falciparum*), followed by cryptoheptine (IC_{50}: 0.8 μM and 1.2 μM against K1 and T996, respectively) (Paulo et al. 2000).

Extracts of the roots have exhibited high efficacy *in vitro* against *P. falciparum* (92.7% to 100% growth inhibition and 65% to 100% growth inhibition for ethanol and dichloromethane extracts, respectively, at concentrations of 6 to 600 μg/ml) and *P. berghei in vivo* (62.9% and 75.07% suppression from dichloromethane and ethanol extracts, respectively) (Benoit-Vical et al. 2003). Studies on mice, rats, and rabbits showed no significant adverse effects in toxicity studies (Wright 2005).

In an open-label study, patients with uncomplicated malaria who were treated with Phyto-laria (one teabag 3 times daily for 5 days) had a mean clearance of 82.3 hours, with a mean fever clearance time of 25.2 hours (Bugyei, Boye, and Addy 2010). Overall, the rate of cure was 93.5%. There was complete clearance of parasitemia in more than 50% of patients within 72 hours, with two patients with late recrudescence. Significant elevations in biochemical markers were observed in alkaline phosphatase and uric acid levels, although the latter level normalized after 28 days. One patient experienced jaundice, which persisted posttreatment.

4.2.4 PR 259 CT1

PR 259 CT1 is a standardized 80% ethanol extract containing 5.6% w/w strictosamide of the stem bark of *Nauclea pobeguinii*, which is in clinical development in the Democratic Republic of Congo (DR Congo). An aqueous decoction is used to treat fever in malaria in DR Congo and tribes in Nigeria (Mesia, Tona, Mampunza, Ntamabyaliro, Muanda, Muyembe, Musuamba et al. 2012).

The glycoalkaloid strictosamide was isolated from the 80% ethanol extract of the stem bark and identified as the putative active compound (Mesia et al. 2010). Other compounds isolated include (5S)-5-carboxystrictosidine, 19-O-methylangustoline, 3-O-beta-fucosylquinovic acid, and 3-ketoquinovic acid.

In vitro studies on the chloroquine-sensitive Ghana strain of *P. falciparum* were done using the aqueous and ethanol extracts of the stem bark with the respective IC_{50} of 44 and 32 μg/ml. There was no apparent cytotoxicity on MRC-5 cells (human fetal lung fibroblast cells) (CC_{50} > 64 μg/ml). A prolonged oral dosing of the ethanol extract with a dose of 300 mg/kg showed 92% reduction of parasitemia and a mean survival time of 17 days (Mesia et al. 2010). No significant abnormalities were observed in acute and subacute toxicity studies.

A Phase IIA trial, which was an open cohort study in 11 adult patients with uncomplicated falciparum malaria, showed that all 11 patients who were treated with PR 259 CT1 had complete fever and parasitemia clearance, with one recurrence after 1 week (Mesia, Tona, Mampunza, Ntamabyaliro, Muanda, Muyembe, Cimanga et al. 2012). It was also well tolerated with mild adverse effects such as headache and fatigue, which self-resolved after 1 and 4 days, respectively.

A Phase IIB, single-blind, prospective trial of 65 patients with uncomplicated falciparum malaria showed significant reduction in parasitemia at day 14 (87.9% versus 96.9% for artesunate-amodiaquine combination). PR 259 CT1 was better tolerated than artesunate-amodiaquine (Mesia, Tona, Mampunza, Ntamabyaliro, Muanda, Muyembe, Musuamba et al. 2012).

REFERENCES

Abosi, A.O., and B.H. Raseroka. 2003. "In vivo antimalarial activity of *Vernonia amygdalina*." *British Journal of Biomedical Science* 60 (2):89–91.

Achan, J., A.O. Talisuna, A. Erhart, A. Yeka, J.K. Tibenderana, F.N. Baliraine, P.J. Rosenthal, and U. D'Alessandro. 2011. "Quinine, an old anti-malarial drug in a modern world: Role in the treatment of malaria." *Malaria Journal* 10:144.

Adebayo, J.O., and A.U. Krettli. 2011. "Potential antimalarials from Nigerian plants: A review." *Journal of Ethnopharmacology* 133:289–302.

Aguiar, A.C.C., E.M.M.D. Rocha, N.B.D. Souza, T.C.C. França, and A.U. Krettli. 2012. "New approaches in antimalarial drug discovery and development: A review." *Memórias do Instituto Oswaldo Cruz* 107 (7):831–845.

Asase, A., G.A. Akwetey, and D.G. Achel. 2010. "Ethnopharmacological use of herbal remedies for the treatment of malaria in the Dangme West District of Ghana." *Journal of Ethnopharmacology* 129:367–376.

Ashton, M., T.N. Hai, N.D. Sy, D.X. Huong, N. Van Huong, N.T. Niêu, and L.D. Công. 1998. "Artemisinin pharmacokinetics is time-dependent during repeated oral administration in healthy male adults." *Drug Metabolism and Disposition: The Biological Fate of Chemicals* 26:25–27.

Attah, A.F., M. O'Brien, J. Koehbach, M.A. Sonibare, J.O. Moody, T.J. Smith, and C.W. Gruber. 2012. "Uterine contractility of plants used to facilitate childbirth in Nigerian ethnomedicine." *Journal of Ethnopharmacology* 143:377–382.

Avello, C.S. 2009. "Investigation of antiplasmodial compounds from various plant extracts." Thesis, University of Geneva.

Benoit-Vical, F., A. Valentin, M. Mallié, J.M. Bastide, and J.M. Bessière. 1999. "In vitro antimalarial activity and cytotoxicity of *Cochlospermum tinctorium* and *C. planchonii* leaf extracts and essential oils." *Planta Medica* 65:378–381.

Benoit-Vical, F., A. Valentin, B. Da, Z. Dakuyo, L. Descamps, and M. Mallié. 2003. "N'Dribala (*Cochlospermum planchonii*) versus chloroquine for treatment of uncomplicated *Plasmodium falciparum* malaria." *Journal of Ethnopharmacology* 89:111–114.

Bloland, P.B. 2001. "Drug resistance in malaria." Accessed May 11, 2014. http://www.who.int/csr/resources/publications/drugresist/malaria.pdf.

Borris, R.P., and J.M. Schaeffer. 1992. "Antiparasitic agents from plants." *Phytochemical Resources for Medicine and Agriculture*, 117–158. US: Springer.

Brown, G.D. 2010. "The biosynthesis of artemisinin (Qinghaosu) and the phytochemistry of *Artemisia annua* L. (Qinghao)." *Molecules* 15:7603–7698.

Bugyei, K.A., G.L. Boye, and M.E. Addy. 2010. "Clinical efficacy of a tea-bag formulation of *Cryptolepis sanguinolenta* root in the treatment of acute uncomplicated falciparum malaria." *Ghana Medical Journal* 44:3–9.

Challand, S., and M. Willcox. 2009. "A clinical trial of the traditional medicine *Vernonia amygdalina* in the treatment of uncomplicated malaria." *Journal of Alternative and Complementary Medicine* 15:1231–1237.

Cheng, G.G., X.H. Cai, B.H. Zhang, Y. Li, J. Gu, M.F. Bao, Y.P. Liu, and X.D. Luo. 2014. "Cinchona alkaloids from *Cinchona succirubra* and *Cinchona ledgeriana*." *Planta Medica* 80 (2-3):223–230.

Chiang, W. 1961. "The mechanism of the emetic action of β-dichroine in dog." *Acta Physiologica Sinica* 24:180–186.

Chinese Pharmacopoeia Commission. 2010. *Pharmacopoeia of the People's Republic of China (2010)*, Volume 1 (English edition). Beijing, China: China Medical Science Press.

Chu, T.L., Q. Guan, C.Y. Nguan, and C. Du. 2013. "Halofuginone suppresses T cell proliferation by blocking proline uptake and inducing cell apoptosis." *International Immunopharmacology* 16 (4):414–423.

De Smet, P.A. 1997. "The role of plant-derived drugs and herbal medicines in healthcare." *Drugs* 54 (6):801–840.

Derbyshire, E.R., R. Mazitschek, and J. Clardy. 2012. "Halofuginone inhibition of liver stage malaria." *ChemMedChem* 7:844–849.

Duran-Reynals, M.L. 1946. *The Fever Bark Tree: The Pageant of Quinine*. Garden City, New York: Doubleday.

Eckstein-Ludwig, U., R.J. Webb, I.D. Van Goethem, J.M. East, A.G. Lee, M. Kimura, P.M. O'Neill, P.G. Bray, S.A. Ward, and S. Krishna. 2003. "Artemisinins target the SERCA of *Plasmodium falciparum*." *Nature* 424 (6951):957–961.

Elfawal, M.A., M.J. Towler, N.G. Reich, D. Golenbock, P.J. Weathers, and S.M. Rich. 2012. "Dried whole plant *Artemisia annua* as an antimalarial therapy." *PLoS ONE* 7 (12):e52746.

Ericsson, T., J. Sundell, A. Torkelsson, K.-J. Hoffmann, and M. Ashton. 2014. "Effects of artemisinin antimalarials on Cytochrome P450 enzymes in vitro using recombinant enzymes and human liver microsomes: Potential implications for combination therapies." *Xenobiotica; The Fate of Foreign Compounds in Biological Systems* 8254:1–12.

Eseyin, A.O., C.A. Igboasoiyi, C. Igbo, A. Igboasoiyi, J. Ekarika, and B. Dooka. 2012. "Effects of the leaf extract of *Vernonia amygdalina* on the pharmacokinetics of dihydroartemisinin in rat." *Pharmacologia* 3 (12):713–718.

Fairhurst, R.M., G.M.L. Nayyar, J.G. Breman, R. Hallett, J.L. Vennerstrom, S. Duong, P. Ringwald, T.E. Wellems, C.V. Plowe, and A.M. Dondorp. 2012. "Artemisinin-resistant malaria: Research challenges, opportunities, and public health implications." *The American Journal of Tropical Medicine and Hygiene* 87 (2):231–241.

Flannery, E.L., A.K. Chatterjee, and E.A. Winzeler. 2013. "Antimalarial drug discovery—Approaches and progress towards new medicines." *Nature Reviews Microbiology* 11: 849–862.

Geary, T.G., A.A. Divo, and J.B. Jensen. 1989. "Stage specific actions of antimalarial drugs on *Plasmodium falciparum* in culture." *American Journal of Tropical Medicine and Hygiene* 40 (3):240–244.

GlaxoSmithKline. 2015. "A multi-centre, double-blind, randomised, parallel-group, active controlled study to evaluate the efficacy, safety and tolerability of tafenoquine (SB-252263, WR238605) in subjects with *Plasmodium vivax* malaria. NLM Identifier: NCT01376167." National Library of Medicine. Accessed June 5, 2015. https://clinicaltrials.gov/ct2/show/NCT01376167.

Golenser, J., J.H. Waknine, M. Krugliak, N.H. Hunt, and G.E. Grau. 2006. "Current perspectives on the mechanism of action of artemisinins." *International Journal of Parasitology* 36 (14):1427–1441.

Gordi, T., R. Xie, N.V. Huong, D.X. Huong, M.O. Karlsson, and M. Ashton. 2005. "A semi-physiological pharmacokinetic model for artemisinin in healthy subjects incorporating autoinduction of metabolism and saturable first-pass hepatic extraction." *British Journal of Clinical Pharmacology* 59:189–198.

Graz, B., M.L. Willcox, C. Diakite, J. Falquet, F. Dackuo, O. Sidibe, S. Giani, and D. Diallo. 2010. "*Argemone mexicana* decoction versus artesunate-amodiaquine for the management of malaria in Mali: Policy and public-health implications." *Transactions of the Royal Society of Tropical Medicine and Hygiene* 104 (1):33–41.

Hempen, C.H., and T. Fischer. 2009. "Herbs that expel parasites." In *A Materia Medica for Chinese Medicine Plants, Minerals and Animal Products*, chap. 17. Edinburgh: Churchill Livingstone.

Huang, K.C. 1998. *The Pharmacology of Chinese Herbs*. 2nd ed. Boca Raton, FL: CRC Press.

Ijeh, I.I., and C.E.C.C. Ejike. 2011. "Current perspectives on the medicinal potentials of *Vernonia amygdalina* Del." *Journal of Medicinal Plants Research* 5 (7):1051–1061.

Jain, V., H. Kikuchi, Y. Oshima, A. Sharma, and M. Yogavel. 2014. "Structural and functional analysis of the anti-malarial drug target prolyl-tRNA synthetase." *Journal of Structural and Functional Genomics* 15 (4):181–190.

Jain, V., M. Yogavel, Y. Oshima, H. Kikuchi, B. Touquet, M.A. Hakimi, and A. Sharma. 2015. "Structure of prolyl-tRNA synthetase-halofuginone complex provides basis for development of drugs against malaria and toxoplasmosis." *Structure* 23 (5):819–829.

Jambou, R., E. Legrand, M. Niang, N. Khim, P. Lim, B. Volney, M.T. Ekala et al. 2005. "Resistance of *Plasmodium falciparum* field isolates to in-vitro artemether and point mutations of the SERCA-type PfATPase6." *Lancet* 366 (9501):1960–1963.

Jang, C.S., F.Y. Fu, C.Y. Wang, K.C. Huang, G. Lu, and T.C. Chou. 1946. "Ch'ang Shan, a Chinese antimalarial herb." *Science* 103 (2663):59.

Jang, C.S., F.Y. Fu, K.C. Huang, and W. Cy. 1948. "Pharmacology of Ch'ang Shan (*Dichroa febrifuga*), a Chinese antimalarial herb." *Nature* 161:400–401.

Jiang, S., Q. Zeng, M. Gettayacamin, A. Tungtaeng, S. Wannaying, A. Lim, P. Hansukjariya, C.O. Okunji, S. Zhu, and D. Fang. 2005. "Antimalarial activities and therapeutic properties of febrifugine analogs." *Antimicrobial Agents and Chemotherapy* 49:1169–1176.

Jin, M.L., S.Y. Park, Y.H. Kim, G. Park, and S.J. Lee. 2014. "Halofuginone induces the apoptosis of breast cancer cells and inhibits migration via downregulation of matrix metalloproteinase-9." *International Journal of Oncology* 44 (1):309–318.

Kacprzak, K.M. 2013. "Chemistry and Biology of *Cinchona* Alkaloids." In *Natural Products*, edited by K.G. Ramawat and J.-M. Mérillon, 605–641. Berlin: Springer.

Keller, T.L., D. Zocco, M.S. Sundrud, M. Hendrick, M. Edenius, J. Yum, Y.-J. Kim et al. 2012. "Halofuginone and other febrifugine derivatives inhibit prolyl-tRNA synthetase." *Nature Chemical Biology* 8:311–317.

Kiang, T.K.L., K.J. Wilby, and M.H.H. Ensom. 2014. "Clinical pharmacokinetic drug interactions associated with artemisinin derivatives and HIV-antivirals." *Clinical Pharmacokinetics* 53 (2):141–153.

Kikuchi, H., H. Tasaka, S. Hirai, Y. Takaya, Y. Iwabuchi, H. Ooi, S. Hatakeyama, H.-S. Kim, Y. Wataya, and Y. Oshima. 2002. "Potent antimalarial febrifugine analogues against the *Plasmodium* malaria parasite." *Journal of Medicinal Chemistry* 45:2563–2570.

Kikuchi, H., S. Horoiwa, R. Kasahara, N. Hariguchi, M. Matsumoto, and Y. Oshima. 2014. "Synthesis of febrifugine derivatives and development of an effective and safe tetrahydroquinazoline-type antimalarial." *European Journal of Medicinal Chemistry* 76:10–19.

Lee, M.R. 2002. "Plants against malaria. Part 1: *Cinchona* or the Peruvian bark." *Journal of the Royal College of Physicians of Edinburgh* 32:189–196.

Liang, J., B. Zhang, R.W. Shen, J.B. Liu, M.H. Gao, Y. Li, Y.Y. Li, and W. Zhang. 2013. "Preventive effect of halofuginone on concanavalin A-induced liver fibrosis." *PLoS ONE* 8 (12):e82232.

Liu, K.C.-S.C., S.L. Yang, M.F. Roberts, B.C. Elford, and J.D. Phillipson. 1992. "Antimalarial activity of *Artemisia annua* flavonoids from whole plants and cell cultures." *Plant Cell Reports* 11:637–640.

Mai, H.D.T., G.V. Thanh, V.H. Tran, V.N. Vu, V.L. Vu, B.N. Truong, T.D. Phi, V.M. Chau, and V.C. Pham. 2014. "Synthesis of febrifuginol analogues and evaluation of their biological activities." *Tetrahedron Letters* 55 (52):7226–7228.

Mannan, A., I. Ahmed, W. Arshad, M.F. Asim, R.A. Qureshi, I. Hussain, and B. Mirza. 2010. "Survey of artemisinin production by diverse *Artemisia* species in northern Pakistan." *Malaria Journal* 9:310.

McCalley, D.V. 2002. "Analysis of the *Cinchona* alkaloids by high-performance liquid chromatography and other separation techniques." *Journal of Chromatography A* 967:1–19.

Meshnick, S.R. 2002. "Artemisinin: Mechanisms of action, resistance and toxicity." *International Journal for Parasitology* 32 (13):1655–1660.

Mesia, K., R.K. Cimanga, L. Dhooghe, P. Cos, S. Apers, J. Totté, G.L. Tona, L. Pieters, A.J. Vlietinck, and L. Maes. 2010. "Antimalarial activity and toxicity evaluation of a quantified *Nauclea pobeguinii* extract." *Journal of Ethnopharmacology* 131 (1):10–16.

Mesia, K., L. Tona, M.M. Mampunza, N. Ntamabyaliro, T. Muanda, T. Muyembe, K. Cimanga et al. 2012. "Antimalarial efficacy of a quantified extract of *Nauclea pobeguinii* stem bark in human adult volunteers with diagnosed uncomplicated falciparum malaria. Part 1: A clinical Phase IIA trial." *Planta Medica* 78 (3):211–218.

Mesia, K., L. Tona, M. Mampunza, N. Ntamabyaliro, T. Muanda, T. Muyembe, T. Musuamba et al. 2012. "Antimalarial efficacy of a quantified extract of *Nauclea pobeguinii* stem bark in human adult volunteers with diagnosed uncomplicated falciparum malaria. Part 2: A clinical Phase IIB trial." *Planta Medica* 78 (9):853–860.

Mojab, F. 2012. "Antimalarial natural products: A review." *Avicenna Journal of Phytomedicine* 2 (2):52–62.

Mueller, M.S., N. Runyambo, I. Wagner, S. Borrmann, K. Dietz, and L. Heide. 2004. "Randomized controlled trial of a traditional preparation of *Artemisia annua* L. (annual wormwood) in the treatment of malaria." *Transactions of the Royal Society of Tropical Medicine and Hygiene* 98:318–321.

Müller, I.B., and J.E. Hyde. 2010. "Antimalarial drugs: Modes of action and mechanisms of parasite resistance." *Future Microbiology* 5:1857–1873.

Murata, K., F. Takano, S. Fushiya, and Y. Oshima. 1998. "Enhancement of NO production in activated macrophages in vivo by an antimalarial crude drug, *Dichroa febrifuga*." *Journal of Natural Products* 61:729–733.

Na-Bangchang, K., J. Karbwang, P. Molunto, V. Banmairuroi, and A. Thanavibul. 1995. "Pharmacokinetics of mefloquine, when given alone and in combination with artemether, in patients with uncomplicated falciparum malaria." *Fundamental & Clinical Pharmacology* 9:576–582.

Njan, A.A., B. Adzu, A.G. Agaba, D. Byarugaba, S. Díaz-Llera, and D.R. Bangsberg. 2008. "The analgesic and antiplasmodial activities and toxicology of *Vernonia amygdalina*." *Journal of Medicinal Food* 11 (3):574–581.

O'Neill, P.M., V.E. Barton, and S.A. Ward. 2010. "The molecular mechanism of action of artemisinin—The debate continues." *Molecules (Basel, Switzerland)* 15:1705–1721.

Oga, E.F., S. Sekine, and T. Horie. 2013. "Ex vivo and in vivo investigations of the effects of extracts of *Vernonia amygdalina, Carica papaya* and *Tapinanthus sessilifolius* on digoxin transport and pharmacokinetics: Assessing the significance on rat intestinal P-glycoprotein efflux." *Drug Metabolism and Pharmacokinetics* 28 (4):314–320.

Ohigashi, H., M.A. Huffman, D. Izutsu, K. Koshimizu, M.S.H. Kawanaka, G.C. Kirby, D.C. Warhurst et al. 1994. "Toward the chemical ecology of medicinal plant use in chimpanzees: The case of *Vernonia amygdalina*, a plant used by wild chimpanzees possibly for parasite-related diseases." *Journal of Chemical Ecology* 20:541–553.

Omoregie, E.S., A. Pal, and B. Sisodia. 2011. "In vitro antimalarial and cytotoxic activities of leaf extracts of *Vernonia amygdalina* (Del.)." *Nigerian Journal of Basic and Applied Science* 19 (1):121–126.

Owolabi, M.A., E.A. Adeniji, O.O. Oribayo, and O.E. Akindehin. 2013. "Effects of *Vernonia amygdalina* aqueous leaf extract on the pharmacokinetics of nifedipine in rabbits." *Journal of Pharmacognosy and Phytochemistry* 2 (1):55–65.

Paulo, A., E.T. Gomes, J. Steele, D.C. Warhurst, and P.J. Houghton. 2000. "Antiplasmodial activity of *Cryptolepis sanguinolenta* alkaloids from leaves and roots." *Planta Medica* 66:30–34.

Petersen, I., R. Eastman, and M. Lanzer. 2011. "Drug-resistant malaria: Molecular mechanisms and implications for public health." *FEBS Letters* 585 (11):1551–1562.

Poonam, K., and G.S. Singh. 2009. "Ethnobotanical study of medicinal plants used by the Taungya community in Terai Arc Landscape, India." *Journal of Ethnopharmacology* 123:167–176.

Rajani, M., and N.S. Kanaki. 2008. "Phytochemical standardization of herbal drugs and polyherbal formulations." In *Bioactive Molecules and Medicinal Plants,* edited by K.G. Ramawat and J.M. Mérillon, 349–369. Berlin: Springer-Verlag.

Rajapakse, S., C. Rodrigo, and S.D. Fernando. 2015. "Tafenoquine for preventing relapse in people with *Plasmodium vivax* malaria." *Cochrane Database Systematic Reviews*, doi: 10.1002/14651858.CD010458.pub2.

Sabchareon, A., T. Chongsuphajaisiddhi, V. Sinhasivanon, P. Chanthavanich, and P. Attanath. 1988. "In vivo and in vitro responses to quinine and quinidine of *Plasmodium falciparum.*" *Bulletin of the World Health Organization* 66 (3):347–352.

Sanogo, R. 2014. "Development of phytodrugs from indigenous plants: The Mali experience." In *Novel Plant Bioresources: Applications in Food, Medicine and Cosmetics*, edited by A. Gurib-Fakim. Chichester, UK: John Wiley & Sons.

Schrader, F.C., M. Barho, I. Steiner, R. Ortmann, and M. Schlitzer. 2012. "The antimalarial pipeline—An update." *International Journal of Medical Microbiology* 302 (4–5):165–171.

Sen, D., A. Banerjee, A.K. Ghosh, and T.K. Chatterjee. 2010. "Synthesis and antimalarial evaluation of some 4-quinazolinone derivatives based on febrifugine." *Journal of Advanced Pharmaceutical Technology & Research* 1:401–415.

Simonsson, U.S.H., B. Jansson, T.N. Hai, D.X. Huong, G. Tybring, and M. Ashton. 2003. "Artemisinin autoinduction is caused by involvement of cytochrome P450 2B6 but not 2C9." *Clinical Pharmacology and Therapeutics* 74:32–43.

Singh, S., T.D. Singh, V.P. Singh, and V.B. Pandey. 2010. "Quaternary alkaloids of *Argemone mexicana.*" *Pharmaceutical Biology* 48:158–160.

Singh, S., V.B. Pandey, and T.D. Singh. 2012. "Alkaloids and flavonoids of Argemone mexicana." *Natural Products Research* 26:16–21.

Solary, E., L. Mannone, D. Moreau, D. Caillot, R.O. Casasnovas, H. Guy, M. Grandjean et al. 2000. "Phase I study of cinchonine, a multidrug resistance reversing agent, combined with the CHVP regimen in relapsed and refractory lymphoproliferative syndromes." *Leukemia* 14 (12):2085–2094.

Sowunmi, A., L.A. Salako, O.J. Laoye, and A.F. Aderounmu. 1990. "Combination of quinine, quinidine and cinchonine for the treatment of acute falciparum malaria: Correlation with the susceptibility of *Plasmodium falciparum* to the cinchona alkaloids in vitro." *Transactions of the Royal Society of Tropical Medicine and Hygiene* 84 (5):626–629.

Svensson, U., and M. Ashton. 1999. "Identification of the human cytochrome P450 enzymes involved in the in vitro metabolism of artemisinin." *British Journal of Clinical Pharmacology* 48:528–535.

Tang, W., and G. Eisenbrand. 1992. *Chinese Drugs of Plant Origin*. Berlin: Springer-Verlag.

Titulaer, H.A., J. Zuidema, P.A. Kager, J.C. Wetsteyn, C.B. Lugt, and F.W. Merkus. 1990. "The pharmacokinetics of artemisinin after oral, intramuscular and rectal administration to volunteers." *Journal of Pharmacy and Pharmacology* 42 (11):810–813.

Tor-anyiin, T.A., R. Sha'ato, and H.O.A. Oluma. 2003. "Ethnobotanical survey of anti-malarial medicinal plants amongst the Tiv people of Nigeria." *Journal of Herbs, Spices & Medicinal Plants* 10:61–74.

Toyang, N.J., and R. Verpoorte. 2013. "A review of the medicinal potentials of plants of the genus *Vernonia* (Asteraceae)." *Journal of Ethnopharmacology* 146:681–723.

Tu, Y. 2011. "The discovery of artemisinin (qinghaosu) and gifts from Chinese medicine." *Nature Medicine* 17 (10):1217–1220.

Unschuld, P.U. 1992. *Medicine in China: A History of Ideas*. Berkeley, CA: University of California Press.

van Valkenburg, J.L.C.H. 2001. "*Dichroa febrifuga* Lour." Accessed May 14, 2014. http://proseanet.org/prosea/e-prosea_detail.php?frt=&id=1062.

Weathers, P.J., and M.J. Towler. 2012. "The flavonoids casticin and artemetin are poorly extracted and are unstable in an *Artemisia annua* tea infusion." *Planta Medica* 78 (10):1024–1026.

Weathers, P.J., P.R. Arsenault, P.S. Covello, A. McMickle, K.H. Teoh, and D.W. Reed. 2011. "Artemisinin production in *Artemisia annua*: Studies in planta and results of a novel delivery method for treating malaria and other neglected diseases." *Phytochemistry Reviews* 10 (2):173–183.

Wesche, D.L., and J. Black. 1990. "A comparison of the antimalarial activity of the cinchona alkaloids against *Plasmodium falciparum* in vitro." *Journal of Tropical Medicine and Hygiene* 93 (3):153–159.

Willcox, M. 2009. "*Artemisia* species: From traditional medicines to modern antimalarials—And back again." *Journal of Alternative and Complementary Medicine* 15:101–109.

Willcox, M. 2011. "Improved traditional phytomedicines in current use for the clinical treatment of malaria." *Planta Medica* 77:662–671.

Willcox, M., G. Bodeker, and P. Rasoanaivo. 2005. *Traditional Medicinal Plants and Malaria*. Vol. 59. Washington, D.C.: CRC Press.

Willcox, M.L., B. Graz, J. Falquet, O. Sidibé, M. Forster, and D. Diallo. 2007. "*Argemone mexicana* decoction for the treatment of uncomplicated falciparum malaria." *Transactions of the Royal Society of Tropical Medicine and Hygiene* 101:1190–1198.

Willcox, M.L., B. Graz, J. Flaquet, C. Diakite, S. Giani, and D. Diallo. 2011. "A 'reverse pharmacology' approach for developing an anti-malarial phytomedicine." *Malaria Journal* 10 (S1):S8.

World Health Organization (WHO). 2010. "Guidelines for the treatment of malaria. Second edition." Accessed March 4, 2015. http://whqlibdoc.who.int/publications/2010/9789241547925_eng.pdf?ua=1.

Wright, C.W. 2005. "Traditional antimalarials and the development of novel antimalarial drugs." *Journal of Ethnopharmacology* 100 (1–2):67–71.

Xing, J., B.J. Kirby, D. Whittington, Y. Wan, and D.R. Goodlett. 2012. "Evaluation of P450 inhibition and induction by artemisinin antimalarials in human liver microsomes and primary human hepatocytes." *Drug Metabolism and Disposition* 40:1757–1764.

Yan, W., W. Fischer, and J. Fratkin. 1997. *Practical therapeutics of traditional Chinese medicine*. Paradigm Publications.

Yeap, S.K., W.Y. Ho, B.K. Beh, W.S. Liang, H. Ky, A.H.N. Yousr, and N.B. Alitheen. 2010. "*Vernonia amygdalina*, an ethnoveterinary and ethnomedical used green vegetable with multiple bioactivities." *Journal of Medicinal Plants Research* 4 (25):2787–2812.

Zhou, H., L. Sun, X.L. Yang, and P. Schimmel. 2013. "ATP-directed capture of bioactive herbal-based medicine on human tRNA synthetase." *Nature* 494 (7435):121–124.
Zhu, S., G. Chandrashekar, L. Meng, K. Robinson, and D. Chatterji. 2012. "Febrifugine analogue compounds: Synthesis and antimalarial evaluation." *Bioorganic and Medicinal Chemistry* 20:927–932.

CHAPTER 5

Conclusion

Malaria is a life-threatening disease with about half the world's population at risk. Ironically, it is preventable and largely curable (except for very severe cases), if given the appropriate resources. A multifaceted approach is required to effectively curb this devastating disease. The World Health Organization's (WHO's) strategies for malaria control include vector control (i.e., the use of long-lasting insecticide-treated nets and indoor residual spraying), vaccine development, implementation of rapid diagnostic testing, chemoprevention, and treatment. Indeed, there is no magic bullet against malaria. International funding is still unable to fully support global malaria targets and remains a key limiting factor in malaria control.

Each of the approaches has its own challenges *per se*. Resistance of vectors to common insecticides has been reported, owing to misuse or overuse of insecticides during farming, and this reduces the options available, as these have to be efficacious, safe, and inexpensive, as well as environmentally friendly. In addition, as with drugs, the development of novel insecticidal agents takes time. Mosquirix (RTS,S/AS01) has recently received a "positive scientific opinion" from the European Medicines Agency as a malaria vaccine, although it displayed partial efficacy. Recommendations by WHO are expected to be made by the end of 2015. Nevertheless, many resources are needed in this area, with the goal of developing a vaccine that is effective, safe, and affordable. In the meantime, efforts to discover effective and safe antimalarial drugs or remedies must continue.

In this book, Chapter 1 gives a broad overview of the disease malaria, epidemiology and implications on public health, the *Plasmodium* life cycle, and signs and symptoms associated with malaria.

Chapter 2 details currently available antimalarial agents, listed according to the chemical classes with information on the modes of action, and examples of drugs and combination therapies. Country-specific antimalarial prevention recommendations from the Centers for Disease Control and Prevention (CDC), as well as details of antimalarial prophylaxis regimens compiled from Public Health England (PHE), WHO, and CDC are presented. In addition, useful emergency standby treatment for travelers from the PHE completes the information that healthcare professionals and the general public will find useful and handy.

Chapter 3 provides an overview to research carried out on medicinal plants used for malaria, in particular ethnobotanical surveys, and an extensive compilation of more than 1800 medicinal plants reportedly used for malaria. Details of plant usage are also included whenever available. Recommendations for prioritizing plant species for further research are discussed. Challenges abound and they include conservation of medicinal plant species, language barriers, incomplete data, limited phytochemical and safety data, and the need for harmonized methodological guidelines, as well as the need for continuous and sustainable funding. For drug discovery, bioassay-guided fractionation involving extraction of plant materials and antimalarial drug assays is presented. Finally, the active antimalarial compounds isolated from plants are listed, together with the IC_{50} values.

Chapter 4 elaborates on four selected antimalarial herbs, namely, *Artemisia annua*, *Cinchona* bark, *Dichroa febrifuga* (Changshan), and *Vernonia amygdalina* (African leaf). The mode of action, phytochemistry, reported pharmacological studies (for *D. febrifuga* and *V. amygdalina* as they are less well-researched compared to the other two), clinical trials carried out, chemical derivatives, place in therapy, reported mechanism of resistance (if any), and safety information are collated. They represent four of the most well-studied antimalarial herbs to date. Halofuginone, a synthetic halogenated derivative of febrifugine that is isolated from *Dichroa febrifuga*, may be the next blockbuster drug to look out for. More important, febrifugine derivatives may have implications for the treatment of inflammation, a cornerstone of many chronic conditions including cardiovascular diseases, autoimmune diseases, and cancer.

It is also of interest to know that several herbal preparations are in clinical use, either as government-approved preparations or are undergoing clinical development. They are Sumafoura Tiemoko Bengaly (*Argemone mexicana*), N'Dribala (*Cochlospermum planchonii*), Phyto-laria (*Cryptolepis sanguinolenta*), and PR 259 CT1 (*Nauclea pobeguinii*). Information on the results of clinical trials, active antimalarial components, and efficacy is presented wherever available.

It is clearly evident that medicinal plants have a vital role to play in malaria control. Medicinal plants are important sources of novel chemical scaffolds, as seen from how *Artemisia annua* and *Cinchona* bark have revolutionized malaria chemotherapy and prophylaxis. The recent progress on *Dichroa febrifuga* as well as the continued interest in identifying plants that are used for malaria in various communities worldwide only goes to show how nature has much more to offer in healthcare. There is certainly potential for further research in antimalarial drug discovery. Plants traditionally used for malaria have been studied extensively for their antimalarial properties, both *in vitro* and *in vivo*, some with promising results. However, very few are further developed. The concept of standardized herbal preparations is already in place in some endemic areas, such as Ghana and Mali, although much remains to be done.

Cocktail therapy has proven useful and effective in various conditions, including cancer, HIV, and tuberculosis, likewise, for malaria. In herbal medicine, multiple components may act on multiple targets in the parasite and the human body. More efforts are needed to understand how the various components work together, perhaps in synergy, or whether the antimalarial activity is due to only one active component. In any case, the prospects look promising. In addition, many plants remain understudied. It is hoped that the information in this book will help to jump-start those who are interested in this field and inspire them to take up the challenge to better understand medicinal plants.

There are many challenges in the discovery and development of new antimalarial prophylaxis and treatment. As with any type of research, well-designed ethnobotanical studies will certainly result in more substantial data that will better guide researchers. Conservation of antimalarial herbs is also critical for sustainable development of these invaluable medicinal sources for future generations. More funding for research to evaluate and develop promising drugs and remedies is warranted. In the face of the relentless emergence of parasitic resistance to currently available antimalarial drugs, nature is still a promising resource to tap on. The various challenges and limitations need to be addressed for further advancement in the field.

The trend toward identifying novel scaffolds from nature and chemically modifying the structures to develop safer and more efficacious derivatives or analogues remains important. It serves to complement other drug discovery approaches such as high throughput screening and combinatorial chemistry, which are limited by chemical diversity. As one soldiers on, and as a global community, we look toward Mother Nature and invaluable traditional knowledge to once again point us toward useful medicinal plants to harness their potential benefits.

Appendix

Table A.1 Medicinal Plants Reported to Be Used for Malaria

Plant Species	Common Name	Family	Part Used	Preparation	Indication	References
Abarema laeta	NA	Fabaceae	Bark	Infusion	Malaria	Milliken 1997a
***Abrus precatorius* L.**	Rosary pea	Fabaceae	Leaf, stem, root, seed	Decoction	Fever, malaria	Adebayo and Krettli 2011; Duke 1996; Gathirwa et al. 2011; Gessler et al. 1995; Milliken 1997a; Rasoanaivo et al. 1992; Wee 1992
Abuta grandifolia (Mart.) Sandwith	NA	Menispermaceae	Not stated	Not stated	Fever, malaria	Milliken 1997a
Abuta rufescens Aubl.	NA	Menispermaceae	Stem, cortex	Decoction, infusion	Malaria	Kvist et al. 2006; Milliken 1997a,b; Roumy et al. 2007; Ruiz et al. 2011
Abuta sp.	NA	Menispermaceae	Stem	Decoction	Malaria	Milliken 1997a
Acacia catechu	Black cutch	Fabaceae	Leaf	Maceration	Malaria	Duke 1996; Karunamoorthi and Tsehaye 2012
***Acacia concinna* DC.**	Soap pod	Fabaceae	Leaf	Infusion	Malaria	Sharma, Chhangte, and Dolui 2001
Acacia dudgeoni Craib. ex Holl	NA	Fabaceae	Not stated	Decoction	Malaria	Nadembega et al. 2011
Acacia farnesiana (L.) Willd	Sweet acacia	Fabaceae	Bark, root	Decoction	Fever, malaria	Milliken 1997a; Shankar, Sharma, and Deb 2012
Acacia foetida Kunth	NA	Fabaceae	Not stated	Not stated	Malaria	Milliken 1997a
Acacia hockii de Wild.	NA	Fabaceae	Bark, root bark	Decoction	Malaria	Geissler et al. 2002; Koch et al. 2005

(*Continued*)

Table A.1 (Continued) Medicinal Plants Reported to Be Used for Malaria

Plant Species	Common Name	Family	Part Used	Preparation	Indication	References
Acacia macrostachya Reich.	NA	Fabaceae	Fruit	Not stated	Malaria, fever	Nadembega et al. 2011
Acacia mellifera (Vahe) Benth.	Black thorn	Fabaceae	Stem bark, bark	Decoction, maceration	Malaria	Koch et al. 2005; Norscia and Borgognini-Tarli 2006; Samuelsson et al. 1992
***Acacia nilotica* (L.) Willd. ex Delile**	Egyptian thorn, red thorn, Egyptian mimosa	Fabaceae	Fruit	Maceration	Malaria	Musa et al. 2011
Acacia pennata Willd.	Rusty mimosa	Fabaceae	Leaf	Decoction	Malaria	Allabi et al. 2011; Nadembega et al. 2011
Acacia robusta Burch.	Splendid acacia, ankle thorn	Fabaceae	Root	Not stated	Malaria	Belayneh et al. 2012
Acacia senegal Willd.	Gum acacia, gum arabic tree	Fabaceae	Leaf	Decoction	Fever, malaria	Milliken 1997a; Nadembega et al. 2011
Acacia seyal Del.	White-galled acacia, white whistling thorn, thirsty thorn, buffalo thorn	Fabaceae	Root	Decoction	Malaria	Nguta et al. 2010b
Acacia sieberiana DC.	White thorn, African laburnum	Fabaceae	Root	Not stated	Malaria	Titanji, Zofou, and Ngemenya 2008
Acacia sp.	NA	Fabaceae	Bark, leaf	Decoction	Malaria	Geissler et al. 2002; Rasoanaivo et al. 1992
Acacia tortilis (Forssk.) Hayne	Umbrella thorn	Fabaceae	Root	Infusion, maceration	Malaria	Ellena, Quave, and Pieroni 2012; Koch et al. 2005
***Acalypha indica* L.**	Indian copperleaf	Euphorbiaceae	Aerial part	Decoction	Malaria	Mesfin et al. 2012

(*Continued*)

Table A.1 (Continued) Medicinal Plants Reported to Be Used for Malaria

Plant Species	Common Name	Family	Part Used	Preparation	Indication	References
***Acanthospermum australe* (Loefl.) Kuntze**	Paraguay starbur, sheep bur, spiny bur	Compositae	Whole plant, leaf	Decoction	Fever, malaria	Brandão et al. 1992; Elisabetsky and Shanley 1994; Milliken 1997a
Acanthospermum hispidum DC.	Hispid starbur	Compositae	Leaf, whole plant, stem	Decoction, bath, consumed	Fever, malaria, jaundice	Asase et al. 2005; Koudouvo et al. 2011; Milliken 1997a; Sanon et al. 2003; Scarpa 2004; Yetein et al. 2013; Zirihi et al. 2005
Acanthospermum hispidus L.	NA	Compositae	Whole plant	Decoction	Malaria	Asase, Akwetey, and Achel 2010
Acanthus polystachyus Delile	NA	Acanthaceae	Root	Not stated	Malaria	Giday et al. 2007
Achyranthes bidentata	Ox knee	Amaranthaceae	Not stated	Not stated	Malaria	Duke 1996
Acokanthera schimperi (A.D.C.) Schweeinf.	Arrow poison tree	Apocynaceae	Leaf, stem	Juice, fumes	Malaria, mosquito repellent	Belayneh et al. 2012; Mesfin et al. 2012
Aconitum deinorhizum	NA	Ranunculaceae	Not stated	Not stated	Malaria	Duke 1996
Aconitum heterophyllum	Atis	Ranunculaceae	Not stated	Not stated	Malaria	Duke 1996
***Acorus calamus* L.**	Calamus	Araceae	Root	Infusion, decoction	Fever, malaria	Milliken 1997a; Shankar, Sharma, and Deb 2012
Acorus gramineus	Grassleaf sweetflag	Acoraceae	Not stated	Not stated	Malaria	Duke 1996

(Continued)

Table A.1 (Continued) Medicinal Plants Reported to Be Used for Malaria

Plant Species	Common Name	Family	Part Used	Preparation	Indication	References
Acosmium panamense (Benth.) Yakovlev	NA	Fabaceae	Bark	Decoction	Malaria	Milliken 1997a
Acosmium sp.	NA	Fabaceae	Bark	Decoction	Malaria	Milliken 1997b
***Acrostichum aureum* L.**	Swamp fern	Pteridaceae	Stem	Not stated	Malaria	Hout et al. 2006
***Adansonia digitata* L.**	Baobab, sour gourd, cream of tartar tree, monkey-bread tree	Bombacaceae	Leaf, root, bark, fruit	Decoction, sauce	Malaria, fever	Adebayo and Krettli 2011; Ajibesin et al. 2007; Duke 1996; Koudouvo et al. 2011; Musa et al. 2011; Nguta et al. 2010a,b; Rasoanaivo et al. 1992; Yetein et al. 2013
Adenanthera microsperma Teijsm. & Binn.	Acacia coral, red bead tree	Fabaceae	Stem bark	Paste	Malaria	Prasad et al. 2008
Adenia cissampeloides (Planch ex Hook) Harms	Monkey rope	Passifloraceae	Root	Cold or hot infusion	Prevention only	Ngarivhume et al. 2015
Adhatoda schimperiana	NA	Acanthaceae	Not stated	Not stated	Malaria	Seid and Tsegay 2011

(Continued)

Table A.1 (Continued) Medicinal Plants Reported to Be Used for Malaria

Plant Species	Common Name	Family	Part Used	Preparation	Indication	References
Adhatoda zeylanica Medicus [Syn: *Adhatoda vasica* Nees, *Justicia adhatoda* L.]	Malabar nut	Acanthaceae	Leaf, root	Decoction, powdered, pounded, and boiled	Malarial fever, malaria (prevention and cure)	Ghimire and Bastakoti 2009; Nagendrappa, Naik, and Payyappallimana 2013; Namsa, Mandal, and Tangjang 2011; Poonam and Singh 2009; Shankar, Sharma, and Deb 2012; Sharma, Chhangte, and Dolui 2001
Aegle mamelos (L.) Correa ex. Roxb	Bael	Rutaceae	Bark	Juice	Malaria	Duke 1996
Afraegle paniculata	Nigerian powder-flask fruit	Rutaceae	Root	Consumed	Malaria	Asase et al. 2005
Aframomum citratum	Mbongo spice	Zingiberaceae	Fruit	Maceration	Malaria	Nadembega et al. 2011; Tetik, Civelek, and Cakilcioglu 2013
Aframomum latifoilum	Grape-seeded amomum, large amomum	Zingiberaceae	Not stated	Not stated	Malaria	Tetik, Civelek, and Cakilcioglu 2013
Aframomum melegueta K. Schum	Melegueta pepper	Zingiberaceae	Leaf, stem, fruit, seed, root	Maceration, decoction, infusion	Malaria, fever	Asase, Hesse, and Simmonds 2012; Duke 1996; Milliken 1997a; Tetik, Civelek, and Cakilcioglu 2013; Traore et al. 2013; Yetein et al. 2013
Aframomum sceptrum	Black amomum, guinea grains	Zingiberaceae	Not stated	Not stated	Malaria	Tetik, Civelek, and Cakilcioglu 2013

(*Continued*)

Table A.1 (Continued) Medicinal Plants Reported to Be Used for Malaria

Plant Species	Common Name	Family	Part Used	Preparation	Indication	References
Aframomum zambesiacum	NA	Zingiberaceae	Not stated	Not stated	Malaria	Tetik, Civelek, and Cakilcioglu 2013
Afzelia afranica Sm.	African mahogany	Fabaceae	Stem, leaf, stem bark	Decoction	Malaria	Adebayo and Krettli 2011; Asase et al. 2005; Traore et al. 2013
Aganosma marginata (Roxb) G. Don	NA	Apocynaceae	Stem, stem bark	Not stated	Malaria	Hout et al. 2006
Aganthesanthemum bojeri Klotzsch.	NA	Rubiaceae	Not stated	Not stated	Malaria	Gathirwa et al. 2011
Agathisanthenum globosum (A. Rich) Hiern	NA	Rubiaceae	Root	Decoction	Malaria	Nguta et al. 2010b
***Ageratum conyzoides* L.**	Tropical whiteweed	Compositae	Whole plant, essential oil of leaves, leaf, aerial part	Decoction	Fever, malaria	Adebayo and Krettli 2011; de Madureira et al. 2002; Kaou et al. 2008; Milliken 1997a; Rasoanaivo et al. 1992
Ageratum echioides Hemsl.	NA	Compositae	Whole plant	Infusion	Malaria	Milliken 1997a
Ailanthus altissima	Tree-of-heaven, shumac	Simaroubaceae	Not stated	Not stated	Malaria	Duke 1996
Ailanthus excelsa Roxb.	Indian tree-of-heaven	Simaroubaceae	Leaflets	Decoction	Malaria	El-Kamali and El-Khalifa 1999

(*Continued*)

Table A.1 (Continued) Medicinal Plants Reported to Be Used for Malaria

Plant Species	Common Name	Family	Part Used	Preparation	Indication	References
Ajuga remota Benth.	NA	Labiatae	Leaf, whole plant	Decoction, infusion	Malaria	Gakuya et al. 2013; Githinji and Kokwaro 1993; Muregi et al. 2007; Muthaura et al. 2007; Suleman et al. 2009
Albizia amara (Roxb.) Boiv.	NA	Fabaceae	Stem bark	Decoction	Malaria	Muthaura et al. 2007; Rukunga et al. 2009
Albizia anthelmintica Brongn	Mucenna albizia	Fabaceae	Stem bark, root	Not stated	Malaria	Muthee et al. 2011; Nanyingi et al. 2008; Rukunga et al. 2009
Albizia chevalieri Harms.	NA	Fabaceae	Root	Decoction	Malaria	Nadembega et al. 2011
Albizia coriaria Welw. ex Oliv.	NA	Fabaceae	Bark	Not stated	Malaria	Geissler et al. 2002
Albizia ferruginea (Guill. and Perr.) Benth.	West African albizia	Fabaceae	Leaf	Decoction	Malaria	Guédé et al. 2010; Zirihi et al. 2005
Albizia grandibracteata Taub.	Red nongo	Fabaceae	Leaf	Decoction	Malaria	Lacroix et al. 2011
Albizia gummifera J.F. Gmel. [Syn: *Albizzia fastigiata* Oliv.]	Peacock flower	Fabaceae	Stem bark, leaf	Decoction	Malaria, fever	Gakuya et al. 2013; Muregi et al. 2007; Muthaura et al. 2007; Rasoanaivo et al. 1992
Albizia zygia (DC.) Macbr.	Albizia	Fabaceae	Bark, stem bark	Decoction	Malaria	Tetik, Civelek, and Cakilcioglu 2013; Titanji, Zofou, and Ngemenya 2008; Traore et al. 2013

(*Continued*)

Table A.1 (Continued) Medicinal Plants Reported to Be Used for Malaria

Plant Species	Common Name	Family	Part Used	Preparation	Indication	References
***Albizzia lebbek* Benth.**	Woman's tongue	Fabaceae	Aerial part	Decoction	Malaria	Rasoanaivo et al. 1992
Alchornea cordifolia (Schumach. & Thonn.) Müll. Arg	Christmas bush	Euphorbiaceae	Root, leaf, stem, root bark	Decoction (also in combination), maceration	Malaria, fever	Adebayo and Krettli 2011; Asase, Hesse, and Simmonds 2012; Duke 1996; Guédé et al. 2010; Koudouvo et al. 2011; Magassouba et al. 2007; Mesia et al. 2008; Okpekon et al. 2004; Tetik, Civelek, and Cakilcioglu 2013; Tor-anyiin, Sha'ato, and Oluma 2003; Traore et al. 2013; Zirihi et al. 2005
Alchornea triplinervia (Spreng.) Mull.Arg.	NA	Euphorbiaceae	Not stated	Not stated	Malaria	Ruiz et al. 2011
Alchromanes difformis	NA	Euphorbiaceae	Not stated	Not stated	Malaria	Tetik, Civelek, and Cakilcioglu 2013
Allanblackia floribunda Oliv.	Vegetable tallow tree	Guttiferae	Leaf, stem	Not stated	Malaria*	Adebayo and Krettli 2011
Allanblackia monticola	NA	Clusiaceae	Not stated	Not stated	Malaria	Tetik, Civelek, and Cakilcioglu 2013
***Allium cepa* L.**	Onion	Amaryllidaceae	Bulb	Consumed	Malaria (in children too)	Betti and Yemefa 2011; Ibrahim et al. 2010; Parveen et al. 2007; Titanji, Zofou, and Ngemenya 2008

(*Continued*)

Table A.1 (Continued) Medicinal Plants Reported to Be Used for Malaria

Plant Species	Common Name	Family	Part Used	Preparation	Indication	References
Allium sativum **L.**	Garlic	Amaryllidaceae	Fruit, bulblets, rhizome, root (with *Gardinia diversifolia*), tuber	Pulverized paste (rhizome), consumed (bulb), infusion, intranasal, maceration (of tuber with roots of *Zyziphus micronata* [oral and bath])	Malaria (cure and prevention) (Prevention: bulblets)	Adebayo and Krettli 2011; Diarra et al. 2015; Duke 1996; Karunamoorthi and Tsehaye 2012; Mesfin, Demissew, and Teklehaymanot 2009; Milliken 1997a; Nadembega et al. 2011; Seid and Tsegay 2011; Suleman et al. 2009; Teklehaymanot et al. 2007; Tetik, Civelek, and Cakilcioglu 2013; Zerabruk and Yirga 2012
Allophylus pervillei Blume.	NA	Sapindaceae	Not stated	Not stated	Malaria	Gathirwa et al. 2011
Alnus glutinosa (L.) Gaertn.	Common alder, black alder	Betulaceae	Bark	Decoction	Fever, malaria	Milliken 1997a
Alnus serrulata	Hazel alder	Betulaceae	Not stated	Not stated	Malaria	Duke 1996
Aloe deserti Berger.	NA	Xanthorrhoeaceae	Leaf	Dissolved in water, decoction	Malaria	Nguta et al. 2010a,b
Aloe kedongensis Reynolds	NA	Xanthorrhoeaceae	Not stated	Infusion	Malaria	Jeruto et al. 2008

(*Continued*)

Table A.1 (Continued) Medicinal Plants Reported to Be Used for Malaria

Plant Species	Common Name	Family	Part Used	Preparation	Indication	References
Aloe macrosiphon Bak.	NA	Xanthorrhoeaceae	Leaf	Dissolved in water, infusion	Malaria	Nguta et al. 2010a,b
Aloe pirottae Berger	NA	Xanthorrhoeaceae	Leaf	Gel	Malaria	Belayneh et al. 2012
Aloe secundiflora Engl.	NA	Xanthorrhoeaceae	Root, leaf	Infusion (hot/cold water)	Malaria	Gakuya et al. 2013; Muthaura et al. 2007; Muthee et al. 2011; Nguta et al. 2010b
Aloe sp.	NA	Xanthorrhoeaceae	Leaf	Decoction, juice, infusion	Malaria	Giday et al. 2007; Jamir, Sharma, and Dolui 1999; Mesfin et al. 2012; Stangeland et al. 2011
Aloe vera **(L.) Burm. F. [Syn: *Aloe humilis, Aloe barbadensis*]**	Barbados aloe	Xanthorrhoeaceae	Leaf, leaf sap	Dissolved in water, decoction, infusion	Malaria	de Madureira et al. 2002; Gakuya et al. 2013; Milliken 1997a; Namukobe et al. 2011; Nguta et al. 2010a,b
Aloe volkensii Engl.	NA	Xanthorrhoeaceae	Leaf	Infusion	Malaria	Ssegawa and Kasenene 2007
Alpinia officinarum	Lesser galangal	Zingiberaceae	Not stated	Not stated	Malaria	Duke 1996
Alpinia zerumbet **[Syn: *Alpinia nutans* Rosc.]**	Shell plant, pink porcelain lily	Zingiberaceae	Leaf	Decoction	Malaria (in combination with *Citrus aurantiifolia*), fever	Cano and Volpato 2004; Ruiz et al. 2011

(*Continued*)

Table A.1 (Continued) Medicinal Plants Reported to Be Used for Malaria

Plant Species	Common Name	Family	Part Used	Preparation	Indication	References
Alstonia boonei De. Wild.	Cheesewood/stool wood/pattern wood	Apocynaceae	Leaf, root, stem, stem bark, bark	Decoction	Malaria, fever	Adebayo and Krettli 2011; Ajibesin et al. 2007; Akendengue 1992; Asase, Hesse, and Simmonds 2012; Asase and Oppong-Mensah 2009; Guédé et al. 2010; Idowu et al. 2010; Koudouvo et al. 2011; Mesia et al. 2008; Muganza et al. 2012; Okpekon et al. 2004; Zirihi et al. 2005
Alstonia congensis Engl.	Cheesewood/stool wood/pattern wood	Apocynaceae	Leaf, stem, bark	Maceration	Malaria	Dike, Obembe, and Adebiyi 2012; Mesia et al. 2008; Tetik, Civelek, and Cakilcioglu 2013
Alstonia constricta	Bitter bark, quinine tree, quinine bush, Australian fever bark tree, Peruvian bark	Apocynaceae	Not stated	Not stated	Malaria	Duke 1996
Alstonia macrophylla	Broad-leaved Pulai	Apocynaceae	Not stated	Not stated	Malaria	Duke 1996

(Continued)

Table A.1 (Continued) Medicinal Plants Reported to Be Used for Malaria

Plant Species	Common Name	Family	Part Used	Preparation	Indication	References
***Alstonia scholaris* (L.) R. Br.**	Blackboard tree, milkwood pine, white cheesewood	Apocynaceae	Root, bark, leaf, sap	Infusion, decoction, exudate (bark), pill (bark)	Malaria, fever, malarial fever	Asase, Hesse, and Simmonds 2012; Duke 1996; Elliott and Brimacombe 1987; Islam et al. 2014; Jorim et al. 2012; Kadir et al. 2014; Leaman et al. 1995; Namsa, Mandal, and Tangjang 2011; Pushpangadan and Atal 1984; Sharma, Chhangte, and Dolui 2001; Wiart 2000; Willcox 2011
Alternanthera brasiliana (L.) Kuntze	Brazilian joyweed	Amaranthaceae	Not stated	Not stated	Malaria	Ruiz et al. 2011
Alternanthera pungens Kunth [Syn: *Alternanthera repens* (Linn.) Link]	Khaki weed	Amaranthaceae	Whole plant	Decoction, maceration	Malaria	Guédé et al. 2010; Tchacondo et al. 2012; Zirihi et al. 2005
***Alternanthera sessilis* (L.) R. Br. ex DC.**	Sessile joyweed, dwarf copperleaf	Amaranthaceae	Not stated	Not stated	Malaria	Duke 1996
Alternanthera tenella Colla var. *bettzickiana* (Regel) Veldk.	NA	Amaranthaceae	Leaf	Juice	Malaria	Gessler et al. 1995
Althaea rosea	Hollyhock	Malvaceae	Not stated	Not stated	Malaria	Duke 1996
Alyxia spp.	NA	Apocynaceae	Not stated	Not stated	Malaria	Duke 1996

(*Continued*)

Table A.1 (Continued) Medicinal Plants Reported to Be Used for Malaria

Plant Species	Common Name	Family	Part Used	Preparation	Indication	References
Amanoa sp.	NA	Euphorbiaceae	Not stated	Not stated	Malaria	Milliken 1997a
Amaranthus hybridus L.	Slim amaranth	Amaranthaceae	Leaf	Decoction	Malaria	Nguta et al. 2010a
Amaranthus viridis	Slender amaranth	Anacardiaceae	Leaf, stem	Not stated	Malaria	Tetik, Civelek, and Cakilcioglu 2013
Amberboa ramosa (Roxb.) Jafri.	NA	Compositae	Whole plant	Decoction	Malaria	Qureshi and Bhatti 2008
Ambrosia artemisiifolia L.	Annual ragweed	Compositae	Leaf	Juice	Malaria	Milliken 1997a
Amorphophallus angolensis N.E. Br.	NA	Araceae	Tuber	Maceration	Malaria	Chifundera 2001
Ampelocissus bombycina (Bak.) Planch.	NA	Vitaceae	Leaf	Decoction	Malaria	Yetein et al. 2013
Ampelozizyphus amazonicus Ducke	NA	Rhamnaceae	Root, bark	Infusion, maceration (bark, prevention)	Malaria, fever	Brandão et al. 1992; Milliken 1997a,b
Amygdalus davidiana (Carr.) C. de Vos.	NA	Rosaceae	Leaf	Decoction	(Malaria)	Au et al. 2008
***Anacardium occidentale* L.**	Cashew tree	Anacardiaceae	Bark, stem, leaf	Decoction	Malaria	Adebayo and Krettli 2011; Ajibesin et al. 2007; Allabi et al. 2011; Diarra et al. 2015; Dike, Obembe, and Adebiyi 2012; Milliken 1997a; Ruiz et al. 2011; Traore et al. 2013

(*Continued*)

Table A.1 (Continued) Medicinal Plants Reported to Be Used for Malaria

Plant Species	Common Name	Family	Part Used	Preparation	Indication	References
Anadendrum montanum	NA	Araceae	Not stated	Not stated	Malaria	Duke 1996
***Ananas comosus* (L.) Merr.**	Pineapple	Bromeliaceae	Fruit, pericarp of fruit, leaf	Decoction	Malaria (also in combination), jaundice	Allabi et al. 2011; Asase, Hesse, and Simmonds 2012; Oni 2010; Traore et al. 2013; Yetein et al. 2013
Ancistrocladus extensus Wall.	NA	Ancistrocladaceae	Not stated	Not stated	Malaria	Duke 1996
***Andira inermis* (Wright) DC.**	Cabbage tree, brown heart	Fabaceae	Bark, Seed	Not stated	Fever, malaria	Duke 1996; Milliken 1997a
Andira surinamensis (Bondt) Splitg. ex Pulle	NA	Fabaceae	Bark	Decoction	Malaria	Milliken 1997b
***Andrographis paniculata* Wall. ex Nees**	Indian snake grass	Acanthaceae	Whole plant, leaf, stem	Decoction, infusion, paste, powder	Malaria (prevention and cure), malarial fever, fever	Duke 1996; Gurib-Fakim et al. 1993; Milliken 1997a; Nagendrappa, Naik, and Payyappallimana 2013; Namsa, Mandal, and Tangjang 2011; Nazar, Ravikumar, and, Williams 2008; Pushpangadan and Atal 1984; Rahman et al. 1999; Shankar, Sharma, and Deb 2012; Sharma, Chhangte, and Dolui 2001; Singh, Raghubanshi, and Singh 2002

(*Continued*)

Table A.1 (Continued) **Medicinal Plants Reported to Be Used for Malaria**

Plant Species	Common Name	Family	Part Used	Preparation	Indication	References
Andropogon schireusis	NA	Gramineae	Root	Not stated	Malaria	Adamu et al. 2005
Andropogon schoenanthus/ nardis L.	NA	Gramineae	Leaf	Decoction, fumes (febrifuge)	Malaria, fever	Randrianarivelojosia et al. 2003
Anemone vitifolia	NA	Ranunculaceae	Not stated	Not stated	Malaria	Duke 1996
Anethum graveolens L.	Dill	Apiaceae	Leaf, root	Decoction	Malaria	Mesfin et al. 2012
Angelica archangelica L.	Angelica	Umbelliferae	Not stated	Not stated	Malaria	Milliken 1997a
Angelica sylvestris L.	Woodland angelica	Umbelliferae	Flower, leaf, root	Decoction (root), infusion	Malaria	Milliken 1997a
Angostura spp.	NA	Rutaceae	Bark	Decoction	Fever, malaria	Milliken 1997a
***Anisomeles indica* Kuntze.**	Catmint	Labiatae	Not stated	Not stated	Malaria	Duke 1996
Anisophyllea laurina R. Br. ex Sabine	English monkey apple	Anisophylleaceae	Leaf	Decoction	Malaria	Traore et al. 2013
Anneslea fragrans Wall.	Spitting plant	Theaceae	Leaf	Not stated	Malaria*	Nguyen-Pouplin et al. 2007
Annickia affinis (Excell) Versteegh & Sosef	Yellowwood	Annonaceae	Not stated	Not stated	Malaria	Towns et al. 2014
Annickia chlorantha (Oliv.)	African yellowwood	Annonaceae	Stem	Decoction	Malaria	Tsabang et al. 2012

(Continued)

Table A.1 (Continued) Medicinal Plants Reported to Be Used for Malaria

Plant Species	Common Name	Family	Part Used	Preparation	Indication	References
***Annona muricata* L.**	Soursop	Annonaceae	Fruit, leaf, seed	Infusion (leaf), decoction (fruit, leaf), juice (fruit), maceration (leaf)	Malaria, fever	Asase, Hesse, and Simmonds 2012; Koudouvo et al. 2011; Milliken 1997a; Nguyen-Pouplin et al. 2007; Novy 1997; Tetik, Civelek, and Cakilcioglu 2013; Tsabang et al. 2012
***Annona reticulata* L.**	Custard apple	Annonaceae	Stem, leaf	Decoction	Malaria, fever	Milliken 1997a; Namsa, Mandal, and Tangjang 2011
Annona senegalensis Pers.	Wild custard apple	Annonaceae	Leaf, root	Decoction	Malaria, fever	Adebayo and Krettli 2011; Nadembega et al. 2011; Tor-anyiin, Sha'ato, and Oluma 2003; Traore et al. 2013; Tsabang et al. 2012
***Annona squamosa* L.**	Sugar apple	Annonaceae	Leaf	Decoction (with *Ocimum americanum* and *O. gratissimum*)	Malaria	Kaou et al. 2008; Nguyen-Pouplin et al. 2007
Anogeissus leiocarpa Guill. & Perr.; *Anogeissus leiocarpus* Guill. & Perr.	African birch	Combretaceae	Stem, leaf, bark	Decoction	Malaria	Adebayo and Krettli 2011; Diarra et al. 2015; Nadembega et al. 2011; Okpekon et al. 2004; Tetik, Civelek, and Cakilcioglu 2013

(*Continued*)

Table A.1 (Continued) Medicinal Plants Reported to Be Used for Malaria

Plant Species	Common Name	Family	Part Used	Preparation	Indication	References
Anplectrum glaucum Tr.	NA	Melastomataceae	Not stated	Not stated	Malaria	Duke 1996
Anthocleista amplexicaulus Bak.	NA	Loganiaceae	Aerial part	Decoction	Malaria	Rasoanaivo et al. 1992
Anthocleista djalonensis A. Chev.	NA	Loganiaceae	Root, leaf, stem, stem bark	Decoction, infusion, maceration	Malaria	Adebayo and Krettli 2011; Ajibesin et al. 2007; Diarra et al. 2015; Guédé et al. 2010; Tchacondo et al. 2012; Togola et al. 2005; Zirihi et al. 2005
Anthocleista nobilis G. Don.	Cabbage palm, cabbage tree	Loganiaceae	Stem, stem bark, root	Decoction	Malaria	Asase and Oppong-Mensah 2009; Traore et al. 2013
Anthocleista rhizophoroides Bak.	NA	Loganiaceae	Root, leaf	Decoction	Malaria	Rasoanaivo et al. 1992
Anthocleista schweinfurthii	NA	Loganiaceae	Not stated	Not stated	Malaria	Tetik, Civelek, and Cakilcioglu 2013
Anthocleista vogelii Planch.	Cabbage tree	Loganiaceae	Leaf	Not stated	Malaria	Adebayo and Krettli 2011; Tetik, Civelek, and Cakilcioglu 2013
Anthonotha macrophylla P. Beauv.	NA	Fabaceae	Stem bark	Decoction	Malaria	Guédé et al. 2010; Zirihi et al. 2005
Antidesma laciniatum	NA	Euphorbiaceae	Essential oils	Not stated	Malaria	Tetik, Civelek, and Cakilcioglu 2013
Antidesma montanum Blume	NA	Euphorbiaceae	Not stated	Not stated	Malaria	Duke 1996

(*Continued*)

Table A.1 (Continued) Medicinal Plants Reported to Be Used for Malaria

Plant Species	Common Name	Family	Part Used	Preparation	Indication	References
Apium graveolens L.	Celery	Umbelliferae	Leaf	Juice	Fever, (Malaria)	Milliken 1997a
Apocynum cannabinum	Indian hemp	Apocynaceae	Not stated	Not stated	Malaria	Duke 1996
Aquilaria agallocha Roset.	Agarwood	Thymelaeaceae	Root, stem	Infusion or decoction: thrice daily	Malaria	Kichu et al. 2015
***Arachis hypogaea* L.**	Peanut, groundnut	Fabaceae	Leaf, shell	Juice, decoction (Shell: oral and as bath)	Malaria	Allabi et al. 2011; Diarra et al. 2015; Nadembega et al. 2011; Stangeland et al. 2011
Araliopsis tabuensis	NA	Rutaceae	Stem	Not stated	Malaria	Tetik, Civelek, and Cakilcioglu 2013
Arcangelisia flava (L.) Merr.	Yellow fruit moonseed	Menispermaceae	Stem	Not stated	Malaria*	Nguyen-Pouplin et al. 2007
Arcypteris difformis Moore	NA	Dryopteridaceae	Not stated	Not stated	Malaria	Duke 1996
Ardisia sp.	NA	Myrsinaceae	Not stated	Not stated	Malaria	Duke 1996
Ardisia virens Kurz	NA	Myrsinaceae	Root	Decoction	Malaria	Ghorbani et al. 2011
Areca catechu	Betel nut	Arecaceae	Not stated	Not stated	Malaria	Duke 1996
Arenga pinnata (Wurmbe) Merr.	Red sugar cane	Palmae	Not stated	Not stated	Malaria, fever	Elliott and Brimacombe 1987
Argemone mexicana L.	Mexican poppy	Papaveraceae	Leaf, aerial part, flower, fruit, latex with lemon juice, leaf	Infusion, decoction, juice	Malaria, fever, jaundice	Allabi et al. 2011; Diarra et al. 2015; Idowu et al. 2010; Kosalge and Fursule 2009; Milliken 1997a; Poonam and Singh 2009; Willcox 2011; Yetein et al. 2013

(*Continued*)

Table A.1 (Continued) Medicinal Plants Reported to Be Used for Malaria

Plant Species	Common Name	Family	Part Used	Preparation	Indication	References
Argyrovernonia martii (DC.) MacLeish	NA	Compositae	Leaf	Infusion	Malaria	Milliken 1997a
Arisaema sp.	NA	Araceae	Not stated	Not stated	Malaria	Duke 1996
***Aristolochia acuminata* Lamk.**	Indian birthwort	Aristolochiaceae	Root, stem, leaf	Decoction	Malaria	Rasoanaivo et al. 1992
Aristolochia albida Duch.	NA	Aristolochiaceae	Tuber	Infusion	Malaria and prevention	Ngarivhume et al. 2015
Aristolochia bracteata Retz.	NA	Aristolochiaceae	Root	Maceration	Malaria	Musa et al. 2011
Aristolochia brasiliensis Mart. & Zucc.	Rooster flower	Aristolochiaceae	Root	Infusion	Fever, malaria	Milliken 1997a
Aristolochia brevipes Benth.	NA	Aristolochiaceae	Root	Decoction	Malaria	Dimayuga, Murillo, and Pantoja 1987
***Aristolochia elegans* Master**	Elegant Dutchman's pipe	Aristolochiaceae	Root, seed	Infusion, maceration, decoction	Fever, malaria	Hamill et al. 2000; Milliken 1997a; Stangeland et al. 2011
Aristolochia heppii Merxm	NA	Aristolochiaceae	Root	Infusion	Malaria	Ngarivhume et al. 2015
Aristolochia indica L.	Indian birthwort	Aristolochiaceae	Root, leaf, bark	Decoction	Malaria	Duke 1996; Jorim et al. 2012; Namsa, Mandal, and Tangjang 2011
Aristolochia serpentaria	Virginia snakeroot	Aristolochiaceae	Not stated	Not stated	Malaria	Duke 1996
Aristolochia spp.	NA	Aristolochiaceae	Root	Not stated	Fever, malaria	Milliken 1997a
Aristolochia stahelii O.C. Schmidt	NA	Aristolochiaceae	Stem	Decoction	Fever, malaria	Milliken 1997a; Vigneron et al. 2005

(Continued)

Table A.1 (Continued) Medicinal Plants Reported to Be Used for Malaria

Plant Species	Common Name	Family	Part Used	Preparation	Indication	References
Aristolochia trilobata L.	Bejuco de santiago	Aristolochiaceae	Root, leaf	Infusion, decoction	Fever, malaria (treatment and prevention in French Guiana, in combination with other herbs)	Milliken 1997a; Vigneron et al. 2005
Artabotrys odoratissimus R.	NA	Annonaceae	Root, stem	Not stated	Malaria	Hout et al. 2006; Ranganathan, Vijayalakshmi, and Parameswari 2012
Artemisia absinthium L.	Absinthium	Compositae	Leaf, shoot	Infusion, decoction, consumed	Fever, malaria	Kültür 2007; Milliken 1997a
Artemisia afra Jack. ex Wild	African wormwood	Compositae	Leaf	Not stated	Malaria	Mesfin, Demissew, and Teklehaymanot 2009
Artemisia annua L.	Sweet wormwood, sweet sagewort	Compositae	Aerial part, root	Infusion, juice, decoction	Malaria	Au et al. 2008; Stangeland et al. 2011; Willcox 2011
Artemisia apiacea	NA	Compositae	Not stated	Not stated	Malaria	Duke 1996
Artemisia argyi	Mugwort	Compositae	Not stated	Not stated	Malaria	Ghorbani et al. 2011
Artemisia capillaris	Oriental wormwood	Compositae	Not stated	Not stated	Malaria	Duke 1996
Artemisia nilagirica (C.B. Clarke) Pamp.	Indian wormwood	Compositae	Leaf	Decoction	Malaria	Shankar, Sharma, and Deb 2012; Sharma, Chhangte, and Dolui 2001

(Continued)

Table A.1 (Continued) Medicinal Plants Reported to Be Used for Malaria

Plant Species	Common Name	Family	Part Used	Preparation	Indication	References
***Artemisia scoparia* Waldst and Kit.**	Redstem wormwood	Compositae	Not stated	Not stated	Malaria	Mahmood et al. 2013
Artemisia spp.	NA	Compositae	Leaf	Not stated	Malaria (prevention, taken with other herbs)	Vigneron et al. 2005
Artemisia tridentata Nutt.	Big sagebrush	Compositae	Leaf	Decoction	Malarial fever (in combination with root of *Leptotaenia multifida*)	Shemluck 1982
***Artemisia vulgaris* L.**	Common wormwood	Compositae	Leaf, fruit, flower, root	Decoction, juice, infusion	Malaria	Milliken 1997a,b; Namsa, Mandal, and Tangjang 2011; Neves et al. 2009; Wiart 2000
Arthrostemma macrodesmum	NA	Melastomataceae	Aerial part	Decoction	Malaria	Milliken 1997a
Arthrostemma volubile (Bonpl. Ex Naud.) Triana	NA	Melastomataceae	Aerial part	Decoction	Fever, malaria	Milliken 1997a
Asparagus africanus Lam.	African asparagus	Asparagaceae	Leaf	Juice	Malaria	Mesfin et al. 2012
Aspidosperma discolor	NA	Apocynaceae	Not stated	Not stated	Malaria	Duke 1996; Schultes 1979
Aspidosperma excelsum Benth. [Syn: *Aspidosperma nitidum* Benth.]	NA	Apocynaceae	Bark, cortex	Decoction	Fever, malaria	Duke 1996; Kvist et al. 2006; Milliken 1997a,b; Ruiz et al. 2011; Schultes 1979

(*Continued*)

Table A.1 (Continued) Medicinal Plants Reported to Be Used for Malaria

Plant Species	Common Name	Family	Part Used	Preparation	Indication	References
Aspidosperma illustris (Vell.) Kuhlm & Piraja [Syn: *Coutinia illustris* Velloso; *Dipladenia ilustris* Mull. Arg.]	NA	Apocynaceae	Not stated	Not stated	Quinine substitute	Cosenza et al. 2013
Aspidosperma quebracho-blanco Schltr.	Quebracho	Apocynaceae	Bark	Not stated	Malaria (also in combination with "quemadillo")*	Duke 1996; Milliken 1997a; Scarpa 2004
Aspidosperma rigidum Rusby	NA	Apocynaceae	Cortex	Not stated	Malaria	Kvist et al. 2006
Aspilia africana (P. Beauv.) C.D. Adams	NA	Compositae	Stem, leaf	Decoction	Malaria	Adebayo and Krettli 2011; Stangeland et al. 2011; Tetik, Civelek, and Cakilcioglu 2013
Asplenium adiantoides C. Chr.	NA	Aspleniaceae	Whole plant	Not stated	Malaria	Shankar, Sharma, and Deb 2012
Aster amellus L.	European Michaelmas daisy, Italian aster, Italian starwort	Compositae	Root	Not stated	Malaria	Shankar, Sharma, and Deb 2012
Aster trinervius	Purple aster	Compositae	Not stated	Not stated	Malaria	Duke 1996
Astragalus hoantchy	NA	Fabaceae	Not stated	Not stated	Malaria	Duke 1996
Astripomoea malvacea (Klotzsch) A. Meeuse.	Common star creeper	Convolvulaceae	Root	Not stated	Malaria	Gakuya et al. 2013
Astrocaryum chonta C. Martius	NA	Arecaceae	Fruit	Not stated	Malaria	Kvist et al. 2006
Atractylis sinensis	NA	Compositae	Not stated	Not stated	Malaria	Duke 1996

(*Continued*)

Table A.1 (Continued) Medicinal Plants Reported to Be Used for Malaria

Plant Species	Common Name	Family	Part Used	Preparation	Indication	References
Autranella congolensis A. Chev. & De Wild.	NA	Sapotaceae	Stem bark	Decoction	Malaria	Muganza et al. 2012
Avicennia basilicum L.	NA	Avicenniaceae	Aerial part	Decoction	Malaria	Rasoanaivo et al. 1992
***Avicennia marina* (Forsk) Vierh.**	Grey mangrove, api-api jambu	Avicenniaceae	Aerial part	Decoction	Malaria	Rasoanaivo et al. 1992
***Axonopus compressus* (Sw.) P. Beauv.**	Broadleaf carpet grass	Poaceae	Whole plant	Not stated	Malaria	Dike, Obembe, and Adebiyi 2012
Ayapana lanceolata R.M. King & H. Rob.	NA	Compositae	Not stated	Not stated	Malaria	Ruiz et al. 2011
Ayapana triplinervis (Vahl) R.M. King & H. Rob.	NA	Compositae	Leaf, whole plant	Infusion, decoction	Fever, (Malaria), Malaria (with other herbs)	Milliken 1997a; Vigneron et al. 2005
***Azadirachta indica* (A. Juss) L. [Syn: *Melia azadiracha* L.]**	Neem tree	Meliaceae	Leaf, root, twig, bark, fruit, root bark, stem bark, seed	Decoction, infusion, juice (also as enema), maceration, fumigation, consumed (seed), powder (root)	Malaria (also in combination) (prevention and cure) (adult and pediatric), malarial fever, repellent	Adamu et al. 2005; Adebayo and Krettli 2011; Al-adhroey et al. 2010; Asase, Akwetey, and Achel 2010; Asase, Hesse, and Simmonds 2012; Asase and Kadera 2014; Asase and Oppong-Mensah 2009; Asase et al. 2005; Belayneh et al. 2012; Betti and Yemefa 2011; Beverly and Sudarsanam 2011;

(*Continued*)

Table A.1 (Continued) Medicinal Plants Reported to Be Used for Malaria

Plant Species	Common Name	Family	Part Used	Preparation	Indication	References
						Dike, Obembe, and Adebiyi 2012; Duke 1996; El-Kamali and El-Khalifa 1999; Gathirwa et al. 2011; Gessler et al. 1995; Giday et al. 2007; Hout et al. 2006; Idowu et al. 2010; Katewa, Chaudhary, and Jain 2004; Mesfin et al. 2012; Milliken 1997a; Muregi et al. 2007; Musa et al. 2011; Muthee et al. 2011; Nadembega et al. 2011; Nagendrappa, Naik, and Payyappallimana 2013; Namsa, Mandal, and Tangjang 2011; Nanyingi et al. 2008; Nguta et al. 2010a,b; Prabhu et al. 2014; Singh, Raghubanshi, and Singh 2002; Ssegawa and Kasenene 2007; Stangeland et al. 2011; Tchacondo et al. 2012; Tetik, Civelek, and Cakilcioglu 2013; Titanji, Zofou, and Ngemenya 2008; Tor-anyiin, Sha'ato, and Oluma 2003; Yetein et al. 2013

(*Continued*)

Table A.1 (Continued) Medicinal Plants Reported to Be Used for Malaria

Plant Species	Common Name	Family	Part Used	Preparation	Indication	References
Baccharis genistelloides Pers.	NA	Compositae	Leaf	Not stated	Fever, malaria	Milliken 1997a
Baccharis lanceolata Kunth	NA	Compositae	Not stated	Not stated	Malaria	Milliken 1997a
Baccharis trimera (Less.) DC. [Syn: *Baccharis genistelloides* subsp. *crispa* (Spreng.) Joch. Mull]	NA	Compositae	Not stated	Not stated	Quinine substitute	Cosenza et al. 2013
Bacopa monniera (L.) Wettst.	Water hyssop	Scrophulariaceae	Whole plant	Infusion	Fever, malaria	Prasad et al. 2008
Bactris gasipaes Kunth	Peach palm	Arecaceae	Root or leaf	Decoction	Malaria	Giovannini 2015
Baissea multiflora A.DC.	NA	Apocynaceae	Root bark	Decoction	Malaria	Traore et al. 2013
Balanites aegyptiaca (L.) Delile	Desert date	Balanitaceae	Leaf	Decoction: oral and bath	Malaria	Belayneh et al. 2012; Diarra et al. 2015; Duke 1996; Geissler et al. 2002; Nadembega et al. 2011
Balanites rotundifolia (van Tieghem) Blatter	NA	Balanitaceae	Leaf, fruit	Decoction, maceration in water	Malaria	Hassan-Abdallah et al. 2013; Mesfin et al. 2012
***Bambusa vulgaris* Schrad ex J.C. Wendl.**	Common bamboo	Gramineae	Leaf	Decoction (oral and as bath)	Malaria	Asase, Akwetey, and Achel 2010; Diarra et al. 2015; Milliken 1997a

(*Continued*)

Table A.1 (Continued) Medicinal Plants Reported to Be Used for Malaria

Plant Species	Common Name	Family	Part Used	Preparation	Indication	References
Banisteriopsis caapi (Spruce ex Grises) Morton	Soulvine	Malpighiaceae	Leaf	Decoction	Malaria	Giovannini 2015; Ruiz et al. 2011
Barleria acanthoides Vahl	NA	Acanthaceae	Not stated	Not stated	Malarial fever	Yaseen et al. 2015
***Barringtonia acutangula* (L.) Gaertn.**	Freshwater mangrove, mangopine	Lecythidaceae	Bark	Not stated	Malaria*	Nguyen-Pouplin et al. 2007
Basella alba	Ceylon spinach	Basellaceae	Not stated	Not stated	Malaria	Tetik, Civelek, and Cakilcioglu 2013
Bauhinia guianensis Aubl. var. *kunthiana* (Vogel) Wunderlin	NA	Fabaceae	Not stated	Not stated	Malaria	Milliken 1997a
Bauhinia rufescens Lam.	NA	Fabaceae	Leafy stem	Decoction	Malaria	Jansen et al. 2010
Bauhinia sp.	NA	Fabaceae	Sap, stem	Not stated, decoction (stem)	Fever, malaria	Milliken 1997a; Odonne et al. 2013
Bauhinia ungulata L.	Pata de vaca	Fabaceae	Root	Decoction	Malaria	Milliken 1997b
Begonia inflata Clarke	NA	Begoniaceae	Rhizome	Decoction	Malaria	Sharma, Chhangte, and Dolui 2001
Begonia parviflora Poepp. & Endl	NA	Begoniaceae	Leaf	Infusion	Malaria, fever	Valadeau et al. 2010
Belamcanda chinensis	Blackberry lily	Iridaceae	Not stated	Not stated	Malaria	Duke 1996
Berberis aristata D.C.	Indian barberry, tree turmeric	Berberidaceae	Root	Not stated	Malaria	Duke 1996; Shankar, Sharma, and Deb 2012
Berberis goudotii Triana & Planch.	NA	Berberidaceae	Stem, root	Infusion	Fever, (Malaria)	Milliken 1997a

(Continued)

Table A.1 (Continued) Medicinal Plants Reported to Be Used for Malaria

Plant Species	Common Name	Family	Part Used	Preparation	Indication	References
Berberis lutea Ruiz & Pav.	NA	Berberidaceae	Stem, root	Infusion	Fever, (Malaria)	Milliken 1997a
Berberis rigidifolia Kuth	NA	Berberidaceae	Stem, root	Infusion	(Malaria)	Milliken 1997a
Berberis ruscifolia Lam.	NA	Berberidaceae	Not stated	Not stated	Malaria	Tabuti 2008
Berberis sp.	Barberry	Berberidaceae	Not stated	Decoction	Malaria	Milliken 1997a
Berberis vulgaris L.	Common barberry	Berberidaceae	Root	Decoction	Malaria	Milliken 1997a
Bergia suffruticosa Fenzl.	NA	Elatinaceae	Whole plant	Decoction	Malaria (Children)	Jansen et al. 2010
Bersama abyssinica Fresen.	Winged bersama	Melianthaceae	Leaf, stem, root bark	Decoction	Malaria	Guédé et al. 2010; Karunamoorthi and Tsehaye 2012; Zirihi et al. 2005
Bersama engleriana	NA	Melianthaceae	Leaf	Not stated	Malaria	Tetik, Civelek, and Cakilcioglu 2013
***Bertholletia excelsa* Humb. & Bonpl.**	Brazil nut	Lecythidaceae	Not stated	Not stated	Malaria	Milliken 1997a
Bidens bipinata	Spanish needles	Compositae	Not stated	Not stated	Malaria	Tetik, Civelek, and Cakilcioglu 2013
Bidens cynapiifolia Kunth	NA	Compositae	Leaf, root	Infusion, decoction	Fever, malaria	Milliken 1997a,b
Bidens grantii	NA	Compositae	Leaf, flower	Decoction	Malaria	Stangeland et al. 2011
Bidens pilosa L.	Hairy beggartick	Compositae	Leaf	Infusion	Fever, malaria	Milliken 1997a; Tetik, Civelek, and Cakilcioglu 2013; Valadeau et al. 2010; Wiart 2000

(*Continued*)

Table A.1 (Continued) Medicinal Plants Reported to Be Used for Malaria

Plant Species	Common Name	Family	Part Used	Preparation	Indication	References
Biophytum petersianum Klotzsch.	African sensitive plant	Oxalidaceae	Not stated	Not stated	Malaria	Grønhaug et al. 2008
Biophytum umbraculum Welw.	NA	Oxalidaceae	Leaf	Decoction: oral and bath	Malaria	Diarra et al. 2015
***Bixa orellana* L.**	Lipstick tree	Bixaceae	Root, seed, shoot, leaf	Decoction	Fever, malaria	Bertani et al. 2005; Brandão et al. 1992; Dike, Obembe, and Adebiyi 2012; Duke 1996; Milliken 1997a,b; Nguyen-Pouplin et al. 2007; Ruiz et al. 2011
***Blighia sapida* K. Konig**	Ackee	Sapindaceae	Aril, stem, fruit, leaf	Eaten, powder, decoction	Fever, malaria	Adebayo and Krettli 2011; Diarra et al. 2015; Milliken 1997a; Tchacondo et al. 2012
Blighia unijugata Bak.	Triangle tops	Sapindaceae	Stem bark	Decoction	Malaria	Namukobe et al. 2011
Blumea balsamifera	Ngai camphor	Compositae	Not stated	Not stated	Malaria	Duke 1996
Bocia senegalensis Lam.	NA	Capparaceae	Not stated	Powder	Malaria	Nadembega et al. 2011
***Boerhaavia diffusa* L.**	Red spiderling	Nyctaginaceae	Aerial part, root	Decoction	Malaria, fever	Duke 1996; Rasoanaivo et al. 1992
Boerhavia elegans Choisy	NA	Nyctaginaceae	Flower, thin branch	Infusion	Malaria	Sadeghi et al. 2014
Boerhaavia erecta	NA	Nyctaginaceae	Not stated	Not stated	Malaria	Duke 1996
Boerhavia coccinea Mill.	Scarlet spiderling	Nyctaginaceae	Root, leaf	Decoction	Malaria, (Malaria)	Milliken 1997a

(Continued)

Table A.1 (Continued) Medicinal Plants Reported to Be Used for Malaria

Plant Species	Common Name	Family	Part Used	Preparation	Indication	References
Boerhavia hirsuta Willd.	NA	Nyctaginaceae	Leaf	Not stated	Malaria	Brandão et al. 1992
Bombax buonopozense P.Beauv. [Syn: *Bombax flammeum*]	Gold Coast bombax	Bombacaceae	Stem	Decoction	Malaria	Asase and Oppong-Mensah 2009; Tetik, Civelek, and Cakilcioglu 2013
Bombax costatum Pellegr. & Vuillet	Cotton tree	Bombacaceae	Leaf	Decoction: oral and bath	Malaria	Diarra et al. 2015
Bonnetia paniculata Spruce ex Benth	NA	Theaceae	Not stated	Not stated	Fever, malaria	Milliken 1997a
Borassus aethiopum Mart.	Agobeam, fan palm, desert palm, elephant palm	Arecaceae	Bulb, root, leaf, pulp	Maceration, powder, decoction (leaf, pulp: oral and as bath)	Malaria	Diarra et al. 2015; Nadembega et al. 2011; Tchacondo et al. 2012
Borassus flabellifer **L.**	Toddy palm	Arecaceae	Leaf, fruit	Decoction, juice (fruit)	Malaria	Ellena, Quave, and Pieroni 2012; Koudouvo et al. 2011
Borreria ocimoides (Burm. F.) DC.	NS	Rubiaceae	Whole plant	Infusion, decoction	Fever, malaria	Milliken 1997a
Boscia angustifolia A. Rich	NA	Rubiaceae	Stem bark, leaf	Decoction, juice	Malaria	Diarra et al. 2015; Koch et al. 2005; Muthaura et al. 2007
Boscia coriacea Pax	NA	Capparaceae	Root	Not stated	Malaria	Norscia and Borgognini-Tarli 2006
Boswellia dalzielii Hutch.	NA	Burseraceae	Leaf, bark	Decoction	Malarial fever	Jansen et al. 2010

(Continued)

Table A.1 (Continued) Medicinal Plants Reported to Be Used for Malaria

Plant Species	Common Name	Family	Part Used	Preparation	Indication	References
Bothriocline longipes (Oliv. & Hiern)	NA	Compositae	Leaf	Decoction, juice	Malaria	Jamir, Sharma, and Dolui 1999; Stangeland et al. 2011
Bouchea prismatica Kuntze	NA	Verbenaceae	Whole plant	Decoction	Fever, malaria	Milliken 1997a
Bowdichia virgilioides Kunth	Alcornoque	Fabaceae	Bark, seed	Infusion, decoction	Malaria, fever	Bourdy, Dewalt, and Cha 2000; Deharo et al. 2001; Milliken 1997a
Brachylaena huillensis O. Hoffm.	Silver oak	Compositae	Root	Infusion	Malaria	Ngarivhume et al. 2015
Brachylaena ramiflora (DC.) H. Humb.	NA	Compositae	Aerial part	Decoction	Malaria	Rasoanaivo et al. 1992
Brassica nigra (L.) Koch.	Black mustard	Brassicaceae	Root	Juice	Malaria	Karunamoorthi and Tsehaye 2012
Breonadia salicina (Vahl) Hepper	NA	Rubiaceae	Root, bark	Not stated	Malaria	Gakuya et al. 2013
Breynia vitis-idaea (Burm.f.) C.E.C. Fischer	NA	Euphorbiaceae	Leaf	Not stated	Malaria	Chander et al. 2014
Brickellia cavanillesi	NA	Compositae	Not stated	Not stated	Malaria	Duke 1996
Bridelia ferruginea Benth.	NA	Euphorbiaceae	Leaf, stem, root	Decoction (leaf: oral and as bath), powder	Malaria	Adebayo and Krettli 2011; Diarra et al. 2015; Magassouba et al. 2007; Okpekon et al. 2004; Tchacondo et al. 2012; Tor-anyiin, Sha'ato, and Oluma 2003; Yetein et al. 2013

(*Continued*)

Table A.1 (Continued) Medicinal Plants Reported to Be Used for Malaria

Plant Species	Common Name	Family	Part Used	Preparation	Indication	References
***Bridelia micrantha* (Hochst.) Baill.**	Bridelia	Euphorbiaceae	Leaf, root bark, stem bark	Decoction	Malaria	Adebayo and Krettli 2011; Namukobe et al. 2011; Nguta et al. 2010b; Tetik, Civelek, and Cakilcioglu 2013; Traore et al. 2013
Brillantaisia patula T. Anders.	NA	Acanthaceae	Whole plant	Not stated	Malaria	Mbatchi et al. 2006
Briquetia spicata (Kunth) Fryxell	NA	Malvaceae	Leaf, root	Infusion	Malaria	Milliken 1997a
Brosimum lactescens S. Moore	Lechoso	Moraceae	Cortex	Not stated	Malaria	Kvist et al. 2006
Brosimum rubescens Taub.	Bloodwood cacique	Moraceae	Not stated	Not stated	Malaria	Kvist et al. 2006; Ruiz et al. 2011
Brucea antidysenterica	NA	Scrophulariaceae	Leaf, stem, seed	Decoction	Malaria	Karunamoorthi and Tsehaye 2012; Suleman et al. 2009; Tetik, Civelek, and Cakilcioglu 2013
***Brucea javanica* (Linn.) (Merr.)**	Macassar kernel	Simaroubaceae	Fruit, root, leaf	Infusion	Malaria	Al-adhroey et al. 2010; Duke 1996; Hout et al. 2006; Nguyen-Pouplin et al. 2007; Shankar, Sharma, and Deb 2012; Zheng and Xing 2009

(*Continued*)

Table A.1 (Continued) Medicinal Plants Reported to Be Used for Malaria

Plant Species	Common Name	Family	Part Used	Preparation	Indication	References
Brunfelsia grandiflora D. Don ssp. *Schultesii* Plowman	Largeflower brunfelsia	Solanaceae	Root	Not stated	Fever, malaria	Kvist et al. 2006; Milliken 1997a; Ruiz et al. 2011
Brunfelsia sp.	NA	Solanaceae	Not stated	Not stated	Fever, malaria	Milliken 1997a
Brunsvigia littoralis	NA	Amaryllidaceae	Not stated	Not stated	Malaria	Wiart 2000
Buchholzia coriacea	Musk tree	Capparaceae	Seed	Not stated	Malaria	Tetik, Civelek, and Cakilcioglu 2013
Buchholzia macrophylla Pax	NA	Capparaceae	Leaf	Not stated	Malaria	Mbatchi et al. 2006
Buddleja asiatica	Dogtail	Loganiaceae	Not stated	Not stated	Malaria	Duke 1996
Bupleurum chinense	Thorowax, northern bupleurum	Apiaceae	Not stated	Not stated	Malaria	Duke 1996
Bupleurum falcatum	Chinese thoroughwax, sickle-leaf hare's ear	Apiaceae	Not stated	Not stated	Malaria	Duke 1996
Bupleurum longicaule Wall. ex DC.	NA	Apiaceae	Whole plant	Powder	Malaria	Kayani et al. 2015
Bupleurum scorzoneraefolium	Southern bupleurum	Apiaceae	Not stated	Not stated	Malaria	Duke 1996
Burasaia australis Sc. Elliot	NA	Menispermaceae	Root bark	Decoction	Malaria, adjunct to chloroquine and quinine	Rasoanaivo et al. 1992
Burasaia congesta Decne	NA	Menispermaceae	Root bark	Decoction	Malaria, adjunct to chloroquine and quinine	Rasoanaivo et al. 1992

(*Continued*)

Table A.1 (Continued) Medicinal Plants Reported to Be Used for Malaria

Plant Species	Common Name	Family	Part Used	Preparation	Indication	References
Burasaia gracilis Decne	NA	Menispermaceae	Root bark	Decoction	Malaria, adjunct to chloroquine and quinine	Rasoanaivo et al. 1992
Burasaia madagascariensis Thouars	NA	Menispermaceae	Root, root bark	Infusion, decoction	Malaria, adjunct to quinine and chloroquine	Novy 1997
Burasaia nigrescens R. Cap.	NA	Menispermaceae	Root bark	Decoction	Malaria, adjunct to chloroquine and quinine	Rasoanaivo et al. 1992
Burdachia prismatocarpa A. Juss.	NA	Malpighiaceae	Cortex	Not stated	Malaria	Kvist et al. 2006
Bursera simaruba (L.) Sarg.	Gumbo limbo	Burseraceae	Bark, leaf	Infusion, decoction	Malaria, fever	Milliken 1997a
Buxus hyrcana Pojark.	NA	Buxaceae	Not stated	Not stated	Malaria	Esmaeili et al. 2009
Buxus sempervirens L.	Boxwood	Buxaceae	Bark, leaf, flower	Decoction	Fever, malaria	Duke 1996; Milliken 1997a; Neves et al. 2009
Byrsonima crassa Nied.	NA	Malpighiaceae	Not stated	Not stated	Malaria	Milliken 1997a
Byrsonima spp.	NA	Malpighiaceae	Bark	Decoction	Malaria	Milliken 1997b
Cabralea canjerana (Vell.) Mart.	NA	Meliaceae	Bark	Infusion	Malaria	Milliken 1997a

(Continued)

Table A.1 (Continued) **Medicinal Plants Reported to Be Used for Malaria**

Plant Species	Common Name	Family	Part Used	Preparation	Indication	References
Cadaba farinosa Forsk.	Cadaba	Capparaceae	Leaf	Decoction	Malaria	Nadembega et al. 2011; Tetik, Civelek, and Cakilcioglu 2013
***Caesalpinia bonduc* (L.) Roxb. [Syn: *Caesalpinia bonducella* (L.) Fleming]**	Nickernut	Fabaceae	Seed, root, leaf, stem	Infusion, powder/cold drink of seeds; decoction (root), powdered (seed pulp)	Malarial fever, malaria	Duke 1996; Milliken 1997a; Novy 1997; Poonam and Singh 2009; Rasoanaivo et al. 1992; Willcox 2011; Yetein et al. 2013
***Caesalpinia coriaria* Willd.**	Divi divi	Fabaceae	Pods	Not stated	Fever, malaria	Duke 1996; Milliken 1997a
***Caesalpinia ferrea* Mart ex Tul.**	Brazilian ironwood, leopard tree	Fabaceae	Fruit	Alcoholic extract	Malaria	Milliken 1997b
***Caesalpinia pulcherrima* (L.) Sw.**	Pride-of-Barbados	Fabaceae	Root, fruit, flower, leaf	Decoction (root, fruit, leaf), pill (root)	Fever, malaria	Chander et al. 2014; Milliken 1997a; Nguyen-Pouplin et al. 2007; Shil, Dutta Choudhury, and Das 2014; Yetein et al. 2013
Caesalpinia sepiara	Mysore thorn, Mauritius thorn	Fabaceae	Not stated	Not stated	Malaria	Duke 1996
Caesalpinia volkensii Harms	NA	Fabaceae	Leaf, seed	Not stated	Malaria	Muregi et al. 2007

(*Continued*)

Table A.1 (Continued) Medicinal Plants Reported to Be Used for Malaria

Plant Species	Common Name	Family	Part Used	Preparation	Indication	References
Cajanus cajan (L.) Millsp.	Pigeonpea	Fabaceae	Leaf, root	Infusion (bath), maceration, decoction	Fever, malaria	Adebayo and Krettli 2011; Allabi et al. 2011; Chander et al. 2014; Idowu et al. 2010; Kaou et al. 2008; Milliken 1997a; Stangeland et al. 2011; Titanji, Zofou, and Ngemenya 2008; Tetik, Civelek, and Cakilcioglu 2013; Yetein et al. 2013
Calamus salicifolius Becc.	NA	Arecaceae	Root	Not stated	Malaria*	Nguyen-Pouplin et al. 2007
Calea berteriana DC.	NA	Compositae	Not stated	Not stated	Malaria	Milliken 1997a
Calea zacatechichi	NA	Compositae	Not stated	Not stated	Malaria	Duke 1996
Calliandra anomala	NA	Fabaceae	Not stated	Not stated	Malaria	Duke 1996
Calliandra clavellina Karst.	NA	Fabaceae	Root	Not stated	Malaria, fever	Milliken 1997a
Calliandra houstoniana	NA	Fabaceae	Not stated	Not stated	Malaria	Duke 1996
Calotropis gigantea	Milkweed, swallow wort	Asclepiadaceae	Leaf, stem	Decoction	Malaria	Duke 1996; Kadir et al. 2014
Calotropis procera (Aiton) W.T. Aiton	Roostertree	Asclepiadaceae	Leaf, root, root bark, root, flower	Maceration, decoction (leaf: oral and bath)	Malaria, (malaria), malarial fever	Diarra et al. 2015; Duke 1996; Nadembega et al. 2011; Parveen et al. 2007; Upadhyay et al. 2010; Yaseen et al. 2015; Yetein et al. 2013

(*Continued*)

Table A.1 (Continued) Medicinal Plants Reported to Be Used for Malaria

Plant Species	Common Name	Family	Part Used	Preparation	Indication	References
Calycophylum acreanum Ducke	NA	Rubiaceae	Bark	Decoction	Malaria	Milliken 1997a
Calycopteris floribunda Lam.	NA	Combretaceae	Leaf	Decoction	Malaria	Kadir et al. 2014
Camellia sinensis	Tea	Theaceae	Not stated	Not stated	Malaria	Duke 1996
Campomanesia aromatica (Aubl.) Griseb.	NA	Myrtaceae	Leaf	Not stated	Malaria (with other herbs)	Vigneron et al. 2005
Campomanesia grandiflora (Aubl.) Sagot.	NA	Myrtaceae	Leaf	Not stated	Malaria (with other herbs)	Vigneron et al. 2005
Campsiandra angustifolia Spruce ex Benth.	NA	Fabaceae	Cortex	Infusion	Malaria	Kvist et al. 2006; Milliken 1997a
Campsiandra comosa Benth.	NA	Fabaceae	Not stated	Not stated	Malaria	Tetik, Civelek, and Cakilcioglu 2013
Campsiandra comosa Benth. var. *laurifolia* (Benth.) Cowan	NA	Fabaceae	Not stated	Not stated	Malaria	Milliken 1997a
***Cananga odorata* (Lam.) Hook. f. & Thomson**	Kenanga	Annonaceae	Flower, leaf	Massage (direct application, leaf), infusion	Malaria, fever	Duke 1996; Elliott and Brimacombe 1987; Waruruai et al. 2011; Wee 1992
Canarium schweinfurthii Engl.	Incense tree, bush candle tree	Burseraceae	Root bark	Decoction	Malaria, fever	Hamill et al. 2000
Canna bidentata L.	NA	Cannaceae	Root	Not stated	Malaria*	de Madureira et al. 2002

(*Continued*)

Table A.1 (Continued) Medicinal Plants Reported to Be Used for Malaria

Plant Species	Common Name	Family	Part Used	Preparation	Indication	References
***Canna indica* L.**	Wild canna lily	Cannaceae	Leaf, root	Not stated	Malaria	Adebayo and Krettli 2011; Dike, Obembe, and Adebiyi 2012; Tetik, Civelek, and Cakilcioglu 2013
Canna odorata (Lam.) Hook.f. et Thoms	NA	Annonaceae	Leaf, bark	Not stated	Malaria*	Nguyen-Pouplin et al. 2007
Canthium glaucum Hiern.	NA	Rubiaceae	Fruit	Decoction	Malaria	Nguta et al. 2010a,b
Cantua quercifolia Juss.	NA	Polemoniaceae	Whole plant	Infusion	Malaria	Milliken 1997a
Capparis decidua	Karira	Capparaceae	Not stated	Not stated	Malaria	Duke 1996
Capparis sepiaria	Indian caper	Capparaceae	Root	Decoction	Malaria	Nadembega et al. 2011
Capparis sp.	NA	Capparaceae	Root	Decoction	Malaria	Geissler et al. 2002
Capsella bursa-pastoris (L.) Medic.	Shepherd's purse	Cruciferae	Leaf, aerial part	Infusion, decoction	Fever, malaria	Milliken 1997a
Capsiandra angustifolia Spring ex Benth	NA	Fabaceae	Not stated	Not stated	Malaria	Ruiz et al. 2011
***Capsicum annuum* L.**	Chilli pepper	Solanaceae	Fruit	Eaten	Fever, malaria (in combination) (cure and prevention)	Giday et al. 2007; Milliken 1997a; Ngarivhume et al. 2015
***Capsicum frutescens* L.**	Red pepper, cayenne pepper	Solanaceae	Fruit	Not stated	Fever, malaria	Betti and Yemefa 2011; Duke 1996; Milliken 1997a,b; Nadembega et al. 2011

(Continued)

Table A.1 (Continued) Medicinal Plants Reported to Be Used for Malaria

Plant Species	Common Name	Family	Part Used	Preparation	Indication	References
***Carallia brachiata* (Lour.) Merr.**	Freshwater mangrove	Rhizophoraceae	Leaf	Not stated	Malaria*	Nguyen-Pouplin et al. 2007
Caralluma dalzielii N.E. Br.	NA	Asclepiadaceae	Tubercle	Maceration	Malaria	Nadembega et al. 2011
***Carapa guianensis* Aubl.**	Crabwood	Meliaceae	Leaf, bark, cortex	Decoction	Fever, malaria	Duke 1996; Kvist et al. 2006; Milliken 1997a
Carapa procera DC.	English tallicoonah oil tree	Meliaceae	Leaf	Decoction	Malaria	Traore et al. 2013
Carica papaya L.	Papaya	Caricaceae	Leaf, root, twig, seed, fruit, flower + leaf, root bark	Decoction, maceration (fruit, leaf + root), decoction (fruit + leaf + root), infusion, decoction (also with *O. americanum* and *O. gratissimum*), added to water with other herbs, latex of immature fruit, juice of leaves	Malaria, fever (leaf + root), jaundice (fruit + leaf + root), suspension of fruit latex	Adebayo and Krettli 2011; Asase, Akwetey, and Achel 2010; Asase, Hesse, and Simmonds 2012; Asase and Oppong-Mensah 2009; Asase et al. 2005; Bertani et al. 2005; Dike, Obembe, and Adebiyi 2012; Ellena, Quave, and Pieroni 2012; Elliott and Brimacombe 1987; Giday et al. 2007; Kaou et al. 2008; Karunamoorthi and Tsehaye 2012; Koudouvo et al. 2011; Leaman et al. 1995; Milliken 1997a,b; Nadembega et al. 2011; Ngarivhume et al. 2015; Ong and Nordiana 1999; Ruiz et al. 2011;

(*Continued*)

Table A.1 (Continued) Medicinal Plants Reported to Be Used for Malaria

Plant Species	Common Name	Family	Part Used	Preparation	Indication	References
						Seid and Tsegay 2011; Shankar, Sharma, and Deb 2012; Stangeland et al. 2011; Suleman et al. 2009; Tor-anyiin, Sha'ato, and Oluma 2003; Traore et al. 2013; Valadeau et al. 2010; Vigneron et al. 2005; Waruruai et al. 2011; Yetein et al. 2013
Carissa carandas	Bengal currant	Apocynaceae	Not stated	Not stated	Malaria	Duke 1996
Carissa edulis (Forsk.) Vahl. [Syn: *Carrisa spinarum*]	Egyptian carissa	Apocynaceae	Root, root bark	Decoction, boiled in meat bone broth	Malaria, fever	Gakuya et al. 2013; Jeruto et al. 2008; Koch et al. 2005; Muthaura et al. 2007; Nanyingi et al. 2008; Rasoanaivo et al. 1992; Stangeland et al. 2011; Teklehaymanot et al. 2007; Titanji, Zofou, and Ngemenya 2008
***Carludovica palmata* Ruiz & Pav.**	Panama hat palm	Cyclanthaceae	Shoot	Decoction (internally, and bath)	Fever, malaria	Milliken 1997a
Casearia aff. *spruceana* Benth. ex Eichl.	NA	Flacourtiaceae	Leaf	Infusion	Malaria	Milliken 1997a
Casearia sp.	NA	Flacourtiaceae	Bark	Decoction	Malaria	Milliken 1997b

(*Continued*)

Table A.1 (Continued) Medicinal Plants Reported to Be Used for Malaria

Plant Species	Common Name	Family	Part Used	Preparation	Indication	References
Cassia aff. *abbreviata* Oliv.	NA	Fabaceae	Root, leaf, root bark	Decoction, powdered, juice	Malaria	Gessler et al. 1995; Rukunga et al. 2009
Cassia abbreviata Oliv.	Long-pod cassia	Fabaceae	Root, bark	Cold or hot infusion	Malaria (cure and prevention)	Ngarivhume et al. 2015.
Cassia arereh Del.	NA	Fabaceae	Stem bark	Decoction	Malaria	Musa et al. 2011
***Cassia fistula* L. [Syn: *Cassia fistulosa*]**	Cascara, Golden shower	Fabaceae	Flower, root, leaf, stem bark	Not stated	Fever, malaria	Idowu et al. 2010; Milliken 1997a; Ranganathan, Vijayalakshmi, and Parameswari 2012
***Cassia mimosoides* L.**	NA	Fabaceae	Stem bark	Decoction	Malaria	Allabi et al. 2011; Nadembega et al. 2011
Cassia nigricans Vahl. [Syn: *Chamaecrista nigricans* (Vahl) Greene]	NA	Fabaceae	Whole plant, stem bark, leaf	Decoction	Malaria	Nadembega et al. 2011; Traore et al. 2013
Cassia sieberiana DC.	NA	Fabaceae	Leaf, root, root bark	Decoction	Malaria	Adebayo and Krettli 2011; Asase et al. 2005; Diarra et al. 2015; Nadembega et al. 2011; Traore et al. 2013
Cassinopsis madagascariensis (Baill.) H. Bn.	NA	Icacinaceae	Leaf, stem bark	Decoction	Malaria	Rasoanaivo et al. 1992
***Cassytha filiformis* Linn.**	Devil's gut	Lauraceae	Twig, aerial part	Decoction	Malaria*, fever	Adebayo and Krettli 2011; Rasoanaivo et al. 1992

(*Continued*)

Table A.1 (Continued) Medicinal Plants Reported to Be Used for Malaria

Plant Species	Common Name	Family	Part Used	Preparation	Indication	References
***Catharanthus roseus* (L.) G. Don [Syn: *Vinca rosea*]**	Madagascar periwinkle	Apocynaceae	Root, whole plant, leaf	Infusion, decoction (whole plant, root)	Fever, malaria	Beverly and Sudarsanam 2011; Milliken 1997a; Ngarivhume et al. 2015; Wee 1992
Cecropia peltata L.	Trumpet tree	Moraceae	Leaf	Decoction	Malaria	Milliken 1997a
Cecropia spp.	NA	Moraceae	Leaf	Not stated	Fever, malaria	Milliken 1997a,b
Cedrela fissilis Vell.	NA	Meliaceae	Bark	Decoction	Malaria	Muñoz, Sauvain, Bourdy, Callapa, Bergeron et al. 2000
Cedrela odorata L.	Spanish cedar	Meliaceae	Bark, leaf, cortex	Infusion (also alcoholic), decoction (consumed, and as bath)	Fever, malaria	Alfaro 1984; de Madureira et al. 2002; Duke 1996; Kvist et al. 2006; Milliken 1997a,b; Tetik, Civelek, and Cakilcioglu 2013
Cedrela toona	NA	Meliaceae	Not stated	Not stated	Malaria	Duke 1996
***Ceiba pentandra* (L.) Gaertn.**	Kapok tree	Bombacaceae	Bark, leaf	Decoction (bark, external; leaf: oral and bath)	Fever, malaria	Diarra et al. 2015; Milliken 1997a; Nguyen-Pouplin et al. 2007; Ruiz et al. 2011
Celtis cf. *tessmannii*	NA	Ulmaceae	Not stated	Not stated	Malaria	Tetik, Civelek, and Cakilcioglu 2013
Celtis durandii Engl.	NA	Ulmaceae	Root	Not stated	Malaria*	Adebayo and Krettli 2011
Cenchrus echinatus L.	Southern sandbur	Gramineae	Whole plant	Infusion	Fever, malaria (in combination)	Milliken 1997a,b

(Continued)

Table A.1 (Continued) Medicinal Plants Reported to Be Used for Malaria

Plant Species	Common Name	Family	Part Used	Preparation	Indication	References
Centaurea calcitrapa L.	Red starthistle	Compositae	Aerial part	Decoction	Malaria	Maxia et al. 2008
Centaurea solstitialis L. ssp. *solstitialis*	Yellow starthistle	Compositae	Capitulum	Decoction, infusion	Malaria	Altundag and Ozturk 2011; Bulut and Tuzlaci 2013
***Centella asiatica* L.**	Gotu kola, asiatic pennywort, spadeleaf	Umbelliferae	Whole plant, stem, leaf, root	Decoction	Malaria	Namsa, Mandal, and Tangjang 2011
Centipeda minima	Spreading sneezeweed	Compositae	Not stated	Not stated	Malaria	Duke 1996
Centipeda minuta	NA	Compositae	Not stated	Not stated	Malaria	Duke 1996
Centratherum punctatum Cass. Ssp. *benedictum*	Larkdaisy	Compositae	Leaf	Infusion	Malaria	Milliken 1997a
Cephalanthus occidentalis	Common buttonbush	Rubiaceae	Not stated	Not stated	Malaria	Duke 1996
Cephalanthus spathelliferus Bak.	NA	Rubiaceae	Leaf	Decoction	Malaria	Rasoanaivo et al. 1992
Cephalostachyum sp.	NA	Bambusaceae	Leaf	Infusion	Malaria	Norscia and Borgognini-Tarli 2006
Ceratotheca sesamoides Endl.	False sesame	Pedaliaceae	Leaf	Decoction: oral and bath	Malaria	Diarra et al. 2015
Ceriops tagal	Yellow mangrove, spurred mangrove	Rhizophoraceae	Not stated	Not stated	Malaria	Duke 1996
Cestrum euanthes Schltdl [Syn: *Cestrum pseudoquina* Mart.]	NA	Solanaceae	Not stated	Not stated	Quinine substitute	Cosenza et al. 2013

(Continued)

Table A.1 (Continued) Medicinal Plants Reported to Be Used for Malaria

Plant Species	Common Name	Family	Part Used	Preparation	Indication	References
Cestrum laevigatum Schlecht. var. *puberuleum* Sendth.	Inkberry	Solanaceae	Leaf	Not stated	Malaria*	de Madureira et al. 2002
Cestrum megalophyllum Dunal	NA	Solanaceae	Not stated	Not stated	Malaria	Ruiz et al. 2011
Ceterach officinarum DC. [Syn: *Asplenium ceterach* L.]	NA	Aspleniaceae	Leaf	Tablet	Malaria	Guarrera, Salerno, and Caneva 2005
Chamaecrista rotundifolia (Pers.) Greene	Round-leaved cassia	Fabaceae	Leaf	Decoction	Malaria	Yetein et al. 2013
Chartoloma chartacea Craib	NA	Brassicaceae	Not stated	Not stated	Malaria	Duke 1996
Chasallia chartacea	Lado-lado	Rubiaceae	Not stated	Not stated	Malaria	Duke 1996
Chasmanthera dependens Hochst.	NA	Menispermaceae	Leaf, root	Decoction	Malaria	Allabi et al. 2011; Hamill et al. 2000
Chasmanthera uviformis Baill. [Syn: *Hyalosepalum uviforme* Troupin]	NA	Menispermaceae	Stem bark	Decoction	Malaria	Rasoanaivo et al. 1992
Cheiloclinium cognatum (Miers) A.C. Sm [Syn: *Kippistia cognata* Miers]	Corocito	Celastraceae	Not stated	Not stated	Quinine substitute	Cosenza et al. 2013

(Continued)

Table A.1 (Continued) Medicinal Plants Reported to Be Used for Malaria

Plant Species	Common Name	Family	Part Used	Preparation	Indication	References
Chenopodium ambrosioides L.	Mexican tea	Chenopodiaceae	Leaf (also in combination), whole plant, shoot	Decoction, direct application (leaf), inhalation	Fever, malaria, jaundice, splenomegaly	Kvist et al. 2006; Milliken 1997a; Odonne et al. 2013; Rasoanaivo et al. 1992; Titanji, Zofou, and Ngemenya 2008; Yetein et al. 2013
Chenopodium opulifolium	Seaport goosefoot	Chenopodiaceae	Leaf	Decoction (also in combination), juice	Malaria	Namukobe et al. 2011; Titanji, Zofou, and Ngemenya 2008
Chimarrhis turbinata D.C.	NA	Rubiaceae	Bark, leaf	Decoction	Malaria	Bertani et al. 2005
Chiococcca alba (L.) Hitsch. [Syn: *Chiococca anguifuga* Mart.]	West Indian snowberry	Rubiaceae	Not stated	Not stated	Quinine substitute	Cosenza et al. 2013
Chionanthus virginicus	White fringetree	Oleaceae	Not stated	Not stated	Malaria	Duke 1996
Chlorocardium rodiaei (Schomb.) Rowher, Richter & van der Werff	Greenheart	Lauraceae	Bark, seed	Decoction	Fever, malaria	Milliken 1997a
Chloroxylon falcatum Capuron	NA	Rutaceae	Bark	Infusion	Malaria	Norscia and Borgognini-Tarli 2006
Chondrodendron platyphyllum (A. St.-Hil.) Miers	NA	Menispermaceae	Root	Not stated	Fever, malaria	Milliken 1997a
Chretia cymosa	NA	Boraginaceae	Not stated	Not stated	Malaria	Tetik, Civelek, and Cakilcioglu 2013

(*Continued*)

Table A.1 (Continued) **Medicinal Plants Reported to Be Used for Malaria**

Plant Species	Common Name	Family	Part Used	Preparation	Indication	References
***Christia vespertilionis* (L.f.) Bakh.f.**	NA	Fabaceae	Whole plant	Not stated	Malaria*	Nguyen-Pouplin et al. 2007
***Chromolaena odorata* Linn. King & Robinson [Syn: *Eupatorium odoratum* L.]**	Jack-in-the-bush, Christmas bush	Compositae	Leaf, twig, whole plant, root, aerial part	Infusion, decoction	Malaria, fever	Adebayo and Krettli 2011; Ajibesin et al. 2007; Idowu et al. 2010; Milliken 1997a; Tor-anyiin, Sha'ato, and Oluma 2003
Chrysochlamys sp.	NA	Guttiferae	Not stated	Not stated	Malaria	Milliken 1997a
Chrysophyllum albidum G. Don	White star apple	Sapotaceae	Leaf	Decoction	Malaria	Yetein et al. 2013
Chrysophyllum perpulchrum	Monkey star apple	Sapotaceae	Stem bark	Decoction	Malaria	Guédé et al. 2010
Chuquiraga jussieui J.F. Gmel.	NA	Compositae	Whole plant	Decoction	Malaria	Milliken 1997a; Tene et al. 2007
Cichorium intybus L.	Chicory	Compositae	Not stated	Juice	Malaria	Milliken 1997a
Cimicifuga foetida	Foetid bugbane	Ranunculaceae	Not stated	Not stated	Malaria	Duke 1996
Cimicifuga racemosa	Black cohosh	Ranunculaceae	Not stated	Not stated	Malaria	Duke 1996
Cinchona barbacoensis H. Karst.	NA	Rubiaceae	Bark	Infusion	Malaria	Milliken 1997a
Cinchona henleana H. Karst.	NA	Rubiaceae	Bark	Infusion	Malaria	Milliken 1997a
Cinchona ledgeriana Muens [Syn: *Cinchona calisaya*]	Quinine	Rubiaceae	Stem bark	Decoction	Malaria	Duke 1996; Rasoanaivo et al. 1992; Tetik, Civelek, and Cakilcioglu 2013

(Continued)

Table A.1 (Continued) Medicinal Plants Reported to Be Used for Malaria

Plant Species	Common Name	Family	Part Used	Preparation	Indication	References
Cinchona officinalis Linn. F.	Quinine	Rubiaceae	Bark, stem bark	Decoction, infusion, powder, alcoholic extract	Fever, malaria	Duke 1996; Milliken 1997a; Rasoanaivo et al. 1992; Shankar, Sharma, and Deb 2012; Tene et al. 2007
Cinchona pitayensis Wedd.	NA	Rubiaceae	Bark	Infusion	Malaria	Milliken 1997a
Cinchona pubescens M. Vahl [Syn: *Cinchona cordifolia; Cinchona succirubra* Pavon et Kiutzsch]	Red cinchona	Rubiaceae	Bark, stem bark	Decoction, infusion, tincture, powder	Fever, malaria	de Madureira et al. 2002; Duke 1996; Hanlidou et al. 2004; Milliken 1997a; Rasoanaivo et al. 1992; Tetik, Civelek, and Cakilcioglu 2013
Cinchona sp.	NA	Rubiaceae	Bark	Infusion	Malaria (treatment and prevention)	Milliken 1997a
Cinnamomum bejolghota (Buch. Ham)	NA	Lauraceae	Bark, leaf	Decoction	Malaria	Shankar, Sharma, and Deb 2012
***Cinnamomum burmanni* Blume**	Padang cassia, padang cinnamon, batavia cinnamon	Lauraceae	Not stated	Not stated	Malaria	Duke 1996
***Cinnamomum camphora* (L.) Sieb.**	Camphor tree	Lauraceae	Leaf	Inhalation, infusion	Malaria	Rasoanaivo et al. 1992
***Cinnamomum iners* Reinn.**	Wild cinnamon; kayu manis	Lauraceae	Leaf	Not stated	Malaria*	Nguyen-Pouplin et al. 2007

(*Continued*)

Table A.1 (Continued) Medicinal Plants Reported to Be Used for Malaria

Plant Species	Common Name	Family	Part Used	Preparation	Indication	References
Cinnamomum zeylanicum **Breyn.**	NA	Lauraceae	Leaf	Decoction (with *O. americanum* and *O. gratissimum*), infusion, inhalation	Malaria, fever	Kaou et al. 2008; Rasoanaivo et al. 1992
Cinnamosma fragrans H. Bn.	NA	Canellaceae	Leaf, bark	Decoction	Malaria	Randrianarivelojosia et al. 2003
Cissampelos glaberrima A. St.-Hil.	NA	Menispermaceae	Bark, root	Infusion, decoction	Malaria, quinine substitute	Milliken 1997a
Cissampelos mucronata A. Rich	NA	Menispermaceae	Aerial part, leaf, root, tuber	Decoction, powdered, infusion	Malaria	Gessler et al. 1995; Ngarivhume et al. 2015; Tor-anyiin, Sha'ato, and Oluma 2003
Cissampelos ovalifolia DC.	NA	Menispermaceae	Root, rhizome	Infusion, decoction	Fever, malaria	Milliken 1997a,b
Cissampelos pareira L. [Syn: *Cissampelos madagascariensis* (Baill.) Diels]	Velvetleaf, false pareira	Menispermaceae	Root, leaf, whole plant, root bark	Decoction, juice, infusion (leaf, bath), raw (root)	Malaria	Duke 1996; Joly et al. 1990; Kichu et al. 2015; Milliken 1997a; Muthaura et al. 2007; Namsa, Mandal, and Tangjang 2011; Norscia and Borgognini-Tarli 2006; Rasoanaivo et al. 1992; Rukunga et al. 2009; Shankar, Sharma, and Deb 2012; Shinwari and Khan 2000

(Continued)

Table A.1 (Continued) Medicinal Plants Reported to Be Used for Malaria

Plant Species	Common Name	Family	Part Used	Preparation	Indication	References
Cissus aralioides (Baker) Planch.	NA	Vitaceae	Root	Powder, decoction	Malaria	Tchacondo et al. 2012; Traore et al. 2013
Cissus flavicans Planch.	NA	Vitaceae	Tubercle	Decoction	Malaria	Nadembega et al. 2011
Cissus populnea Guill. & Perr.	NA	Vitaceae	Leaf	Not stated	Malaria*	Adebayo and Krettli 2011
Cissus quadrangularis	Veldt grape	Vitaceae	Leaf	Decoction: oral and bath	Malaria	Diarra et al. 2015; Tetik, Civelek, and Cakilcioglu 2013
Cissus rotundifolia (Forssk.) Vahl	Venezuelan treebine	Vitaceae	Root, root bark	Decoction	Malaria	Mesfin et al. 2012
Citronella (cf. *melliodora* or *incanum*)	NA	Cardiopteridaceae	Root	Not stated	Malaria	Odonne et al. 2013
Citrullus lanatus (Thunb.) Masum. & Nakai	Watermelon	Cucurbitaceae	Leaf	Decoction	Malaria	Duke 1996; Milliken 1997a
***Citrus aurantiifolia* L.**	Lime tree	Rutaceae	Leaf, fruit, root, seed, dried peel of fruits	Decoction (also with *O. americanum* and *O. gratissimum*), infusion, juice (bath and per os), powder	Fever, malaria (also in combination)	Asase, Akwetey, and Achel 2010; Asase, Hesse, and Simmonds 2012; Cano and Volpato 2004; Kaou et al. 2008; Koudouvo et al. 2011; Milliken 1997a,b; Odonne et al. 2013; Tchacondo et al. 2012
Citrus aurantium L.	Bitter orange	Rutaceae	Leaf, fruit	Infusion, decoction	Fever, malaria	Milliken 1997a; Traore et al. 2013

(*Continued*)

Table A.1 (Continued) Medicinal Plants Reported to Be Used for Malaria

Plant Species	Common Name	Family	Part Used	Preparation	Indication	References
***Citrus limon* (L.) Burm. F. [Syn: *Citrus limonum* Risso.]**	Lemon	Rutaceae	Fruit, root, leaf	Juice, decoction (also in combination), fruit and leaf also eaten with *C. papaya,* maceration (fresh or powdered roots)	Fever, malaria	Diarra et al. 2015; Ellena, Quave, and Pieroni 2012; Kvist et al. 2006; Milliken 1997a; Namukobe et al. 2011; Ruiz et al. 2011; Semenya et al. 2012; Ssegawa and Kasenene 2007; Tetik, Civelek, and Cakilcioglu 2013; Yetein et al. 2013
***Citrus maxima* (Burm.) Merrill**	Shaddock	Rutaceae	Leaf, fruit	Decoction	Malaria	Yetein et al. 2013
Citrus medica L.	Citron	Rutaceae	Fruit, leaf	Juice (fruit), inhalation, decoction (leaf)	Malaria, fever	Idowu et al. 2010; Rasoanaivo et al. 1992; Ruiz et al. 2011; Shankar, Sharma, and Deb 2012; Traore et al. 2013
Citrus nobilis Lour.	Tangerine	Rutaceae	Leaf	Decoction (with *O. americanum* and *O. gratissimum*)	Malaria	Duke 1996; Kaou et al. 2008
***Citrus paradisi* Macfadyn.**	Grapefruit	Rutaceae	Root	Not stated	Malaria	Kvist et al. 2006; Ruiz et al. 2011

(*Continued*)

Table A.1 (Continued) Medicinal Plants Reported to Be Used for Malaria

Plant Species	Common Name	Family	Part Used	Preparation	Indication	References
***Citrus sinensis* (L.) Osbeck**	Orange	Rutaceae	Leaf, root, bark, stem, fruit peel	Decoction, infusion, juice	Malaria (leaves also used in treatment [with other herbs] and prevention [alone]), fever, repellent (fumigation)	Adebayo and Krettli 2011; Koudouvo et al. 2011; Milliken 1997a,b; Shankar, Sharma, and Deb 2012; Tetik, Civelek, and Cakilcioglu 2013; Titanji, Zofou, and Ngemenya 2008; Vigneron et al. 2005
Citrus sp.	NA	Rutaceae	Root	Decoction	Malaria	Bertani et al. 2005
Clausena anisata Hook.f. De Wild. & Staner	Horsewood	Rutaceae	Leafy stem, leaf, root bark, stem bark	Decoction	Malaria, jaundice (bath)	Koudouvo et al. 2011; Muthaura et al. 2007; Nguta et al. 2010b; Yetein et al. 2013
***Clausena excavata* Burm. F.**	Pink wampee	Rutaceae	Leaf	Juice	(Malaria)	Shankar, Sharma, and Deb 2012
Cleistopholis patens Engl. & Diels.	NA	Annonaceae	Stem, stem bark	Decoction	Malaria	Asase and Oppong-Mensah 2009; Tetik, Civelek, and Cakilcioglu 2013; Traore et al. 2013
Clematis brachiata Thunb	Traveler's-joy	Ranunculaceae	Root bark, root	Decoction	Malaria	Koch et al. 2005; Muthaura et al. 2007
Clematis mauritiana Lamk. var. *normalis*	NA	Ranunculaceae	Aerial part	Decoction	Malaria	Rasoanaivo et al. 1992
Clematis minor [Syn: *Clematis chinensis*]	Chinese clematis	Ranunculaceae	Not stated	Not stated	Malaria	Duke 1996

(*Continued*)

Table A.1 (Continued) Medicinal Plants Reported to Be Used for Malaria

Plant Species	Common Name	Family	Part Used	Preparation	Indication	References
Cleodendrum colebrookianum Walp.	NA	Verbenaceae	Leaf	Decoction	Malaria	Shankar, Sharma, and Deb 2012
Cleodendrum serratum (L.) Moon	NA	Verbenaceae	Root	Not stated	Malaria	Shankar, Sharma, and Deb 2012
Cleome arborea Kunth	NA	Capparaceae	Galls	Decoction	Fever, (Malaria)	Milliken 1997a
Cleome ciliata	NA	Capparaceae	Not stated	Not stated	Malaria	Tetik, Civelek, and Cakilcioglu 2013
Cleome rutidosperma	Fringed spiderflower	Capparaceae	Not stated	Not stated	Malaria	Tetik, Civelek, and Cakilcioglu 2013
Clerodendron indicum (L.) Kuntz	NA	Verbenaceae	Not stated	Not stated	Malaria, fever	Pei 1985
Clerodendron scandens	NA	Verbenaceae	Not stated	Not stated	Malaria	Tetik, Civelek, and Cakilcioglu 2013
Clerodendrum eriophyllum Guerke	NA	Verbenaceae	Root bark, leaf	Decoction	Malaria	Muthaura et al. 2007
***Clerodendrum inerme* (L.) Gaertn.**	Wild jasmine, sorcerers' bush, seaside clerodendrum	Verbenaceae	Leaf	Decoction (also in combination)	Malaria	Mahishi, Srinivasa, and Shivanna 2005; Nguyen-Pouplin et al. 2007
Clerodendrum infortunatum Gaertn. [Syn: *Clerodendrum viscosum* Vent.]	NA	Verbenaceae	Leaf, root	Decoction, paste	Malaria	Bhandary, Chandrashekar, and Kaveriappa 1995; Duke 1996; Kadir et al. 2014; Namsa, Mandal, and Tangjang 2011

(*Continued*)

Table A.1 (Continued) Medicinal Plants Reported to Be Used for Malaria

Plant Species	Common Name	Family	Part Used	Preparation	Indication	References
***Clerodendrum myricoides* (Hoschst.) Vatke**	Butterfly bush	Labiatae	Root, leaf (inhaled nasally), root bark	Crushed, homogenized, and drunk; burnt on charcoal for nasal inhalation; decoction	Malaria	Karunamoorthi and Tsehaye 2012; Koch et al. 2005; Muregi et al. 2007; Muthaura et al. 2007; Wondimu, Asfaw, and Kelbessa 2007
Clerodendrum rotundifolium	NA	Labiatae	Leaf, root	Juice	Malaria	Jamir, Sharma, and Dolui 1999
Clerodendrum serratum var. *amplexifolium* Moldenke	NA	Verbenaceae	Not stated	Not stated	Malaria	Ghorbani et al. 2011; Pei 1985
Clerodendrum splendens	Flaming glorybower vine, bleeding heart vine	Verbenaceae	Whole plant, leaf	Decoction, nasal drops	Malaria (including cerebral)	Akendengue 1992; Guédé et al. 2010
Clerodendrum wallii Moldenke	NA	Verbenaceae	Root	Infusion	Malaria	Hamill et al. 2000
***Clidemia hirta* (L.) D. Don**	Koster's curse	Melastomataceae	Not stated	Not stated	Malaria	Ruiz et al. 2011
Clinopodium laevigatum	NA	Labiatae	Not stated	Not stated	Malaria	Duke 1996
Clinopodium taxifolium (Kunth.) Harley	NA	Labiatae	Not stated	Infusion	Malaria	Tene et al. 2007
Clutia abyssinica Jaub. & Spach	NA	Euphorbiaceae	Leaf, root bark	Infusion, decoction, juice	Malaria	Muthaura et al. 2007; Seid and Tsegay 2011; Stangeland et al. 2011
***Cnestis palala* Merr.**	NA	Connaraceae	Not stated	Not stated	Malaria	Duke 1996

(*Continued*)

Table A.1 (Continued) Medicinal Plants Reported to Be Used for Malaria

Plant Species	Common Name	Family	Part Used	Preparation	Indication	References
Cocculus leaeba [Syn: *Cocculus pendulus*]	NA	Menispermaceae	Leaf, root, bark	Cataplasm, decoction	Malarial fever, malaria	Duke 1996; Sadeghi et al. 2014
Cochlospermum planchonii Hook. F. ex Planch	NA	Bixaceae	Root, tubercle	Decoction, maceration	Malaria	Nadembega et al. 2011; Willcox 2011
Cochlospermum tinctorium A. Rich.	NA	Cochlospemaceae	Root, leaf	Decoction (leaf: ral and as bath), maceration	Malaria	Asase et al. 2005; Diarra et al. 2015; Seid and Tsegay 2011
***Cocos nucifera* L.**	Coconut	Palmae	Root, leaf, fruit	Decoction, juice (fruit), infusion	Fever, malaria (cure and prevention), jaundice (root + leaf)	Al-adhroey et al. 2010; Asase, Akwetey, and Achel 2010; Koudouvo et al. 2011; Milliken 1997a; Odonne et al. 2013; Ruiz et al. 2011; Tetik, Civelek, and Cakilcioglu 2013; Yetein et al. 2013
Coffea arabica L.	Arabian coffee	Rubiaceae	Leaf, fruit, seed	Infusion, decoction	Fever, malaria	Calderón, Simithy-Williams, and Gupta 2012; Duke 1996; Milliken 1997a,b; Ruiz et al. 2011; Tetik, Civelek, and Cakilcioglu 2013
***Coffea canephora* Froehner**	Robusta coffee	Rubiaceae	Leaf	Not stated	Malaria	Titanji, Zofou, and Ngemenya 2008
Cogniauxia podolaena Baill.	NA	Cucurbitaceae	Whole plant, root	Not stated	Malaria	Mbatchi et al. 2006

(*Continued*)

Table A.1 (Continued) Medicinal Plants Reported to Be Used for Malaria

Plant Species	Common Name	Family	Part Used	Preparation	Indication	References
Cola acuminata (P. Beaub.) Schott & Endl.	Abata cola	Malvaceae	Fruit	Not stated	Malaria	Allabi et al. 2011
Cola cordifolia (Cav.) R. Br.	NA	Sterculiaceae	Leaf	Decoction	Malaria	Diarra et al. 2015; Traore et al. 2013
Cola millenii K. Schum.	NA	Malvaceae	Leaf	Decoction	Malaria	Yetein et al. 2013
***Cola nitida* A. Chev.**	Ghanja kola	Malvaceae	Not stated	Not stated	Malaria	Duke 1996; Nadembega et al. 2011
Colubrina glomerata Benth. Hemsl.	NA	Rhamnaceae	Bark	Decoction (in combination)	Malaria, fever	Dimayuga and Agundez 1986
Colubrina guatemalensis	NA	Rhamnaceae	Not stated	Not stated	Malaria	Duke 1996
Combretum adenogonium Stend ex A.Rich.	NA	Combretaceae	Root	Decoction	Malaria	Nadembega et al. 2011
Combretum decandrum [Syn: *Combretum roxburghii* Spreng.]	NA	Combretaceae	Root	Decoction	Malaria, malarial fever	Duke 1996; Shil, Dutta Choudhury, and Das 2014
Combretum ghasalense	NA	Combretaceae	Leaf	Not stated (in combination with other herbs)	Malaria	Asase et al. 2005
Combretum glutinosum Perr.	NA	Combretaceae	Leaf	Decoction	Malaria	Tetik, Civelek, and Cakilcioglu 2013; Traore et al. 2013
Combretum illairii Engl.	NA	Combretaceae	Not stated	Not stated	Malaria	Gathirwa et al. 2011

(*Continued*)

Table A.1 (Continued) Medicinal Plants Reported to Be Used for Malaria

Plant Species	Common Name	Family	Part Used	Preparation	Indication	References
Combretum latiatum	NA	Combretaceae	Not stated	Not stated	Malaria	Nadembega et al. 2011; Tetik, Civelek, and Cakilcioglu 2013
Combretum lecardii Engl. & Diels	NA	Combretaceae	Leaf	Decoction: oral and bath	Malaria	Diarra et al. 2015
Combretum micranthum G. Don.	Opium antidote	Combretaceae	Leaf	Decoction with *Trichilia emetica*: oral and bath	Malaria, malarial fever	Diarra et al. 2015; Gurib-Fakim et al. 1993; Tetik, Civelek, and Cakilcioglu 2013; Traore et al. 2013
Combretum molle R. Br. ex G. Don.	NA	Combretaceae	Leaf	Decoction: oral and bath	Malaria	Diarra et al. 2015; Grønhaug et al. 2008; Titanji, Zofou, and Ngemenya 2008
Combretum nigricans Lepr. var. *elliotii* (Engl. & Diels)	NA	Combretaceae	Leaf	Decoction	Malaria	Traore et al. 2013
Combretum padoides Engl. & Diels	Thicket bushwillow	Combretaceae	Leaf	Decoction	Malaria	Gathirwa et al. 2011; Nguta et al. 2010a,b
Combretum paniculatum Vent.	Burning bush	Combretaceae	Leaf	Decoction	Malaria	Traore et al. 2013
Combretum platysterum	NA	Combretaceae	Not stated	Not stated	Malaria	Tetik, Civelek, and Cakilcioglu 2013
Combretum raimbaulti Heckel	NA	Combretaceae	Leaf	Decoction	Fever, malaria	Rasoanaivo et al. 1992

(*Continued*)

Table A.1 (Continued) Medicinal Plants Reported to Be Used for Malaria

Plant Species	Common Name	Family	Part Used	Preparation	Indication	References
Combretum sp.	NA	Combretaceae	Leaf	Not stated (in combination with other herbs)	Malaria	Asase et al. 2005
Combretum spinesis	NA	Combretaceae	Not stated	Not stated	Malaria	Tetik, Civelek, and Cakilcioglu 2013
***Commelina benghalensis* L.**	Indian dayflower, tropical spiderwort	Commelinaceae	Leaf, aerial part	Decoction	Malaria, jaundice	Rasoanaivo et al. 1992; Tetik, Civelek, and Cakilcioglu 2013; Yetein et al. 2013
Commelina communis	Asiatic dayflower	Commelinaceae	Not stated	Not stated	Malaria	Duke 1996
Commelina erecta L.	Whitemouth dayflower	Commelinaceae	Whole plant, leaf, stem	Decoction, juice	Fever, malaria	Bertani et al. 2005; Milliken 1997a
Commiphora africana Engl. var. *africana*	African myrrh	Burseraceae	Root	Decoction	Malaria	Nadembega et al. 2011
Commiphora schimperi (Berg) Engl.	Glossy-leaved corkwood	Burseraceae	Root, leaf, stem bark	Decoction	Malaria	Nguta et al. 2010b; Koch et al. 2005
Condaminea sp.	NA	Rubiaceae	Not stated	Not stated	Fever, malaria	Milliken 1997b
Conyza aegyptiaca Ait. var. *lineariloba*	NA	Compositae	Aerial part	Decoction	Malaria	Rasoanaivo et al. 1992
***Conyza bonariensis* (L.) Cronquist [Syn: *Conyza floribunda* Kunth; *Erigeron floribundus* (Kunth) Sch. Bip.]**	Asthmaweed	Compositae	Leaf, whole plant, stem	Decoction	Fever, malaria	Milliken 1997a; Stangeland et al. 2011; Zirihi et al. 2005

(*Continued*)

Table A.1 (Continued) Medicinal Plants Reported to Be Used for Malaria

Plant Species	Common Name	Family	Part Used	Preparation	Indication	References
Conyza lyrata Kunth	NA	Compositae	Leaf	Decoction	Malaria	Duke 1996; Milliken 1997a
Conyza sp.	NA	Compositae	Leaf, bark	Not stated	Malaria	Jamir, Sharma, and Dolui 1999
Conyza sumatrensis	NA	Compositae	Leaf	Not stated	Malaria	Tetik, Civelek, and Cakilcioglu 2013; Titanji, Zofou, and Ngemenya 2008
Copaifera pauperi (Herzog) Dwyer	NA	Fabaceae	Resin	Not stated	Malaria	Kvist et al. 2006
Coptis teeta Wall	NA	Ranunculaceae	Root, seed, rhizome	Decoction	Malaria	Duke 1996; Namsa, Mandal, and Tangjang 2011; Shankar, Sharma, and Deb 2012; Tangjang et al. 2011
Cordia curassavica (Jacq.) Roem. & Schult.	Black sage	Boraginaceae	Leaf	Decoction, juice	Fever, malaria	Duke 1996; Milliken 1997a
Cordia riparia Kunth	NA	Boraginaceae	Aerial part	Juice	Malaria	Milliken 1997a
Cordia sinensis Lam	NA	Boraginaceae	Flower	Not stated	Malaria, fever	Nanyingi et al. 2008
Coriandrum sativum L.	Coriander	Umbelliferae	Seed	Eaten	Fever, malaria	Milliken 1997a; Ruiz et al. 2011
Cornus florida	Flowering dogwood	Cornaceae	Not stated	Not stated	Malaria	Duke 1996
Cornus officinalis	Japanese cornel dogwood	Cornaceae	Not stated	Not stated	Malaria	Duke 1996
Cornutia sp.	NA	Labiatae	Leaf	Juice	Malaria	Odonne et al. 2013

(Continued)

Table A.1 (Continued) Medicinal Plants Reported to Be Used for Malaria

Plant Species	Common Name	Family	Part Used	Preparation	Indication	References
Coronopus didymus (L.) Smith	Lesser swinecress	Cruciferae	Whole plant, root	Infusion (whole plant), decoction infusion	Malaria	Duke 1996; Milliken 1997a
***Cordyline fruticosa* (L.) A. Chev.**	Cabbage tree	Asparagaceae	Root	Pill (small balls of paste were made followed by drying)	Malarial fever	Shil, Dutta Choudhury, and Das 2014
Corydalis govaniana	NA	Papaveraceae	Not stated	Not stated	Malaria	Duke 1996
Costus afer Ker Gawl.	Bush cane, ginger lily	Zingiberaceae	Leaf, twig, root	Decoction	Malaria	Yetein et al. 2013
Costus arabicus L.	Kostus	Zingiberaceae	Stem	Not stated	Fever, malaria	Kvist et al. 2006; Milliken 1997a; Ruiz et al. 2011
Costus dubius	African costus	Zingiberaceae	Not stated	Not stated	Malaria	Tetik, Civelek, and Cakilcioglu 2013
Costus spectabilis (Fenzl) K. Schum.	NA	Costaceae	Leaf	Decoction: oral and bath	Malaria	Diarra et al. 2015
***Couroupita guianensis* Aubl.**	Cannonball tree	Lecythidaceae	Cortex, fruit	Not stated	Malaria	Kvist et al. 2006; Ruiz et al. 2011
Coutarea hexandra (Jacq.) K. Schum.	NA	Rubiaceae	Bark, leaf, leafy branches	Decoction, infusion	Fever, malaria, quinine substitute	Brandão et al. 1992; Cosenza et al. 2013; Duke 1996; Milliken 1997a; Muñoz, Sauvain, Bourdy, Callapa, Bergeron et al. 2000
Coutoubea racemosa Aubl.	NA	Gentianaceae	Whole plant	Decoction	Malaria	Bertani et al. 2005

(*Continued*)

Table A.1 (Continued) Medicinal Plants Reported to Be Used for Malaria

Plant Species	Common Name	Family	Part Used	Preparation	Indication	References
Coutoubea spicata (Aubl.) Mart.	NA	Gentianaceae	Root, leaf, whole plant	Decoction	Fever, malaria	Bertani et al. 2005; Milliken 1997a; Vigneron et al. 2005
Crabbea velutina S. Moore.	NA	Acanthaceae	Not stated	Decoction, infusion (bath)	Malaria	Geissler et al. 2002
Crassocephalum biafrae (Oliv. et Hiern) S. Moore	NA	Compositae	Leaf, flower	Infusion	Malaria	Hamill et al. 2000
Crataeva adansonii	NA	Capparaceae	Not stated	Not stated	Malaria	Tetik, Civelek, and Cakilcioglu 2013
Craterispermum laurinum (Poir.) Benth.	NA	Rubiaceae	Leaf	Decoction	Malaria	Traore et al. 2013
Crateva adansonii DC. ssp. *adansonii*	NA	Capparaceae	Leaf	Decoction	Malaria	Betti and Yemefa 2011; Yetein et al. 2013
Crateva nurvala	NA	Capparaceae	Not stated	Not stated	Malaria	Duke 1996
***Crateva religiosa* G. Forst.**	Sacred garlic pear	Capparaceae	Leaf	Decoction	Malaria	Koudouvo et al. 2011
Cremastosperma cauliflorum R. E. Fr.	NA	Annonaceae	Not stated	Decoction	Malaria	Giovannini 2015
***Cresentia cujete* L.**	Common calabash tree	Bignoniaceae	Not stated	Soup	Fever, malaria	Milliken 1997a; Ruiz et al. 2011
Crinum erubescens	NA	Liliaceae	Not stated	Not stated	Malaria	Duke 1996
Crinum zeylanicum L.	Ceylon swamplily	Amaryllidaceae	Tubercle	Not stated	Malaria	Duke 1996; Milliken 1997a; Nadembega et al. 2011
Crocus sativus L.	Saffron	Iridaceae	Not stated	Not stated	Malaria	Milliken 1997a

(*Continued*)

Table A.1 (Continued) Medicinal Plants Reported to Be Used for Malaria

Plant Species	Common Name	Family	Part Used	Preparation	Indication	References
Crossopteryx febrifuga (Afzel. Ex G. Don) Benth.	NA	Rubiaceae	Stem, leaf, stem bark	Decoction (consumed and as bath), maceration, taken with porridge	Malaria	Adebayo and Krettli 2011; Diarra et al. 2015; Gessler et al. 1995; Jansen et al. 2010; Nadembega et al. 2011; Ngarivhume et al. 2015; Sanon et al. 2003; Tetik, Civelek, and Cakilcioglu 2013; Traore et al. 2013
Crotalaria naragutensis Hutch.	NA	Fabaceae	Whole plant	Decoction	Malaria	Nadembega et al. 2011
Crotolaria occulta Grab	NA	Fabaceae	Whole plant	Juice	Malaria	Shankar, Sharma, and Deb 2012
Crotolaria spinosa Hochst.	NA	Fabaceae	Leaf	Decoction	Malaria	Rasoanaivo et al. 1992
Croton spp.	NA	Euphorbiaceae	Bark, stem, leaf, root	Infusion, decoction, inhalation	Fever, malaria	Milliken 1997a,b; Rasoanaivo et al. 1992
Croton cajucara Benth.	NA	Euphorbiaceae	Leaf	Infusion	Fever, malaria	Milliken 1997a,b
Croton eleuteria (L.) Sw.	NA	Euphorbiaceae	Twig, leaf, bark	Infusion	Fever, quinine substitute	Milliken 1997a
Croton goudoti H. Bn.	NA	Euphorbiaceae	Leaf	Inhalation, decoction	Malaria	Rasoanaivo et al. 1992
Croton gratissimus	NA	Euphorbiaceae	Not stated	Not stated	Malaria	Duke 1996
Croton guatemalensis Lotsy	Copalchi	Euphorbiaceae	Bark	Decoction	Fever, malaria	Duke 1996; Milliken 1997a

(*Continued*)

Table A.1 (Continued) Medicinal Plants Reported to Be Used for Malaria

Plant Species	Common Name	Family	Part Used	Preparation	Indication	References
Croton humilis L.	Pepperbush	Euphorbiaceae	Bark	Decoction	Malaria	Duke 1996; Milliken 1997a
Croton lechleri Mull.-Arg.	Dragon's blood croton	Euphorbiaceae	Resin, latex	Not stated	Malaria	Kvist et al. 2006; Odonne et al. 2013
Croton leptostachyus Kunth	NA	Euphorbiaceae	Leaf	Decoction	Fever, malaria	Milliken 1997a
Croton macrostachis	NA	Euphorbiaceae	Leaf	Decoction	Malaria	Suleman et al. 2009
Croton macrostachyus Hochst. ex. Del.	NA	Euphorbiaceae	Leaf, bark, root	Boiled	Malaria (also in combination)	Giday et al. 2007; Karunamoorthi and Tsehaye 2012
Croton megalocarpus Hutch.	NA	Euphorbiaceae	Root, bark	Not stated	Malaria, fever	Nanyingi et al. 2008
Croton niveus Jacq.	NA	Euphorbiaceae	Bark	Decoction	Fever, malaria	Duke 1996; Milliken 1997a
Croton reflexifolius Kunth	NA	Euphorbiaceae	Leaf, bark	Decoction	Fever, malaria	Duke 1996; Milliken 1997a
***Croton tiglium* L.**	Purging croton	Euphorbiaceae	Leaf, flower	Not stated	Malaria	Shankar, Sharma, and Deb 2012
Croton xalapensis Kunth	NA	Euphorbiaceae	Bark	Decoction	Fever, malaria	Milliken 1997a
Croton zambesicus Mull. Arg.	NA	Euphorbiaceae	Leaf	Decoction	Malaria	Adebayo and Krettli 2011; Ajibesin et al. 2007
Cryptolepis sanguinolenta (Lindl.) Schltr.	NA	Apocynaceae	Root	Decoction	Malaria	Adebayo and Krettli 2011; Willcox 2011
Cucumis ficifolius A. Rich.	Cucumis	Cucurbitaceae	Whole plant	Decoction	Malaria	Mesfin et al. 2012

(*Continued*)

Table A.1 (Continued) Medicinal Plants Reported to Be Used for Malaria

Plant Species	Common Name	Family	Part Used	Preparation	Indication	References
Cudrania triloba	Silkworm thorn	Moraceae	Not stated	Not stated	Malaria	Duke 1996
Cuphea carthagenensis (Jacq.) Macbr.	Colombian waxweed	Lythraceae	Whole plant	Decoction	Fever, malaria	Milliken 1997a
Cuphea glutinosa Cham. & Schltdl.	Sticky waxweed	Lythraceae	Not stated	Not stated	Malaria	Tabuti 2008
Curarea sp.	NA	Menispermaceae	Stem	Not stated	Malaria	Odonne et al. 2013
Curarea tecunarum Barneby and Krukoff	NA	Menispermaceae	Stem	Not stated	Malaria	Kvist et al. 2006
Curculigo pilosa Engl.	NA	Hypoxidaceae	Whole plant	Decoction	Malaria	Nadembega et al. 2011
Curcuma caesia Roxb.	Black turmeric	Zingiberaceae	Rhizome	Taken as pills (small balls of paste were made followed by drying)	Malarial fever	Shil, Dutta Choudhury, and Das 2014
***Curcuma longa* L. [Syn: *Curcuma domestica*]**	Turmeric	Zingiberaceae	Whole plant, rhizome, bulb, leaf, root	Infusion, maceration in combination, decoction, inhalation, decoction (external)	Fever, malaria (prevention, with other herbs, and cure)	Adebayo and Krettli 2011; Al-adhroey et al. 2010; Duke 1996; Jaganath and Ng 2000; Kvist et al. 2006; Milliken 1997a; Nagendrappa, Naik, and Payyappallimana 2013; Odonne et al. 2013; Rasoanaivo et al. 1992; Ruiz et al. 2011; Vigneron et al. 2005
Curroria volubilis	NA	Asclepiadaceae	Bark	Decoction	Malaria	Jeruto et al. 2008
Cussonia arborea Hochst. ex A.Rich.	Octopus cabbage tree	Araliaceae	Leaf	Decoction: oral and bath	Malaria	Diarra et al. 2015

(*Continued*)

Table A.1 (Continued) Medicinal Plants Reported to Be Used for Malaria

Plant Species	Common Name	Family	Part Used	Preparation	Indication	References
Cyathula cylindrica Moq [Syn: *Cyathula schimperiana*]	NA	Amaranthaceae	Root	Decoction	Malaria	Jeruto et al. 2008
***Cyathula prostrata* (L.) Blume**	Pastureweed	Amaranthaceae	Leaf, whole plant	Decoction (external wash and consumed)	Fever, malaria	Guédé et al. 2010; Mbatchi et al. 2006; Milliken 1997a
Cylicodiscus gabunensis (Taub.) Harms	African greenheart	Fabaceae	Leaf	Decoction	Malaria	Ajibesin et al. 2007; Tetik, Civelek, and Cakilcioglu 2013
***Cymbopogon citratus* (DC.) Stapf.**	Lemongrass	Gramineae	Aerial part, leaf	Decoction, infusion, maceration	Fever, malaria (Treatment and prevention) (also in combination), Shiver	Adebayo and Krettli 2011; Allabi et al. 2011; Asase, Akwetey, and Achel 2010; Asase, Hesse, and Simmonds 2012; Asase and Oppong-Mensah 2009; Dike, Obembe, and Adebiyi 2012; Duke 1996; Idowu et al. 2010; Koudouvo et al. 2011; Kvist et al. 2006; Mesia et al. 2008; Nadembega et al. 2011; Odonne et al. 2013; Rehecho et al. 2011; Ruiz et al. 2011; Stangeland et al. 2011; Tchacondo et al. 2012; Tetik, Civelek, and Cakilcioglu 2013; Tor-anyiin, Sha'ato, and Oluma 2003; Vigneron et al. 2005; Yetein et al. 2013

(Continued)

Table A.1 (Continued) Medicinal Plants Reported to Be Used for Malaria

Plant Species	Common Name	Family	Part Used	Preparation	Indication	References
Cymbopogon giganteus Chiov.	NA	Poaceae	Leaf	Not stated	Malaria	Adebayo and Krettli 2011; Nadembega et al. 2011
Cymbopogon nardus	Citronella, serai wangi	Poaceae	Not stated	Not stated	Malaria	Duke 1996
Cymbopogon proximus Stapf.	NA	Poaceae	Not stated	Not stated	Malaria	Nadembega et al. 2011
Cyclea peltata Hook. F & Thoms.	Pata root	Menispermaceae	Not stated	Decoction	Malarial fever	Sivasankari, Anandharaj, and Gunasekaran 2014
Cynara cardunculus L.	Cardoon	Compositae	Not stated	Not stated	Fever, malaria	Milliken 1997a
Cynoglossum glochidion Wall.	NA	Bombacaceae	Root	Not stated	Malaria	Shankar, Sharma, and Deb 2012
Cynometra sphaerocarpa	NA	Fabaceae	Not stated	Not stated	Malaria	Duke 1996
Cyperus articulatus L.	Jointed flatsedge	Cyperaceae	Leaf, tuber, rhizome	Nasal drops; infusion (hot/cold water)	Malaria (including cerebral)	Akendengue 1992; Muthaura et al. 2007; Rukunga et al. 2009
Cyperus esculentus L.	Yellow nutsedge	Cyperaceae	Leaf	Decoction (oral and as bath)	Malaria	Diarra et al. 2015; Yetein et al. 2013
***Cyperus papyrus* L.**	Papyrus	Cyperaceae	Root	Not stated	Malaria	Gakuya et al. 2013
Cyphomandra pendula (Ruiz & Pav.) Sendtn.	NA	Solanaceae	Leaf	Decoction (also in combination)	Malaria	Odonne et al. 2013
Cyphomandra sp.	NA	Solanaceae	Leaf	Juice, decoction (also in combination)	Malaria	Milliken 1997a; Odonne et al. 2013

(Continued)

Table A.1 (Continued) Medicinal Plants Reported to Be Used for Malaria

Plant Species	Common Name	Family	Part Used	Preparation	Indication	References
Cyphostemma digitatum (Forssk.) Descoings	NA	Vitaceae	Leaf	Maceration	Malaria	Samuelsson et al. 1993
Dacryodes edulis (G. Don.) H.J. Lam.	African native pear, African plum, bush butter tree	Burseraceae	Leaf, bark	Not stated	Malaria	Dike, Obembe, and Adebiyi 2012
Dalbergia pinnata (Lour.) Merr.	NA	Fabaceae	Bark	Decoction	Malaria	Chander et al. 2014
Danais breviflora Bak.	NA	Rubiaceae	Root	Decoction	Malaria	Rasoanaivo et al. 1992
Danais cernua Bak.	NA	Rubiaceae	Root	Decoction	Malaria	Rasoanaivo et al. 1992
Danais fragrans Gaertn.	NA	Rubiaceae	Root	Decoction	Malaria	Rasoanaivo et al. 1992
Danais gerrardii Bak.	NA	Rubiaceae	Root	Decoction	Malaria	Rasoanaivo et al. 1992
Danais verticillata Bak.	NA	Rubiaceae	Root	Decoction	Malaria	Rasoanaivo et al. 1992
Daniellia ogea (Harrms) Rolfe ex Holl.	Gum copal tree	Fabaceae	Root	Decoction, infusion	Malaria	Adebayo and Krettli 2011; Ajibesin et al. 2007
Daniellia oliveri (Rolfe) Hutch. & Dalz.	African copaiba balsam tree	Fabaceae	Stem, stem bark, leaf	Decoction	Malaria	Adebayo and Krettli 2011; Nadembega et al. 2011; Traore et al. 2013
Daphne genkwa	Lilac daphne	Thymelaeaceae	Not stated	Not stated	Malaria	Duke 1996
Datura inoxia L.	Pricklyburr	Solanaceae	Not stated	Not stated	Malaria	Yaseen et al. 2015
Datura metel **L.**	Sacred thorn-apple	Solanaceae	Seed, leaf, root	Decoction	Fever, malaria (including cerebral)	Milliken 1997a; Shankar, Sharma, and Deb 2012

(Continued)

Table A.1 (Continued) Medicinal Plants Reported to Be Used for Malaria

Plant Species	Common Name	Family	Part Used	Preparation	Indication	References
Datura stramonium L.	Jimsonweed	Solanaceae	Leaf	Decoction	Malaria	Nadembega et al. 2011
Daucus carota	Queen Anne's lace, wild carrot	Apiaceae	Not stated	Not stated	Malaria	Duke 1996
Dedonaea viscosa	NA	Sapindaceae	Not stated	Not stated	Malaria	Tetik, Civelek, and Cakilcioglu 2013
Deinbollia pinnata Schum. & Thonn.	NA	Sapindaceae	Leaf	Decoction	Malaria	Asase, Akwetey, and Achel 2010
Delonix elata	NA	Fabaceae	Not stated	Not stated	Malaria	Duke 1996
Delonix regia	Royal poinciana	Fabaceae	Not stated	Not stated	Malaria	Duke 1996
Denteromallotus acuminatus Pax et Hoffm.	NA	Euphorbiaceae	Stem	Infusion, or chewed	Malaria	Novy 1997
Desmodium gangeticum DC.	NA	Fabaceae	Whole plant	Juice	Malaria	Poonam and Singh 2009
Desmodium hirtum Grill et Perr.	NA	Fabaceae	Aerial part; leaf, bark	Decoction	Splenomegaly; malaria	Randrianarivelojosia et al. 2003; Rasoanaivo et al. 1992
Desmodium incanum (Sw.) DC. [Syn: *Desmodium mauritianum* D.C. de Candolle]	Zarzabacoa comun	Fabaceae	Leaf, whole plant, bark	Infusion, decoction	Fever, malaria	Milliken 1997a; Randrianarivelojosia et al. 2003
Desmodium ramosissimum G. Don.	NA	Fabaceae	Leaf	Decoction	Malaria	Yetein et al. 2013
Detarium microcarpum Harms.	Sweet dattock, tallow tree	Fabaceae	Root bark, fruit pulp, leaf	Decoction, fresh or steamed fruit pulp	Malaria	Diarra et al. 2015; Nadembega et al. 2011; Traore et al. 2013

(*Continued*)

Table A.1 (Continued) Medicinal Plants Reported to Be Used for Malaria

Plant Species	Common Name	Family	Part Used	Preparation	Indication	References
Detarium senegalense J.F. Gmel.	Dattock	Fabaceae	Leaf	Decoction	Malaria	Traore et al. 2013
Dialium guineense Willd.	Velvet tamarind	Fabaceae	Leaf	Decoction	Malaria	Traore et al. 2013; Yetein et al. 2013
Dianthus anatolicus	NA	Caryophyllaceae	Not stated	Not stated	Malaria	Duke 1996
Dichapetalum madagascariense Poir.	NA	Dichapetalaceae	Leaf	Decoction	Malaria	Yetein et al. 2013
Dicliptera paniculata (Forssk.) I. Darbysh.	Panicled foldwing	Acanthaceae	Leaf	Decoction: oral and bath	Malaria	Diarra et al. 2015
Dichorisandra hexandra (Aubl.) Standl.	NA	Commelinaceae	Not stated	Not stated	Fever, malaria	Milliken 1997a
Dichroa febrifuga Lour	Chinese quinine	Saxifragaceae	Root, leaf	Decoction	Fever, malaria	Duke 1996; Nguyen-Pouplin et al. 2007; Sharma, Chhangte, and Dolui 2001; Shankar, Sharma, and Deb 2012
Dichrostachys cinerea (L.) Wight & Arn. [Syn: *Acacia cinerea*]	Aroma	Fabaceae	Leaf, root	Decoction (leaf)	Malaria	Allabi et al. 2011; Milliken 1997a; Nguta et al. 2010b; Traore et al. 2013
Dicoma tomentosa Cassini	NA	Compositae	Whole plant	Decoction	Malaria (children and adults)	Jansen et al. 2010; Nadembega et al. 2011
Dictamnus albus	Gas plant	Rutaceae	Not stated	Not stated	Malaria	Duke 1996

(Continued)

Table A.1 (Continued) Medicinal Plants Reported to Be Used for Malaria

Plant Species	Common Name	Family	Part Used	Preparation	Indication	References
Dieffenbachia sp.	NA	Araceae	Leaf, root	Decoction (root)	Malaria	Giovannini 2015; Kvist et al. 2006
Digitaria abyssinica Stapf.	African couchgrass	Poaceae	Leaf	Decoction (per os and bath)	Malaria	Namukobe et al. 2011
***Dillenia indica* L.**	Elephant apple	Dilleniaceae	Leaf	Not stated	Malaria*	Nguyen-Pouplin et al. 2007
Discaria febrifuga Mart.	NA	Rhamnaceae	Not stated	Not stated	Quinine substitute	Cosenza et al. 2013
Dioclea sp.	NA	Fabaceae	Bark	Infusion	Malaria	Milliken 1997a
Dioscorea dumetorum Kunth	Wild yam, bitter yam	Dioscoreaceae	Leaf, root	Not stated	Malaria	Dike, Obembe, and Adebiyi 2012
Diospyros ebenaster	NA	Ebenaceae	Not stated	Not stated	Malaria	Duke 1996
***Diospyros malabarica* (Desr.) Kostel. [Syn: *Diospyros peregrina*]**	Indian persimmon, mountain ebony	Ebenaceae	Stem bark	Decoction	Malaria	Duke 1996; Upadhyay et al. 2010
Diospyros mespiliformis Hochst.	Ebony diospyros	Ebenaceae	Leaf, root, bark	Decoction (leaf: decocted with *Trichilia emetic*; oral and bath	Malaria	Betti and Yemefa 2011; Diarra et al. 2015; Nadembega et al. 2011; Traore et al. 2013
Diospyros revoluta Poir.	Black apple	Ebenaceae	Leaf	Decoction	Malaria	Milliken 1997a
Diospyros zombensis (B. L. Burtt) F. White	NA	Ebenaceae	Root, leaf	Infusion, decoction	Malaria	Gessler et al. 1995
Diplorhynchus condylocarpon (Muell. Arg.) Pichon	NA	Apocynaceae	Root, leaf, stem bark	Decoction, juice, cold infusion	Malaria	Gessler et al. 1995; Ngarivhume et al. 2015

(Continued)

Table A.1 (Continued) Medicinal Plants Reported to Be Used for Malaria

Plant Species	Common Name	Family	Part Used	Preparation	Indication	References
Disthmonanthus benthamianus	NA	Fabaceae	Not stated	Not stated	Malaria	Tetik, Civelek, and Cakilcioglu 2013
Dodonaea madagascariensis Rdlk.	NA	Sapindaceae	Leaf	Decoction	Malaria	Rasoanaivo et al. 1992
***Dodonaea viscosa* Jacq.**	Hopbush, candlewood	Sapindaceae	Leaf, fruit	Decoction	Fever, malaria	Koudouvo et al. 2011; Milliken 1997a; Rasoanaivo et al. 1992
Dodonea angustifolia L.f.	Florida hopbush	Sapindaceae	Seed	Not stated	Malaria	Giday et al. 2007
Dolichos schweinfurthii Harms	NA	Fabaceae	Bark, leaf	Decoction: oral and bath	Malaria	Diarra et al. 2015
Doliocarpus dentatus (Aub.) Standley	NA	Dilleniaceae	Stem	Juice	(Malaria)	Milliken 1997a
Dombeya shupangae K. Schum.	NA	Malvaceae	Root, leaf	Infusion, decoction	Malaria	Gessler et al. 1995
Doxantha unguis-cati	NA	Bignoniaceae	Not stated	Not stated	Malaria	Duke 1996
***Dracaena reflexa* Lamk.**	Song-of-India	Agavaceae	Aerial part, leaf, bark	Decoction	Fever, malaria	Randrianarivelojosia et al. 2003; Rasoanaivo et al. 1992
***Drymaria cordata* (L.) Willd. ex Schult.**	Whitesnow	Caryophyllaceae	Whole plant	Decoction	Fever, malaria	Calderón, Simithy-Williams, and Gupta 2012; Milliken 1997a
Drynaria quercifolia	Oakleaf fern	Polypodiaceae	Not stated	Not stated	Malaria	Duke 1996

(*Continued*)

Table A.1 (Continued) Medicinal Plants Reported to Be Used for Malaria

Plant Species	Common Name	Family	Part Used	Preparation	Indication	References
Drypetes natalensis (Harv.) Hutch.	NA	Euphorbiaceae	Stem, leaf	Decoction	Malaria	Gessler et al. 1995
Durio oxleyianus Griff.	NA	Bombacaceae	Not stated	Not stated	Malaria	Duke 1996
Dyschoriste perrottetii O. Kuntze	NA	Acanthaceae	Aerial part	Decoction	Malaria (children)	Jansen et al. 2010
Ecballium elaterium (L.) A. Rich.	Squirting cucumber	Cucurbitaceae	Fruit	Juice, cold infusion	Malaria	di Tizio et al. 2012; Pieroni et al. 2002
Ecclinusa ramiflora Mart.	NA	Sapotaceae	Leaf	Decoction	Malaria	Odonne et al. 2013
Echinops hoehnelii Schweinf.	NA	Compositae	Root	Not stated	Malaria	Giday, Asfaw, and Woldu 2010
***Eclipta prostata* L.**	NA	Compositae	Leaf	Not stated	Malarial fever	Bosco and Arumugam 2012
Ehretia buxifolia Willd.	NA	Boraginaceae	Root, bark	Not stated	Malaria	Nanyingi et al. 2008
Ehretia cymosa Thonn. ex Sehum. var. *cymosa* Brenan	NA	Boraginaceae	Leaf	Decoction, infusion	Malaria, fever	Jeruto et al. 2008; Yetein et al. 2013
Ekebergia capensis Sparm.	Cape ash, dog plum	Meliaceae	Bark, leaf, stem bark, root bark	Infusion	Malaria	Koch et al. 2005; Muregi et al. 2007
***Elaeis guineensis* Jacq.**	African oil palm	Palmae	Root, leaf, flower	Decoction, fumigation, infusion	Malaria, fever, repellent	Adebayo and Krettli 2011; Asase, Akwetey, and Achel 2010; Yetein et al. 2013

(*Continued*)

Table A.1 (Continued) Medicinal Plants Reported to Be Used for Malaria

Plant Species	Common Name	Family	Part Used	Preparation	Indication	References
Elaeocarpus kontumensis Gagn.	NA	Elaeocarpaceae	Bark	Not stated	Malaria*	Nguyen-Pouplin et al. 2007
Elephantorrhiza goetzei (Harms) Harms	NA	Fabaceae	Root	Infusion	Malaria	Ngarivhume et al. 2015
Elettaria cardamomum	Cardamom	Zingiberaceae	Not stated	Not stated	Malaria	Duke 1996
***Eleusine indica* Gaertn.**	Indian goosegrass	Poaceae	Whole plant	Decoction	Malaria	Nadembega et al. 2011
Elytraria marginata	NA	Acanthaceae	Whole plant	Decoction	Malaria	Guédé et al. 2010
Emilia coccinea	Scarlet tasselflower	Compositae	Not stated	Not stated	Malaria	Tetik, Civelek, and Cakilcioglu 2013
Enanthia chlorantha Oliv.	NA	Compositae	Bark	Decoction	Malaria	Akendengue 1992
Enantia clorantha Oliv.	NA	Annonaceae	Stem, stem bark	Decoction	Fever, malaria	Adebayo and Krettli 2011; Muganza et al. 2012; Tetik, Civelek, and Cakilcioglu 2013
Enicostema hyssopitolium (Willd.) I.C.	NA	Gentianaceae	Whole plant	Not stated	Malaria, fever	Parveen et al. 2007
Enicostema littorale	NA	Gentianaceae	Not stated	Not stated	Malaria	Duke 1996
Enicostema verticillatum (L.) Engl. ex Gilg	Whitehead	Gentianaceae	Whole Plant, Root	Decoction	Fever, malaria	Duke 1996; Milliken 1997a
Entada abyssinica Steud ex A.Rich	NA	Fabaceae	Leaf	Decoction	Malaria, fever	Ssegawa and Kasenene 2007

(Continued)

Table A.1 (Continued) Medicinal Plants Reported to Be Used for Malaria

Plant Species	Common Name	Family	Part Used	Preparation	Indication	References
Entada africana Guill. & Perr.	NA	Fabaceae	Bark, root, leaf, stem bark	Decoction (root: decoction with leaves of *Trichilia emetic*, oral and bath)	Malaria	Nadembega et al. 2011; Yetein et al. 2013
Entada phaseoloides	St. Thomas bean	Fabaceae	Not stated	Not stated	Malaria	Duke 1996
Enthadrophragma angolense	NA	Meliaceae	Not stated	Not stated	Malaria	Tetik, Civelek, and Cakilcioglu 2013
Ephedra vulgaris	NA	Ephedraceae	Not stated	Not stated	Malaria	Duke 1996
Eremomastax speciosa	NA	Acanthaceae	Leaf	Not stated	Malaria	Tetik, Civelek, and Cakilcioglu 2013
Erioglossum edule Blume	NA	Sapindaceae	Root, stem	Not stated	Malaria	Hout et al. 2006
Eriosema stanerianum Hauman	NA	Fabaceae	Leaf	Decoction	Malaria	Ssegawa and Kasenene 2007
Erlangea cordifolia (Oliv.) S. Moore	NA	Compositae	Leaf, root	Not stated	Malaria	Jamir, Sharma, and Dolui 1999
Erlangea tomentosa S. Moore	NA	Compositae	Stem	Decoction	Malaria	Ssegawa and Kasenene 2007
Eryngium amethystinum L.	Amethyst eryngo, sea holly	Apiaceae	Flower	Consumed	Malaria (prevention)	De Natale and Pollio 2007
***Eryngium foetidum* L.**	Spiritweed	Umbelliferae	Leaf (also in combination), whole plant, root	Infusion, decoction, direct application	Fever, malaria, malaria (prevention, with other herbs)	Duke 1996; Kvist et al. 2006; Milliken 1997a; Odonne et al. 2013; Roumy et al. 2007; Ruiz et al. 2011; Vigneron et al. 2005

(*Continued*)

Table A.1 (Continued) Medicinal Plants Reported to Be Used for Malaria

Plant Species	Common Name	Family	Part Used	Preparation	Indication	References
Erythrina abyssinica Lam. [Syn: *Erythrina tomentosa* Buch. Ham.]	Erythrina, abyssinian coral tree	Fabaceae	Bark	Decoction	Malaria	Lacroix, Prado, Kamoga, Kasenene, Namukobe et al. 2011; Ssegawa and Kasenene 2007
***Erythrina fusca* Lour.**	Coral bean, purple coral tree	Fabaceae	Root, bark	Decoction	Malaria	Milliken 1997a
Erythrina sacleuxii Hua	NA	Fabaceae	Root, leaf	Infusion, decoction	Malaria	Gessler et al. 1995
Erythrina senegalensis DC.	Coral tree, coral flower	Fabaceae	Stem, leaf, stem bark	Decoction: oral and bath	Malaria	Adebayo and Krettli 2011; Diarra et al. 2015; Traore et al. 2013
Erythrina sigmoidea Hua	NA	Fabaceae	Stem bark	Decoction	Malaria	Traore et al. 2013
Erythrocephalum zambesianum Oliv. & Hiern	Red rays	Compositae	Root	Cold infusion	Malaria	Ngarivhume et al. 2015.
Erythrococca anomala	NA	Euphorbiaceae	Leaf	Decoction	Malaria	Guédé et al. 2010
Erythrophleum suaveolens	Ordeal tree, sasswood	Fabaceae	Leaf, root	Not stated	Malaria	Jamir, Sharma, and Dolui 1999
Erythroxylum spp.	NA	Erythroxylaceae	Bark	Not stated	Malaria, quinine substitute	Milliken 1997a
Erythroxylum tortuosum Mart.	NA	Erythroxylaceae	Not stated	Not stated	Malaria	Milliken 1997a
Erythrynia indica Lamk.	NA	Fabaceae	Aerial part	Decoction	Malaria	Rasoanaivo et al. 1992

(Continued)

Table A.1 (Continued) Medicinal Plants Reported to Be Used for Malaria

Plant Species	Common Name	Family	Part Used	Preparation	Indication	References
Esenbeckia febrifuga (A. St.-Hil.) A. Juss. ex Mart.	NA	Rutaceae	Not stated	Not stated	Malaria, quinine substitute	Cosenza et al. 2013; Milliken 1997a
***Eucalyptus camaldulensis* Mehn.**	River redgum	Myrtaceae	Leaf, bulblet	Decoction	Malaria	Nadembega et al. 2011
Eucalyptus citriodora Hook.	Lemonscented gum	Myrtaceae	Leaf	Infusion	Malaria	Milliken 1997a,b
Eucalyptus globulus Labill.	Tasmanian bluegum	Myrtaceae	Leaf	Infusion, decoction, tincture	Fever, malaria	Duke 1996; Maxia et al. 2008; Milliken 1997a; Tetik, Civelek, and Cakilcioglu 2013
Eucalyptus grandis	Grand eucalyptus	Myrtaceae	Not stated	Not stated	Malaria	Tetik, Civelek, and Cakilcioglu 2013
Eucalyptus robusta	Swamp mahogany	Myrtaceae	Leaf	Not stated	Malaria	Tetik, Civelek, and Cakilcioglu 2013
Eucalyptus sp. L'Her.	NA	Myrtaceae	Leaf, flower	Decoction (also in combination), infusion	Fever, malaria	Asase, Hesse, and Simmonds 2012; Milliken 1997a
Euclea divinorum Hiern	NA	Ebenaceae	Not stated	Decoction	Malaria	Nanyingi et al. 2008
Euclea natalensis A. DC.	Natal ebony	Ebenaceae	Root	Eaten with porridge	Malaria	Ngarivhume et al. 2015
Eugenia pitanga Kiaersk.	NA	Myrtaceae	Bark, Leaf	Not stated	Fever, malaria	Milliken 1997a
Eugenia sulcata Spring ex Mart.	NA	Myrtaceae	Leaf	Decoction	Malaria	Milliken 1997a

(*Continued*)

Table A.1 (Continued) Medicinal Plants Reported to Be Used for Malaria

Plant Species	Common Name	Family	Part Used	Preparation	Indication	References
Euonymus thunbergianus	NA	Celastraceae	Not stated	Not stated	Malaria	Duke 1996
Euonynmus europaeus	Common spindle, spindle tree	Celastraceae	Not stated	Not stated	Malaria	Duke 1996
Eupatorium perfoliatum L.	Common boneset	Compositae	Leaf	Decoction	Malaria	Duke 1996; Shemluck 1982
Euphorbia abyssinica Gmel.	Desert candle, Abyssinian spurge	Euphorbiaceae	Nectar	Syrup	Malaria	Belayneh et al. 2012
Euphorbia calyculata	NA	Euphorbiaceae	Not stated	Not stated	Malaria	Duke 1996
Euphorbia helioscopia	Madwoman's milk	Euphorbiaceae	Not stated	Not stated	Malaria	Duke 1996
***Euphorbia heterophylla* L.**	Mexican fireplant	Euphorbiaceae	Leaf	Decoction	Malaria	Traore et al. 2013
***Euphorbia hirta* L. [Syn: *Chamaesyce hirta* (L.) Millsp.]**	Asthmaplant, garden spurge, pill-bearing spurge	Euphorbiaceae	Whole plant, leaf	Powder, infusion, decoction (oral and bath)	Malaria, fever	Diarra et al. 2015; Tchacondo et al. 2012; Tetik, Civelek, and Cakilcioglu 2013; Milliken 1997a; Zirihi et al. 2005
Euphorbia poinsonii	Candle plant	Euphorbiaceae	Leaf	Not stated	Malaria	Nadembega et al. 2011; Tetik, Civelek, and Cakilcioglu 2013
***Eurycoma longifolia* Jack.**	Tongkat ali	Simaroubaceae	Root, aerial part, stem, leaf, bark	Decoction	Malaria (curative, prophylactic), fever	Al-adhroey et al. 2010; Duke 1996; Elliott and Brimacombe 1987; Jaganath and Ng 2000; Mahyar et al. 1991; Nguyen-Pouplin et al. 2007

(*Continued*)

Table A.1 (Continued) Medicinal Plants Reported to Be Used for Malaria

Plant Species	Common Name	Family	Part Used	Preparation	Indication	References
***Euterpe edulis* Mart.**	Assai palm	Palmae	Root	Decoction	Malaria	Milliken 1997a
***Euterpe oleracea* Mart.**	Assai palm	Palmae	Seed	Decoction, infusion	Fever, malaria (with other herbs)	Milliken 1997a; Vigneron et al. 2005
Euterpe precatoria Mart.	Mountain cabbage	Palmae	Root	Decoction, jam (with *Saccharum officinarum*)	Malaria	Bertani et al. 2005; Hajdu and Hohmann 2012; Kvist et al. 2006; Ruiz et al. 2011
Euterpe sp.	NA	Palmae	Root	Decoction	Malaria	Milliken 1997a
Evodia fatraina H. Perr	NA	Rutaceae	Root bark, stem bark	Decoction	Malaria	Rasoanaivo et al. 1992
Excoecaria grahamii Stapf	NA	Euphorbiaceae	Root	Decoction, powder	Malaria	Tchacondo et al. 2012
Exostema australe A. St.-Hil. [Syn: *Bathysa australis* (A. St.-Hil.) K. Schum.]	NA	Rubiaceae	Not stated	Not stated	Quinine substitute	Cosenza et al. 2013
Exostema cuspidatum A. St.-Hil. [Syn: Bathysa cuspidata (A. St.-Hil.) Hook. f. ex K.Schum.; *Schoenleinia cuspidata* (A. St.-Hil.)]	NA	Rubiaceae	Not stated	Not stated	Quinine substitute	Cosenza et al. 2013

(Continued)

Table A.1 (Continued) Medicinal Plants Reported to Be Used for Malaria

Plant Species	Common Name	Family	Part Used	Preparation	Indication	References
Exostema caribaeum (Jacq.) Schult.	Carribean princewood	Rubiaceae	Bark	Infusion	Malaria	Duke 1996; Milliken 1997a
Exostema mexicanum A. Gray	NA	Rubiaceae	Bark	Decoction	Malaria	Milliken 1997a
Exostemma peruviana	NA	Rubiaceae	Not stated	Not stated	Malaria	Duke 1996
Fadogia agrestis (Schweinf. ex Hiern)	NA	Rubiaceae	Leaf	Decoction (consumed and as bath)	Malaria	Sanon et al. 2003
Fagara macrophylla (Oliv.) Engl.	East African Satinwood	Rutaceae	Stem bark	Not stated	Malaria	Tetik, Civelek, and Cakilcioglu 2013; Zirihi et al. 2005
Fagaropsis angolensis (Engl.) Del.	NA	Rutaceae	Leaf, stem bark	Decoction	Malaria	Gakuya et al. 2013; Muthaura et al. 2007; Nguta et al. 2010a,b
Fagraea cochinchinensis	NA	Loganiaceae	Not stated	Not stated	Malaria	Duke 1996
***Fagraea fragans* Roxb.**	Tembusu	Gentianaceae	Leaf	Not stated	Malaria	Duke 1996; Nguyen-Pouplin et al. 2007
Fagus sylvatica L.	Copper beech	Fagaceae	Bark	Infusion	Fever, malaria	Milliken 1997a
Feretia apodenthera Del.	NA	Rubiaceae	Leaf	Decoction: oral and bath	Malaria	Diarra et al. 2015; Nadembega et al. 2011
Fernandoa sp.	NA	Bignoniaceae	Aerial part	Decoction	Malaria	Rasoanaivo et al. 1992
Feroniella lucida (Schaff.) Sw.	Ferioniella	Rutaceae	Leaf	Not stated	Malaria*	Nguyen-Pouplin et al. 2007
***Fibraurea tinctoria* Lour.**	Fibraurea, akar kinching kerbau	Menispermaceae	Root, stem	Not stated	Malaria*	Nguyen-Pouplin et al. 2007

(*Continued*)

Table A.1 (Continued) Medicinal Plants Reported to Be Used for Malaria

Plant Species	Common Name	Family	Part Used	Preparation	Indication	References
Ficus adenosperma	NA	Moraceae	Root	Not stated	Malaria	Mahyar et al. 1991
Ficus bussei Warp ex Mildbr and Burret.	NA	Moraceae	Root, leaf	Decoction	Malaria	Nguta et al. 2010a,b
Ficus carica L. var. *genuina* Boiss.	Edible fig	Moraceae	Not stated	Not stated	Malaria	Esmaeili et al. 2009
Ficus dicranostyla Mildbr	NA	Moraceae	Leaf	Decoction: oral and bath	Malaria	Diarra et al. 2015
Ficus exasperata Vahl	Fig tree	Moraceae	Leaf	Decoction	Malaria	Dike, Obembe, and Adebiyi 2012; Koudouvo et al. 2011; Tetik, Civelek, and Cakilcioglu 2013
Ficus glabrata Kunth	NA	Moraceae	Latex, bark	Not stated	Malaria	Milliken 1997a
Ficus gnaphalocarpa (Miq.) Steud. ex Miq.	NA	Moraceae	Root, leaf	Not stated (with other herbs), decoction (leaf: oral and as bath)	Malaria	Asase et al. 2005; Diarra et al. 2015
Ficus hispida L.f.	NA	Moraceae	Leaf	Not stated	Malaria*	Nguyen-Pouplin et al. 2007
Ficus insipida Willdenow	NA	Moraceae	Resin	Not stated	Malaria	Kvist et al. 2006; Ruiz et al. 2011
Ficus ischnopoda Miq.	NA	Moraceae	Leaf	Decoction: Oral and bath	Malaria	Diarra et al. 2015
Ficus megapoda Bak.	NA	Moraceae	Root, leaf, bark	Decoction	Fever, malaria	Randrianarivelojosia et al. 2003; Rasoanaivo et al. 1992

(Continued)

Table A.1 (Continued) Medicinal Plants Reported to Be Used for Malaria

Plant Species	Common Name	Family	Part Used	Preparation	Indication	References
Ficus ovata Vahl	NA	Moraceae	Stem bark	Decoction	Malaria	Traore et al. 2013
Ficus platyphylla Del.	NA	Moraceae	Stem, bulblet, leaf, stem bark	Decoction (oral and as bath)	Malaria	Adebayo and Krettli 2011; Asase et al. 2005; Diarra et al. 2015; Nadembega et al. 2011; Traore et al. 2013
Ficus polita Vahl	NA	Moraceae	Leaf	Decoction	Malaria	Koudouvo et al. 2011
***Ficus racemosa* L.**	Cluster tree	Moraceae	Leaf	Not stated	Malaria*	Nguyen-Pouplin et al. 2007
Ficus ribes	Walen	Moraceae	Not stated	Not stated	Malaria	Duke 1996
Ficus sur Forssk. [Syn: *Ficus capensis* Thunb.]	Bush fig, cape fig	Moraceae	Leaf, stem bark, root bark	Decoction	Malaria	Muregi et al. 2007; Traore et al. 2013; Zirihi et al. 2005
Ficus sycomorus L.	Sycamore fig	Moraceae	Stem bark, leaf	Decoction (consumed and as bath)	Malaria	Diarra et al. 2015; Nadembega et al. 2011; Sanon et al. 2003
Ficus thonningii Blume	NA	Moraceae	Stem, leaf	Decoction (for leaf: oral and as bath)	Malaria, fever	Adebayo and Krettli 2011; Diarra et al. 2015; Jansen et al. 2010; Tetik, Civelek, and Cakilcioglu 2013
Ficus umbellata Vahl	NA	Moraceae	Leaf	Decoction	Malaria	Koudouvo et al. 2011
Ficus vallis-choudae Del.	False cape fig	Moraceae	Root bark	Decoction	Malaria	Traore et al. 2013
Fischeria cf. *martiana* Decne.	NA	Asclepiadaceae	Leaf	Decoction	Malaria	Milliken 1997a

(*Continued*)

Table A.1 (Continued) Medicinal Plants Reported to Be Used for Malaria

Plant Species	Common Name	Family	Part Used	Preparation	Indication	References
***Flacourtia indica* (Burm.f.) Merr.**	Governor's plum	Flacourtiaceae	Leaf, stem, root, stem bark	Decoction	Malaria	Gakuya et al. 2013; Kaou et al. 2008; Nguta et al. 2010a,b; Yetein et al. 2013
Flueggea microcarpa Blume	NA	Euphorbiaceae	Aerial part	Decoction	Malaria	Rasoanaivo et al. 1992
***Flueggea virosa* Buch.-Ham. ex Wall. [Syn: *Securinega virosa* (Roxb. ex Willd.) Baill.]**	NA	Euphorbiaceae	Leafy stem, leaf, aerial part, root, whole plant	Decoction (leaf: oral and as bath)	Jaundice, malaria	Asase, Akwetey, and Achel 2010; Diarra et al. 2015; Gathirwa et al. 2011; Kaou et al. 2008; Koudouvo et al. 2011; Nadembega et al. 2011; Titanji, Zofou, and Ngemenya 2008; Yetein et al. 2013
Foeniculum vulgare Miller	Sweet fennel	Apiaceae	Root	Infusion	Malaria	Mesfin et al. 2012
Fragaria indica Andr.	NA	Rosaceae	Not stated	Not stated	Malaria	Duke 1996
Fraxinus excelsior L.	European ash	Oleaceae	Bark	Decoction	Fever, malaria	Duke 1996; Milliken 1997a
Fuerstia africana T.C.E. Fries	NA	Labiatae	Whole plant, leaf	Infusion (hot/cold water), decoction	Malaria	Koch et al. 2005; Muthaura et al. 2007
Funtumia africana (Benth.) Stapf	False rubbertree	Apocynaceae	Root	Not stated	Malaria*	Adebayo and Krettli 2011
Funtumia elastica (Preuss)	Silkrubber	Apocynaceae	Stem bark	Decoction	Malaria	Guédé et al. 2010; Zirihi et al. 2005

(*Continued*)

Table A.1 (Continued) Medicinal Plants Reported to Be Used for Malaria

Plant Species	Common Name	Family	Part Used	Preparation	Indication	References
Galipea jasminiflora (A. St.-Hil.) Engl. [Syn: *Galipea multiflora* Schult.; *Ticorea febrifuga* A. St.-Hil.]	NA	Rutaceae	Not stated	Not stated	Quinine substitute	Cosenza et al. 2013
Garcinia kola Heckel	Kola nut tree	Clusiaceae	Leaf, fruit, stem bark	Not stated	Malaria	Dike, Obembe, and Adebiyi 2012; Idowu et al. 2010
Garcinia luzonensis Merr.	NA	Clusiaceae	Root, stem	Boiled	Malaria	Libman et al. 2006
Gardenia aqualla Stapf & Hutch	NA	Rubiaceae	Stem bark; leaf	Decoction (for leaf: decocted with roots and leaves of *Mitragyna inermis, Borassus aethiopum* and *Feretia apodanthera*): oral and bath	Malaria	Diarra et al. 2015; Nadembega et al. 2011
Gardenia erubescens Stapf & Huteh.	NA	Rubiaceae	Leaf, root	Decoction	Malaria	Nadembega et al. 2011; Yetein et al. 2013
Gardenia gummifera Linn.	NA	Rubiaceae	Root, leaf	Infusion	Malaria	Poonam and Singh 2009
Gardenia lutea Fresen.	NA	Rubiaceae	Root	Not stated	Malaria (in combination)	Giday et al. 2007

(*Continued*)

Table A.1 (Continued) Medicinal Plants Reported to Be Used for Malaria

Plant Species	Common Name	Family	Part Used	Preparation	Indication	References
Gardenia sokotensis Hutch.	NA	Rubiaceae	Leaf	Decoction (oral and as bath)	Malaria, fever (both often in combination)	Diarra et al. 2015; Jansen et al. 2010; Nadembega et al. 2011
Gardenia sp.	NA	Rubiaceae	Not stated	Not stated	Malaria	Duke 1996
Gardenia ternifolia Schumach. & Thonn.	NA	Rubiaceae	Root bark	Decoction	Malaria	Traore et al. 2013
Gardenia vogelii Planch.	NA	Rubiaceae	Root	Decoction	Malaria	Gessler et al. 1995
Garrya fremontii	California fever bush, flannel bush, quinine bush	Garryaceae	Not stated	Not stated	Malaria	Duke 1996
Geissospermum argenteum Woodson	NA	Apocynaceae	Bark	Maceration in rum/cognac	Malaria	Bertani et al. 2005; Milliken 1997a
Geissospermum laeve (Vell.) Miers [Syn: *Geissospermum vellosii* Allemao]	NA	Apocynaceae	Bark	Decoction, maceration in rum/cognac	Malaria (treatment and prevention), fever, quinine substitute	Bertani et al. 2005; Cosenza et al. 2013; Duke 1996; Milliken 1997a,b; Muñoz, Sauvain, Bourdy, Callapa, Bergeron et al. 2000; Vigneron et al. 2005
Geissospermum sericeum (Sagot) Benth. & Hook.f.	NA	Apocynaceae	Bark	Decoction, infusion	Malaria (treatment and prevention), fever, quinine substitute	Brandão et al. 1992; Cosenza et al. 2013; Milliken 1997a,b; Vigneron et al. 2005
Gelsemium sempervirens	Evening trumpetflower	Loganiaceae	Not stated	Not stated	Malaria	Duke 1996

(*Continued*)

Table A.1 (Continued) Medicinal Plants Reported to Be Used for Malaria

Plant Species	Common Name	Family	Part Used	Preparation	Indication	References
***Gendarussa vulgaris* Nees [Syn: *Justicia gendarussa* Burm.f.]**	Gandarusa	Acanthaceae	Aerial part	Decoction	Malaria, fever	Duke 1996; Rasoanaivo et al. 1992
Gentiana lutea L. subsp *lutea*	Yellow gentian	Gentianaceae	Root	Infusion	Malarial fever	Idolo, Motti, and Mazzoleni 2010
Gentiana quinquefolia	NA	Gentianaceae	Not stated	Not stated	Malaria	Duke 1996
Geophila obvallata (Schumach.) F.Didr	NA	Rubiaceae	Leaf	Decoction	Malaria	Traore et al. 2013
Gerranthus lobatus (Cogn.) Jeffrey	NA	Cucurbitaceae	Root	Decoction	Malaria	Nguta et al. 2010a,b
Geum urbanum	Herb bennet	Rosaceae	Not stated	Not stated	Malaria	Duke 1996
Glinus oppositifolius (L.) Aug. D.C.	Slender carpetweed	Aizoaceae	Stem, leaf	Not stated	Malaria	Diallo et al. 1999
***Gliricidia sepium* (Jacq.) Walp.**	Quickstick	Fabaceae	Leaf	Decoction	Fever, malaria	Milliken 1997a
Glochidion puberum	Needlebush	Euphorbiaceae	Not stated	Not stated	Malaria	Duke 1996
Glochidion sp.	NA	Euphorbiaceae	Leaf	Decoction	Malaria, fever	Elliott and Brimacombe 1987
Gloriosa superba	Flame lily	Liliaceae	Not stated	Not stated	Malaria	Duke 1996
Glossocalyx brevipes Benth	NA	Monimiaceae	Leaf	Not stated	Malaria	Tetik, Civelek, and Cakilcioglu 2013
Glycine max	Soybean	Fabaceae	Not stated	Not stated	Malaria	Duke 1996
Glycyrrhiza glabra L.	Liquorice	Fabaceae	Not stated	Not stated	Malaria	Esmaeili et al. 2009
Gmelia arborea L.	NA	Verbenaceae	Leaf	Decoction	Malaria	Traore et al. 2013

(Continued)

Table A.1 (Continued) Medicinal Plants Reported to Be Used for Malaria

Plant Species	Common Name	Family	Part Used	Preparation	Indication	References
Gnaphalium leuteo-album L.	Cudweed	Compositae	Whole plant	Decoction	Fever, malaria	Wiart 2000
Gnetum spp.	NA	Gnetaceae	Stem	Not stated	Malaria*	Nguyen-Pouplin et al. 2007
Gnetum ula	Gnemon tree	Gnetaceae	Not stated	Not stated	Malaria	Duke 1996
Gomphostemma parviflora Wall.	NA	Labiatae	Leaf	Not stated	Malaria	Shankar, Sharma, and Deb 2012
Gossypium arboreum L.	Tree cotton	Malvaceae	Leaf	Not stated	Malaria	Adebayo and Krettli 2011; Duke 1996; Idowu et al. 2010
Gossypium barbadense L.	Creole cotton	Malvaceae	Leaf	Decoction	Malaria	Adebayo and Krettli 2011; Duke 1996; Milliken 1997a,b
Gossypium herbaceum L.	Arabian cotton	Malvaceae	Root	Not stated	Fever, malaria	Duke 1996; Milliken 1997a
Gossypium hirsutum L.	Upland cotton	Malvaceae	Leaf	Decoction	Fever, malaria	Adebayo and Krettli 2011; Koudouvo et al. 2011; Milliken 1997a
Gossypium spp.	NA	Malvaceae	Seed	Decoction	Malaria, fever	Milliken 1997a; Nadembega et al. 2011; Tetik, Civelek, and Cakilcioglu 2013
Greenwayodendron sp.	NA	Annonaceae	Leaf	Decoction	Malaria	Asase, Akwetey, and Achel 2010
Grewia cyclea Baill.	NA	Tiliaceae	Leaf	Infusion	Malaria	Norscia and Borgognini-Tarli 2006
Grewia hexamita Burret.	NA	Tiliaceae	Root, leaf	Decoction	Malaria	Nguta et al. 2010a,b

(*Continued*)

Table A.1 (Continued) Medicinal Plants Reported to Be Used for Malaria

Plant Species	Common Name	Family	Part Used	Preparation	Indication	References
Grewia mollis Juss.	NA	Tiliaceae	Leaf	Decoction: oral and bath	Malaria	Diarra et al. 2015
Grewia paniculata Roxb.	Cenderai	Tiliaceae	Leaf	Not stated	Malaria*	Nguyen-Pouplin et al. 2007
Grewia plagiophylla K. Schum	NA	Tiliaceae	Not stated	Not stated	Malaria	Gathirwa et al. 2011
Grewia trichocarpa Hochst ex A.Rich	NA	Tiliaceae	Root	Decoction	Malaria	Nguta et al. 2010b
Grewia villosa Willd.	Mallow raisin	Tiliaceae	Stem bark	Decoction	Malaria	Traore et al. 2013
Grias neuberthii J.F. Macbr.	NA	Lecythidaceae	Stem, bark, seed	Infusion, decoction	(Malaria), malaria	Kvist et al. 2006; Milliken 1997a; Roumy et al. 2007; Ruiz et al. 2011
Grias peruviana Miers	NA	Lecythidaceae	Bark	Decoction	Malaria	Sanz-Biset et al. 2009
Guarea grandifolia (L.) Sleumer	Carapa	Rubiaceae	Bark	Powder	Malaria	Giovannini 2015
Guatteria megalophylla Diels	NA	Annonaceae	Leaf	Decoction	Malaria	Milliken 1997a
Guatteria sp.	NA	Annonaceae	Bark	Not stated	Fever, malaria	Milliken 1997a,b
***Guazuma ulmifolia* Lam. [Syn: *Guazuma tomentosa*]**	Bay cedar, musket tree, pigeon wood	Malvaceae	Bark, latex	Not stated	Fever, malaria	Duke 1996; Milliken 1997a

(Continued)

Table A.1 (Continued) Medicinal Plants Reported to Be Used for Malaria

Plant Species	Common Name	Family	Part Used	Preparation	Indication	References
***Guettarda speciosa* L.**	Ketapang pasir, sea randa, zebrawood	Rubiaceae	Flower	Decoction (with *O. americanum* and *O. gratissimum*), direct application	Malaria	Kaou et al. 2008
Guibourtia tessmannii	NA	Fabaceae	Stem	Not stated	Malaria	Tetik, Civelek, and Cakilcioglu 2013
Guiera senegalensis J.F. Gmel.	NA	Combretaceae	Leaf, root	Decoction (leaf: oral and as bath), maceration (root, with *Combretum micranthum*)	Malaria	Adebayo and Krettli 2011; Diarra et al. 2015; Nadembega et al. 2011; Traore et al. 2013
Guizotia scabra	NA	Compositae	Leaf, flower	Decoction, infusion	Malaria	Stangeland et al. 2011
Gustavia longifolia Poepp. ex O. Berg	NA	Lecythidaceae	Bark	Decoction	Malaria	Sanz-Biset et al. 2009
Gutenbergia cordifolia Benth.	NA	Compositae	Leaf	Infusion	Malaria	Koch et al. 2005
Gutierrezia sarothrae	Broom snakeweed	Compositae	Not stated	Not stated	Malaria	Duke 1996
Gymnopetalum cochinchinensis Kurz	NA	Cucurbitaceae	Root	Decoction	Malaria	Namsa, Mandal, and Tangjang 2011

(*Continued*)

Table A.1 (Continued) Medicinal Plants Reported to Be Used for Malaria

Plant Species	Common Name	Family	Part Used	Preparation	Indication	References
Gynandropsis gynandra [Syn: *Cleome gynandra* L. *Gynandropsis pentaphylla* DC.]	Spiderwisp	Capparaceae	Leaf, root	Decoction	Malaria, fever	Duke 1996; Jeruto et al. 2008; Pushpangadan and Atal 1984; Yetein et al. 2013
Gynura scandens	NA	Compositae	Leaf	Juice, decoction	Malaria	Stangeland et al. 2011
Haematostaphis barteri	NA	Anacardiaceae	Leaf	Not stated (in combination)	Malaria	Asase et al. 2005
Hallea rubrostipulata (K. Schum.) J-F. Leroy	NA	Rubiaceae	Bark	Decoction	Malaria	Ssegawa and Kasenene 2007
Halothamnus somalensis (N.E. Br.) Botsch.	NA	Chenopodiaceae	Root, root bark	Decoction	Malaria	Mesfin et al. 2012
***Hamelia patens* Jacq.**	Scarletbush	Rubiaceae	Leaf, stem	Infusion, decoction	Fever, malaria	Duke 1996; Giovannini 2015; Milliken 1997a; Valadeau et al. 2010
Harrisonia abyssinica Oliv.	NA	Simaroubaceae	Root bark, leaf, root, bark, stem bark	Decoction	Malaria, fever	Gakuya et al. 2013; Maregesi et al. 2007; Muthaura et al. 2007; Nanyingi et al. 2008; Nguta et al. 2010a,b; Norscia and Borgognini-Tarli 2006; Tetik, Civelek, and Cakilcioglu 2013; Titanji, Zofou, and Ngemenya 2008; Traore et al. 2013

(Continued)

Table A.1 (Continued) Medicinal Plants Reported to Be Used for Malaria

Plant Species	Common Name	Family	Part Used	Preparation	Indication	References
Harrisonia perforata (Blanco) Merr.	NA	Simaroubaceae	Leaf, root, stem	Not stated	Malaria	Hout et al. 2006; Nguyen-Pouplin et al. 2007
Harungana madagascariensis Lam. ex Poir	Dragon's blood tree, orange milktree	Guttiferae	Stem, aerial part, stem bark, bark, leaf	Decoction, maceration, infusion	Malaria, splenomegaly, fever	Adebayo and Krettli 2011; Gakuya et al. 2013; Muganza et al. 2012; Rasoanaivo et al. 1992; Tetik, Civelek, and Cakilcioglu 2013; Traore et al. 2013
Hedychium cylindricum Ridley	NA	Zingiberaceae	Stalk	Decoction	Malaria, fever (children)	Leaman et al. 1995
Hedyotis biflora	Twoflower mille graines	Rubiaceae	Not stated	Not stated	Malaria	Duke 1996
Hedyotis diffusa	Snake-needle grass, spreading hedyotis	Rubiaceae	Not stated	Not stated	Malaria	Duke 1996
Hedyotis scandens Roxb.	NA	Rubiaceae	Root, leaf	Infusion	Malaria	Shankar, Sharma, and Deb 2012
Heeria insignis Del.	NA	Anacardiaceae	Stem bark	Not stated	Malaria	Nguta et al. 2010b
***Helianthus annuus* L.**	Common sunflower	Compositae	Flower, leaf, stem, seed	Decoction (flower, infloresence), infusion (stem, leaf), juice (flower, seed)	Malaria, fever	Duke 1996; Milliken 1997a; Shankar, Sharma, and Deb 2012
Helichrysum faradifani Sc. Ell.	NA	Compositae	Leaf, aerial part	Infusion, decoction	Malaria, fever	Novy 1997; Rasoanaivo et al. 1992
Helichrysum sp.	NA	Compositae	Leaf	Decoction	Malaria	Ssegawa and Kasenene 2007

(*Continued*)

Table A.1 (Continued) Medicinal Plants Reported to Be Used for Malaria

Plant Species	Common Name	Family	Part Used	Preparation	Indication	References
Heliotropium arborecens L.	Garden heliotrope	Boraginaceae	Not stated	Not stated	Fever (quinine substitute)	Milliken 1997a
Heliotropium europaeum L.	European heliotrope	Boraginaceae	Not stated	Not stated	Malaria	Yaseen et al. 2015
***Heliotropium indicum* L.**	Indian heliotrope	Boraginaceae	Flower, leaf, root, seed	Infusion, juice (leaf), decoction	Fever, malaria	Milliken 1997a; Yetein et al. 2013
Heliotropium peruvianum	NA	Boraginaceae	Not stated	Not stated	Malaria	Duke 1996
Helminthostachys zeylanica	Flowering fern	Ophioglossaceae	Not stated	Not stated	Malaria	Duke 1996
Hemidesmus indicus	Indian sarsaparilla	Asclepiadaceae	Root, leaf	Juice	Malaria (prevention)	Nagendrappa, Naik, and Payyappallimana 2013
Hernandia voyroni Jum.	NA	Hernandiaceae	Stem bark	Decoction	Splenomegaly, adjuvant to chloroquine or quinine	Rasoanaivo et al. 1992
Heteromorpha trifoliata	NA	Apiaceae	Leaf	Decoction	Malaria	Stangeland et al. 2011
Heterotis rotundifolia (Sm.) Jacq-Fel.	NA	Melastomataceae	Leaf	Decoction	Malaria, fever, shiver	Yetein et al. 2013
Hewittia malabarica (L.) Suresh [Syn: *Hewittia sublobata* (I.f.) Kuntze]	Hweittia	Convolvulaceae	Leaf	Decoction: oral and bath	Malaria	Diarra et al. 2015
Hexalobus crispiflorus	NA	Annonaceae	Not stated	Not stated	Malaria	Tetik, Civelek, and Cakilcioglu 2013

(*Continued*)

Table A.1 (Continued) Medicinal Plants Reported to Be Used for Malaria

Plant Species	Common Name	Family	Part Used	Preparation	Indication	References
Hibiscus asper	NA	Malvaceae	Leaf	Decoction	Malaria	Tetik, Civelek, and Cakilcioglu 2013; Traore et al. 2013
Hibiscus cannabinus L.	Brown Indianhemp	Malvaceae	Not stated	Not stated	Malaria	Nadembega et al. 2011
***Hibiscus esculentus* L. [Syn: *Abelmoschus escentulus*]**	Okra, lady's finger	Malvaceae	Not stated	Not stated	Malaria	Nadembega et al. 2011
***Hibiscus rosa-sinensis* L.**	Shoeblack plant	Malvaceae	Leaf, flower	Decoction, infusion	Fever, malaria	Al-adhroey et al. 2010; Milliken 1997a
***Hibiscus sabdariffa* L.**	Roselle	Malvaceae	Calyx, sepals, whole plant, leaf	Infusion, decoction (for leaf: oral and as bath)	Fever, malaria	Diarra et al. 2015; Milliken 1997a; Nadembega et al. 2011
***Hibiscus surattensis* L.**	Wild sour, shrub althea	Malvaceae	Whole plant, leaf	Decoction	Malaria	Koudouvo et al. 2011; Yetein et al. 2013
***Hibiscus tiliaceus* L.**	Sea hibiscus	Malvaceae	Bark, branch	Infusion	Fever, malaria	Milliken 1997a; Tetik, Civelek, and Cakilcioglu 2013
Himatanthus articulatus (Vahl) Woods.	NA	Apocynaceae	Bark	Decoction	Malaria	Milliken 1997b
Himatanthus sucuuba (Spruce ex Mull. Arg.) Woodson	NA	Apocynaceae	Latex, cortex, leaf	Decoction (leaf)	Fever, malaria	Kvist et al. 2006; Milliken 1997a; Odonne et al. 2013

(*Continued*)

Table A.1 (Continued) Medicinal Plants Reported to Be Used for Malaria

Plant Species	Common Name	Family	Part Used	Preparation	Indication	References
Hintonia latiflora (Sesse & Moc. Ex DC.) Bullock [Syn: *Coutarea pterosperma*, *Coutarea latiflora*]	NA	Rubiaceae	Not stated	Not stated	Malaria	Duke 1996; Milliken 1997a
Hippeastrum puniceum (Lam.) Kuntze	Barbados lily	Amaryllidaceae	Not stated	Not stated	Malaria	Duke 1996; Milliken 1997a
Hippocratea africana (Willd.) Loes ex Engl.	NA	Celastraceae	Root	Decoction, maceration	Malaria, jaundice, hepatitis	Adebayo and Krettli 2011; Ajibesin et al. 2007
Hiptis sidifolia (LiHer.) Briq.	NA	Labiatae	Leaf	Infusion	Malaria	Tene et al. 2007
Holarrhena floribunda T. Durand & Schinz	False rubber tree	Apocynaceae	Leaf, root bark	Decoction	Malaria	Koudouvo et al. 2011; Tetik, Civelek, and Cakilcioglu 2013; Traore et al. 2013; Yetein et al. 2013
Holarrhena pubescens Wall. ex G. Don	Feverpod	Apocynaceae	Root	Decoction	Malaria	Ngarivhume et al. 2015; Sivasankari, Anandharaj, and Gunasekaran 2014
Homalium involucratum (Pers.) Baill.	NA	Flacourtiaceae	Leaf	Decoction	Malaria	Norscia and Borgognini-Tarli 2006
Homalium letestui Pellegr.	NA	Flacourtiaceae	Root	Not stated	Malaria*	Adebayo and Krettli 2011
Homalium sp.	NA	Flacourtiaceae	Leaf	Decoction	Malaria	Rasoanaivo et al. 1992

(*Continued*)

Table A.1 (Continued) Medicinal Plants Reported to Be Used for Malaria

Plant Species	Common Name	Family	Part Used	Preparation	Indication	References
Homonoia riparia Lour.	NA	Euphorbiaceae	Wood, stem	Infusion (wood), not stated (stem)	Malaria	Duke 1996; Nguyen-Pouplin et al. 2007; Shankar, Sharma, and Deb 2012
Hortia sp.	NA	Rutaceae	Bark	Not stated	Malaria	Duke 1996; Milliken 1997a
Hortia brasiliana Vand. ex DC.	NA	Rutaceae	Not stated	Not stated	Quinine substitute	Cosenza et al. 2013
Hoslundia opposita Vahl	NA	Labiatae	Root, flower, leaf	Decoction (also in combination)	Malaria	Asase, Hesse, and Simmonds 2012; Gathirwa et al. 2011; Gessler et al. 1995; Nadembega et al. 2011; Nguta et al. 2010b; Stangeland et al. 2011; Tetik, Civelek, and Cakilcioglu 2013; Yetein et al. 2013
Houttuynia cordata	Chameleon	Saururaceae	Not stated	Not stated	Malaria	Duke 1996
Hua gabonii Pierre ex de Willd.	NA	Celastraceae	Leaf, root	Not stated	Malaria	Mbatchi et al. 2006
Humulus japonicus	Japanese hop	Cannabaceae	Not stated	Not stated	Malaria	Duke 1996
Hura crepitans	Sandbox tree	Euphorbiaceae	Stem bark	Decoction	Malaria	Guédé et al. 2010
Hydnora johannis Becc.	NA	Hydnoraceae	Root	Decoction	Malaria	Mesfin et al. 2012
***Hydrangea macrophylla* (Thunb.) Ser.**	French hydrangea	Hydrangeaceae	Leaf, root, flower	Not stated	Malaria	Shankar, Sharma, and Deb 2012; Tetik, Civelek, and Cakilcioglu 2013; Wee 1992

(*Continued*)

Table A.1 (Continued) Medicinal Plants Reported to Be Used for Malaria

Plant Species	Common Name	Family	Part Used	Preparation	Indication	References
Hydrangea paniculata	Panicled hydrangea	Hydrangeaceae	Not stated	Not stated	Malaria	Duke 1996
Hydrangea strigosa	NA	Hydrangeaceae	Not stated	Not stated	Malaria	Duke 1996
Hydrangea umbellata	NA	Hydrangeaceae	Not stated	Not stated	Malaria	Duke 1996
Hymenachne donacifolia Beauv.	NA	Poaceae	Whole plant	Decoction	Malaria	Muñoz, Sauvain, Bourdy, Arrázola et al. 2000
***Hymenaea courbaril* L.**	NA	Fabaceae	Bark	Decoction	Malaria, quinine substitute	Bertani et al. 2005; Duke 1996; Milliken 1997a,b; Ruiz et al. 2011
Hymenocallis sp.	NA	Amaryllidaceae	Not stated	Not stated	Malaria	Milliken 1997a
Hymenocardia acida Tul.	Heartfruit	Euphorbiaceae	Leaf, stem bark	Decoction	Malaria	Traore et al. 2013
Hymenocardia ulmoides Oliv.	NA	Euphorbiaceae	Bark, leaf	Not stated	Malaria	Mbatchi et al. 2006
Hymenodictyon excelsum	NA	Rubiaceae	Not stated	Not stated	Malaria	Duke 1996
Hymenodyction lohavato Baill.	NA	Rubiaceae	Root bark, stem bark	Decoction	Malaria	Rasoanaivo et al. 1992
Hyptis crenata Pohl ex Benth.	NA	Labiatae	Leaf	Infusion	Malaria	Milliken 1997a
Hyptis pectinata (L.) Poit.	Comb bushmint	Labiatae	Leaf	Decoction, infusion	Malaria (prevention with other herbs, treatment too)	Rasoanaivo et al. 1992; Ssegawa and Kasenene 2007; Vigneron et al. 2005

(Continued)

Table A.1 (Continued) Medicinal Plants Reported to Be Used for Malaria

Plant Species	Common Name	Family	Part Used	Preparation	Indication	References
Hyptis spicigera Lam	Marubio	Labiatae	Leaf	Decoction: oral and bath	Malaria	Asase et al. 2005; Diarra et al. 2015; Nadembega et al. 2011
***Hyptis suaveolens* (L.) Poit.**	Pignut	Labiatae	Root, leaf	Decoction, fumigation	Fever, malaria, repellent	Duke 1996; Koudouvo et al. 2011; Milliken 1997a; Tchacondo et al. 2012
Iboza multiflora (Benth) E.A. Brunce	NA	Labiatae	Leaf	Not stated	Malaria	Muthee et al. 2011
Ilex aquifolium	Common holly	Aquifoliaceae	Not stated	Not stated	Malaria	Duke 1996
Ilex guayusa Loesener	Guayusa	Aquifoliaceae	Leaf	Decoction	Malaria	Milliken 1997a
***Impatiens angustifolia* (Blume)**	Marsh henna	Balsaminaceae	Leaf	Paste	Malaria	Shankar, Sharma, and Deb 2012
***Imperata cylindrica* (L.) P. Beauv.**	Alang-alang	Poaceae	Root, aerial part	Infusion, decoction	Malaria, fever	Rasoanaivo et al. 1992; Yetein et al. 2013
Indigofera arrecta	Natal indigo	Fabaceae	Root	Decoction	Malaria	Stangeland et al. 2011
Indigofera articulata Gouan	NA	Fabaceae	Root	Infusion	Malaria	Mesfin et al. 2012
Indigofera coerulea Roxb.	NA	Fabaceae	Leaf	Decoction	Malaria	Mesfin et al. 2012
Indigofera congesta Welw. ex Bak.f.	NA	Fabaceae	Stem	Decoction	Fever, malaria	Ssegawa and Kasenene 2007
Indigofera emarginella	NA	Fabaceae	Root	Infusion	Malaria	Stangeland et al. 2011
Indigofera erecta Thunb.	NA	Fabaceae	Leaf	Not stated	Malaria	Namukobe et al. 2011

(Continued)

Table A.1 (Continued) Medicinal Plants Reported to Be Used for Malaria

Plant Species	Common Name	Family	Part Used	Preparation	Indication	References
Indigofera lupatana Bak.f.	NA	Fabaceae	Root	Not stated	Malaria	Gakuya et al. 2013
Indigofera nigritana Hook.f.	NA	Fabaceae	Whole plant	Not stated	Malaria	Nadembega et al. 2011
Indigofera pulchra Willd.	NA	Fabaceae	Whole plant	Not stated	Malaria	Adamu et al. 2005; Adebayo and Krettli 2011; Asase et al. 2005
Indigofera sp.	NA	Fabaceae	Leaf	Decoction	Malaria	Asase and Oppong-Mensah 2009
Indigofera spicata Forssk.	NA	Fabaceae	Root	Not stated	Malaria	Giday, Asfaw, and Woldu 2009
***Indigofera suffruticosa* Mill.**	Anil de pasto	Fabaceae	Leaf, root, whole plant	Decoction (whole plant), direct application	Fever, malaria	Milliken 1997a
Inga acreana Harms	NA	Fabaceae	Not stated	Infusion	Malaria	Tene et al. 2007
Inga sessilis Mart.	NA	Fabaceae	Leaf	Not stated	Fever, malaria	Milliken 1997a
Inga setulifera DC.	NA	Fabaceae	Bark	Infusion	Malaria	Milliken 1997a
Inga spp.	NA	Fabaceae	Bark	Decoction	Malaria	Milliken 1997b
Inula perrieri H. Humb.	NA	Compositae	Leaf	Decoction	Malaria, splenomegaly	Rasoanaivo et al. 1992
Inula viscosa	NA	Compositae	Not stated	Not stated	Malaria	Duke 1996
***Ipomoea batatas* (L.) Lam**	Sweet potato	Convolvulaceae	Leaf	Decoction: oral and bath	Malaria	Diarra et al. 2015
***Ipomoea cairica* (L.) Sweet**	Cairo morning glory	Convolvulaceae	Leaf	Decoction: oral and bath	Malaria	Diarra et al. 2015
Ipomoea hederifolia L.	Scarlet creeper	Convolvulaceae	Root	Not stated	Fever, malaria	Duke 1996; Milliken 1997a

(Continued)

Table A.1 (Continued) Medicinal Plants Reported to Be Used for Malaria

Plant Species	Common Name	Family	Part Used	Preparation	Indication	References
***Ipomoea pes-caprae* (L.) Sweet**	Bayhops	Convolvulaceae	Whole plant, leaf	Decoction (whole plant), not stated (leaf)	Fever, malaria	Kaou et al. 2008; Milliken 1997a; Nguyen-Pouplin et al. 2007
Ipomoea spathulala Hall.f.	NA	Convolvulaceae	Not stated	Decoction	Malaria	Geissler et al. 2002
Iresine calea (Ibanez) Standl.	NA	Amaranthaceae	Leaf	Decoction	Fever, malaria	Duke 1996; Milliken 1997a
Iresine diffusa Humb. & Bonpl. ex Willd.	Juba's bush	Amaranthaceae	Aerial part	Maceration, infusion	Malaria	Valadeau et al. 2010
Irlbachia alata (Aubl.) Maas	NA	Gentianaceae	Leaf, root	Decoction, juice, maceration	Malaria	Bertani et al. 2005; Milliken 1997a,b; Valadeau et al. 2010
Irlbachia speciosa (Cham. & Schltr.) Maas	NA	Gentianaceae	Not stated	Not stated	Malaria	Milliken 1997a
Irvingia gabonensis (Aubry-Lecomte ex O'Rorke) Baill.	African wild mango, sweet bush-mango	Irvingiaceae	Stem bark	Decoction	Malaria	Guédé et al. 2010; Zirihi et al. 2005
***Irvingia malayana* Oliv. ex Benn.**	Wild almond	Irvingiaceae	Leaf	Not stated	Malaria*	Nguyen-Pouplin et al. 2007
Isertia haenkeana DC.	NA	Rubiaceae	Leaf, bark	Decoction	Malaria	Milliken 1997a
Isoberlinia doka Craib. et Stapf.	NA	Fabaceae	Leaf	Decoction	Malaria	Nadembega et al. 2011
Isocarpha divaricata	NA	Compositae	Not stated	Not stated	Malaria	Duke 1996

(*Continued*)

Table A.1 (Continued) Medicinal Plants Reported to Be Used for Malaria

Plant Species	Common Name	Family	Part Used	Preparation	Indication	References
Isocarpha microcephala (DC.) S.F. Blake	NA	Compositae	Not stated	Not stated	Malaria	Milliken 1997a
Isolona hexaloba Engl. & Diels	NA	Annonaceae	Stem bark	Decoction	Malaria	Muganza et al. 2012
Jacaranda copaia (Aubl.) D. Don	NA	Bignoniaceae	Leaf	Infusion (per os and bath)	Malaria (preventive and curative), fever	Valadeau et al. 2010
Jacaranda spp.	NA	Bignoniaceae	Not stated	Not stated	Malaria	Milliken 1997a
Jasminum fluminense Vell.	Jazmin de trapo	Oleaceae	Root	Decoction	Malaria	Geissler et al. 2002
Jasminum syringifolia Wall. ex G. Don	NA	Oleaceae	Leaf	Not stated	Malaria	Chander et al. 2014
Jatropha curcas **L.**	Barbados nut	Euphorbiaceae	Leaf, seed, stem, latex, root	Decoction (also with *O. americanum* and *O. gratissimum*), infusion (bath or consumed), eaten (seed), maceration	Fever, malaria (leaf, seed)	Asase et al. 2005; Kaou et al. 2008; Mahishi, Srinivasa, and Shivanna 2005; Mesfin et al. 2012; Milliken 1997a; Musa et al. 2011; Odonne et al. 2013; Rasoanaivo et al. 1992; Titanji, Zofou, and Ngemenya 2008; Traore et al. 2013; Yetein et al. 2013

(Continued)

Table A.1 (Continued) Medicinal Plants Reported to Be Used for Malaria

Plant Species	Common Name	Family	Part Used	Preparation	Indication	References
***Jatropha gossypifolia* L.**	Bellyache bush	Euphorbiaceae	Whole plant, leafy stem, leaf (also in combination), stem	Infusion, decoction, eaten	Fever, malaria	Asase et al. 2005; Jansen et al. 2010; Koudouvo et al. 2011; Milliken 1997a; Ruiz et al. 2011; Tchacondo et al. 2012; Yetein et al. 2013
Jatropha podagrica Hook.	Goutystalk	Euphorbiaceae	Not stated	Not stated	Malaria	Milliken 1997a
Joosia dichotoma (Ruiz & Pav.) H. Karst.	NA	Rubiaceae	Bark	Alcoholic extract	Malaria	Milliken 1997a
Juliania adstringens	NA	Anacardiaceae	Not stated	Not stated	Malaria	Duke 1996
***Justicia betonica* L.**	Squirrel's tail	Acanthaceae	Leaf	Decoction, infusion, juice	Malaria	Namukobe et al. 2011; Stangeland et al. 2011
Justicia flava Vahl.	NA	Acanthaceae	Leaf	Decoction (also in combination)	Malaria	Asase, Hesse, and Simmonds 2012; Namsa, Mandal, and Tangjang 2011; Tetik, Civelek, and Cakilcioglu 2013
Justicia insularis	NA	Acanthaceae	Leaf	Not stated	Malaria	Tetik, Civelek, and Cakilcioglu 2013
Justicia schimperiana Hochst.	NA	Acanthaceae	Leaf	Not stated	Malaria	Zerabruk and Yirga 2012
Kaempferia galanga	Galanga	Zingiberaceae	Not stated	Not stated	Malaria	Duke 1996
Kalanchoe densiflora Rolfe	NA	Crassulaceae	Leaf	Not stated	Malaria	Titanji, Zofou, and Ngemenya 2008

(*Continued*)

Table A.1 (Continued) Medicinal Plants Reported to Be Used for Malaria

Plant Species	Common Name	Family	Part Used	Preparation	Indication	References
***Khaya anthotheca* C.DC.**	White mahogany	Meliaceae	Leaf	Decoction and drank for a week as required with leaves of *Musa paradisiaca* L., whole plant of *Sarcophrynium brachystachys* Schumann., leaves, young stem and roots of *Bambusa vulgaris* Schreber., fruit of *Citrus aurantiifolia* (Christm.) Swingle, leaves and bark of three other unknown species	Malaria (children)	Asase and Kadera 2014
***Khaya grandifoliola* C.DC.**	African mahogany, Benin mahogany, large-leaved mahogany	Meliaceae	Stem, stem bark	Decoction	Malaria	Adebayo and Krettli 2011; Idowu et al. 2010; Tetik, Civelek, and Cakilcioglu 2013

(*Continued*)

Table A.1 (Continued) Medicinal Plants Reported to Be Used for Malaria

Plant Species	Common Name	Family	Part Used	Preparation	Indication	References
***Khaya senegalensis* (Desr.) A. Juss.**	Senegal mahogany	Meliaceae	Bark, stem, leaf, stem bark, root	Decoction (also decoction with *Guiera senegalensis*: oral and bath), maceration, infusion (alcoholic), powder	Malaria	Adebayo and Krettli 2011; Asase, Akwetey, and Achel 2010; Asase and Oppong-Mensah 2009; Asase et al. 2005; Betti and Yemefa 2011; Diarra et al. 2015; El-Kamali and El-Khalifa 1999; Musa et al. 2011; Nadembega et al. 2011; Tchacondo et al. 2012; Tetik, Civelek, and Cakilcioglu 2013; Tor-anyiin, Sha'ato, and Oluma 2003; Traore et al. 2013; Yetein et al. 2013
***Kigelia africana* (Lam.) Benth.**	Sausage tree	Bignoniaceae	Stem, leaf, fruit	Infusion, decoction	Malaria	Gessler et al. 1995
Kigelianthe madagascariensis* Sprague var *hidebrandtii	NA	Bignoniaceae	Leaf	Inhalation, infusion	Malaria	Rasoanaivo et al. 1992
Labisia pumila	Kacip Fatimah	Myrsinaceae	Leaf	Decoction	Malaria (curative, prophylactic)	Al-adhroey et al. 2010
Lacistema aggregatum (Bergius) Rusby	NA	Lacistemataceae	Wood	Maceration in rum, decoction	Malaria, fever	Milliken 1997a; Roumy et al. 2007

(*Continued*)

Table A.1 (Continued) Medicinal Plants Reported to Be Used for Malaria

Plant Species	Common Name	Family	Part Used	Preparation	Indication	References
Ladenbergia cujabensis Klotzsch [Syn: *Cinchona lambertiana* A. Braun ex Mart.]	NA	Rubiaceae	Not stated	Not stated	Quinine substitute	Cosenza et al. 2013
Ladenbergia macrocarpa (M. Vahl) Klotzsch.	NA	Rubiaceae	Bark	Not stated	Malaria	Duke 1996; Milliken 1997a
Ladenbergia magnifolia (Ruiz & Pav.) Klotzsch.	NA	Rubiaceae	Bark	Infusion	Malaria	Milliken 1997a
Ladenbergia malacophylla Standl.	NA	Rubiaceae	Bark	Not stated	Malaria	Milliken 1997a
Ladenbergia moritziana	NA	Rubiaceae	Not stated	Not stated	Malaria	Duke 1996
Ladenbergia oblongifolia (Humb. ex Mutis) L. Andersson [Syn: *Cinchona lutescens* Vell.]	NA	Rubiaceae	Not stated	Not stated	Quinine substitute	Cosenza et al. 2013
Laelia sp.	NA	Orchidaceae	Not stated	Not stated	Malaria	Milliken 1997a

(Continued)

Table A.1 (Continued) Medicinal Plants Reported to Be Used for Malaria

Plant Species	Common Name	Family	Part Used	Preparation	Indication	References
***Lagerstroemia speciosa* Pers.**	Pride-of-India	Lythraceae	Not stated	Not stated	Malaria	Duke 1996
Laggera alata	NA	Compositae	Not stated	Not stated	Malaria	Tetik, Civelek, and Cakilcioglu 2013
Landolphia heudelotii A.DC.	Landolphia rubber	Apocynaceae	Leaf	Decoction: oral and bath	Malaria	Diarra et al. 2015; Traore et al. 2013
Landolphia lanceolata (K. Schum.) Pichon	NA	Apocynaceae	Leaf, root	Not stated	Malaria	Mbatchi et al. 2006
Landolphia sp.	NA	Apocynaceae	Stem	Decoction	Malaria	Asase and Oppong-Mensah 2009
Languas galanga Stuntz.	Lengkuas	Zingiberaceae	Rhizome	Decoction	Malaria	Al-adhroey et al. 2010
Lannea acida A. Rich.	NA	Anacardiaceae	Stem bark, leaf (in combination), bark	Decoction	Malaria	Asase et al. 2005; Diarra et al. 2015; Nadembega et al. 2011
Lannea microcarpa Eng. & Kr.	African grape	Anacardiaceae	Not stated	Decoction	Malaria	Nadembega et al. 2011
Lannea schweinfurthii (Engl.) Engl.	NA	Anacardiaceae	Not stated	Decoction	Malaria	Gathirwa et al. 2011; Geissler et al. 2002
***Lansium domesticum* Corr. Ser.**	Duku, Langsat	Meliaceae	Peel, bark, seed, leaf	Infusion, decoction	Malaria, fever	Al-adhroey et al. 2010; Duke 1996; Elliott and Brimacombe 1987; Leaman et al. 1995; Omar et al. 2003

(*Continued*)

Table A.1 (Continued) Medicinal Plants Reported to Be Used for Malaria

Plant Species	Common Name	Family	Part Used	Preparation	Indication	References
Lantana camara var. *aculeata* (L.) Moldenke	Lantana	Verbenaceae	Whole plant	Not stated	Malaria	Sivasankari, Anandharaj, and Gunasekaran 2014
***Lantana camara* L. [Syn: *Lantana crocea* Jacq.]**	Wild sage, common lantana	Verbenaceae	Whole plant, flower, bark, leaf, root	Decoction, infusion	Malaria, fever	Duke 1996; Gurib-Fakim et al. 1993; Mesia et al. 2008; Milliken 1997a; Nguta et al. 2010a,b; Poonam and Singh 2009; Rasoanaivo et al. 1992; Sharma, Chhangte, and Dolui 2001; Tetik, Civelek, and Cakilcioglu 2013; Titanji, Zofou, and Ngemenya 2008; Traore et al. 2013; Vigneron et al. 2005; Yetein et al. 2013
Lantana involucrata L.	Buttonsage	Verbenaceae	Not stated	Decoction	Fever, malaria	Duke 1996; Milliken 1997a
Lantana rhodesiensis Moldenke	NA	Verbenaceae	Leaf, root	Decoction, fumes	Malaria, insect repellent	Nadembega et al. 2011
Lantana trifolia L.	Threeleaf shrubverbena	Verbenaceae	Stem, leaf, whole plant, root	Infusion, decoction, juice	Fever, malaria	Calderón, Simithy-Williams, and Gupta 2012; Jamir, Sharma, and Dolui 1999; Milliken 1997b; Namukobe et al. 2011; Stangeland et al. 2011

(Continued)

Table A.1 (Continued) Medicinal Plants Reported to Be Used for Malaria

Plant Species	Common Name	Family	Part Used	Preparation	Indication	References
Laudolphia buchananii (Hall.f) Stapf.	NA	Apocynaceae	Leaf	Decoction	Malaria	Nguta et al. 2010b
Launea cornuta (Oliv and Hiern) C.Jeffrey	NA	Compositae	Leaf	Decoction	Malaria	Nguta et al. 2010b
Laurus nobilis L.	Bay tree	Lauraceae	Leaf	Not stated	Malaria	Passalacqua, Guarrera, and De Fine 2007
Lawsonia inermis **L.**	Henna	Lythraceae	Root, root bark, leaf	Decoction, juice	Malaria	Mesfin et al. 2012; Traore et al. 2013
Lecaniodiscus cupanoides	NA	Sapindaceae	Leaf, stem bark	Not stated	Malaria	Idowu et al. 2010; Tetik, Civelek, and Cakilcioglu 2013
Leea sp.	NA	Vitaceae	Leaf	Decoction	Malaria, fever	Elliott and Brimacombe 1987
Lemna minor	Common duckweed	Lemnaceae	Not stated	Not stated	Malaria	Duke 1996
Leonotis nepetifolia (L.) R. Br.	Lion's ear	Labiatae	Whole plant, leaf, flower, aerial part	Infusion, decoction (leaf, flower, aerial part)	Fever, malaria, malaria (prevention, with other herbs)	Duke 1996; Milliken 1997a; Rasoanaivo et al. 1992; Titanji, Zofou, and Ngemenya 2008; Vigneron et al. 2005
Leonurus japonicus Houtt.	Chinese motherwort	Labiatae	Whole plant, flower	Infusion	Fever, malaria	Milliken 1997a
Lepidagathis anobrya Nees	NA	Acanthaceae	Whole plant	Decoction	Malaria	Nadembega et al. 2011
Lepidium sativum L.	Gardencress	Brassicaceae	Seed	Not stated	Malaria	Mesfin, Demissew, and Teklehaymanot 2009

(*Continued*)

Table A.1 (Continued) Medicinal Plants Reported to Be Used for Malaria

Plant Species	Common Name	Family	Part Used	Preparation	Indication	References
Leptadenia hastata (Pers.) Decne.	NA	Asclepiadaceae	Root, root bark	Decoction	Malaria	Mesfin et al. 2012; Nadembega et al. 2011
Leptadenia madagascariensis Decne	NA	Asclepiadaceae	Aerial part	Decoction (with *O. americanum* and *O. gratissimum*)	Malaria, fever	Kaou et al. 2008; Rasoanaivo et al. 1992
Leretia cordata Vell.	NA	Icacinaceae	Root	Not stated	Malaria	Odonne et al. 2013
Leucas martinicensis Jacq. Ait.F.	Whitewort	Labiatae	Leaf, whole plant (in combination)	Charred, boiled, decoction (oral and as bath)	Insect repellent, malaria	Asase et al. 2005; Diarra et al. 2015; Nadembega et al. 2011
Licania hebantha Mart. Ex Hook.f.	NA	Chrysobalanaceae	Fruit	Infusion	(Malaria)	Milliken 1997a
Licania parviflora Benth.	NA	Chrysobalanaceae	Leaf	Infusion	(Malaria)	Milliken 1997a
Limeum pterocarpum (Gay) Heimerl	NA	Aizoaceae	Whole plant	Decoction	Malaria	Diallo et al. 1999
Lindackeria paludosa (Benth.) Gilg	NA	Flacourtiaceae	Bark	Infusion	Malaria	Milliken 1997a
Lindera strychnifolia	Combined spicebush	Lauraceae	Not stated	Not stated	Malaria	Duke 1996
Linociera ramiflora	NA	Oleaceae	Not stated	Not stated	Malaria	Duke 1996
Lippia alba (Mill.) N.E. Br.	Bushy lippia	Verbenaceae	Leaf	Decoction	Fever, malaria (with other herbs)	Milliken 1997a; Vigneron et al. 2005
Lippia chevalieri Mold	NA	Verbenaceae	Leaf	Decoction	Malaria	Diarra et al. 2015; Traore et al. 2013; Willcox 2011

(Continued)

Table A.1 (Continued) Medicinal Plants Reported to Be Used for Malaria

Plant Species	Common Name	Family	Part Used	Preparation	Indication	References
Lippia multiflora Moldenke	NA	Verbenaceae	Leaf, aerial part	Infusion, decoction	Malaria	Adebayo and Krettli 2011; Asase, Akwetey, and Achel 2010; Koudouvo et al. 2011; Okpekon et al. 2004
Lippia schomburgkiana Schum	NA	Verbenaceae	Leaf	Infusion, decoction	Malaria	Milliken 1997b
Liquidambar orientalis	Oriental sweetgum	Hamamelidaceae	Not stated	Not stated	Malaria	Duke 1996
Liriodendron tulipifera	Tuliptree	Magnoliaceae	Not stated	Not stated	Malaria	Duke 1996
Litsea bubeba (Lour.) Pers.	NA	Lauraceae	Not stated	Decoction	Malaria	Ghorbani et al. 2011
Litsea floribunda Gamb.	NA	Lauraceae	Stem	Decoction	Malaria	Lingaraju, Sudarshana, and Rajashekar 2013
Litsea semecarpifolia (Wall. ex Nees)	NA	Lauraceae	Bark	Decoction	Malaria	Ghorbani et al. 2011
Litsea sp.	NA	Lauraceae	Not stated	Not stated	Malaria	Duke 1996
Lobelia chinensis	Chinese lobelia	Campanulaceae	Not stated	Not stated	Malaria	Duke 1996
Lobelia sp.	NA	Campanulaceae	Root	Not stated	Malaria	Giday et al. 2007
Loeselia ciliata L.	NA	Polemoniaceae	Not stated	Decoction	Malaria	Dimayuga and Agundez 1986
Loeseneriella africana (Willd.) N.Halle	NA	Celastraceae	Leaf	Decoction	Malaria (also in pediatrics)	Jansen et al. 2010; Nadembega et al. 2011

(*Continued*)

Table A.1 (Continued) Medicinal Plants Reported to Be Used for Malaria

Plant Species	Common Name	Family	Part Used	Preparation	Indication	References
Lonchocarpus cyanescens (Schum. & Thonn.) Benth.	NA	Fabaceae	Leaf	Not stated	Malaria*	Adebayo and Krettli 2011
Lophira alata Banks ex Gaertn. F.	Ironwood	Ochnaceae	Leaf, stem, stem bark, root, seed	Decoction	Malaria	Adebayo and Krettli 2011; Tetik, Civelek, and Cakilcioglu 2013; Traore et al. 2013
Lophira lanceolata Van Tiegh. ex Keay.	Dwarf red ironwood	Ochnaceae	Leaf	Decoction	Malaria	Adebayo and Krettli 2011; Tchacondo et al. 2012
Loranthus ferrugineus Roxb.	NA	Loranthaceae	Not stated	Not stated	Malaria	Duke 1996
Loranthus kaempferi	NA	Loranthaceae	Not stated	Not stated	Malaria	Duke 1996
Lucuma salicifolia	NA	Sapotaceae	Not stated	Not stated	Malaria	Duke 1996
Ludolphia buchananii (Hall.f) Stapf.	NA	Apocynaceae	Leaf	Decoction	Malaria	Nguta et al. 2010a
Ludwigia erecta (L.) Hara	NA	Onagraceae	Whole plant	Infusion	Malaria	Muthaura et al. 2007
Ludwigia peruviana (L.) Hara.	Primrose Willow	Onagraceae	Whole plant	Not stated	Malaria	Dike, Obembe, and Adebiyi 2012
Luffa acutangula (L.) Roxb.	Chinese okra, ribbed gourd	Cucurbitaceae	Seed	Not stated	Malaria, fever	Elliott and Brimacombe 1987
Luisia teres	NA	Orchidaceae	Not stated	Not stated	Malaria	Duke 1996
Lycopodium crasum Kunth ex Willd.	NA	Lycopodiaceae	Not stated	Not stated	Malaria	Milliken 1997a
Lyopodium saururus Lam.	NA	Lycopodiaceae	Not stated	Infusion	Malaria	Milliken 1997a

(Continued)

Table A.1 (Continued) Medicinal Plants Reported to Be Used for Malaria

Plant Species	Common Name	Family	Part Used	Preparation	Indication	References
Macaranga populifolia Muel.-Ar	NA	Euphorbiaceae	Not stated	Not stated	Malaria	Duke 1996
Machaerium floribunda Benth.	NA	Fabaceae	Stem	Not stated	Malaria	Kvist et al. 2006
Maclura sp.	NA	Moraceae	Leaf	Fresh	Malaria	Jorim et al. 2012
Maclura tinctoria (L.) G. Don ex Steud.	Fustic tree	Moraceae	Wood	Decoction	Malaria	Milliken 1997a
Macroclinidium verticillatum	NA	Compositae	Not stated	Not stated	Malaria	Duke 1996
Macrocnemum roseum (Ruiz & Pav.) Wedd.	NA	Rubiaceae	Bark	Not stated	Malaria, quinine substitute	Cosenza et al. 2013; Milliken 1997a
Macrotyloma axillare	Perennial horsegram	Fabaceae	Leaf	Not stated	Malaria	Stangeland et al. 2011
Maerua oblongifolia (Forssk.) A. Rich.	NA	Capparaceae	Leaf	Decoction	Malaria	Mesfin et al. 2012
Maesa doraena	NA	Myrsinaceae	Not stated	Not stated	Malaria	Duke 1996
Maesa lanceolata Forssk.	NA	Maesaceae	Leaf, bark	Maceration, juice	Malaria	Jamir, Sharma, and Dolui 1999; Stangeland et al. 2011
Magnolia grandiflora **L.**	Southern magnolia	Magnoliaceae	Bark	Not stated	Malaria	Duke 1996
Magnolia officinalis	NA	Magnoliaceae	Not stated	Not stated	Malaria	Duke 1996
Magnolia virginiana	Sweetbay magnolia	Magnoliaceae	Not stated	Not stated	Malaria	Duke 1996
Mahonia aquifolia	NA	Berberidaceae	Not stated	Not stated	Malaria	Duke 1996
Mahonia bealei	Beale's barberry	Berberidaceae	Not stated	Not stated	Malaria	Duke 1996

(*Continued*)

Table A.1 (Continued) Medicinal Plants Reported to Be Used for Malaria

Plant Species	Common Name	Family	Part Used	Preparation	Indication	References
Malachra alceifolia Jacq.	Yellow leafbract	Malvaceae	Whole plant, leaf, shoot	Infusion	Fever, malaria (also in combination)	Kvist et al. 2006; Milliken 1997a; Odonne et al. 2013; Ruiz et al. 2011
Mallotus oppositofolius	NA	Euphorbiaceae	Leaf	Not stated	Malaria	Dike, Obembe, and Adebiyi 2012; Tetik, Civelek, and Cakilcioglu 2013
***Mallotus paniculatus* (Lam.) Muell.-Arg.**	Balek angin	Euphorbiaceae	Leaf	Juice, decoction	Fever, malaria	Duke 1996; Mahyar et al. 1991
Malus sylvestris (L.) Mill.	European crab apple	Rosaceae	Root	Not stated	Malaria, fever	Duke 1996; Milliken 1997a
Mammea africana	NA	Clusiaceae	Not stated	Not stated	Malaria	Tetik, Civelek, and Cakilcioglu 2013
Mammea americana L.	Mammee apple	Guttiferae	Leaf, seed	Decoction	Fever, malaria	Milliken 1997a
Mangifera caesia	Binjai	Anacardiaceae	Leaf	Not stated	Malaria	Tetik, Civelek, and Cakilcioglu 2013

(*Continued*)

Table A.1 (Continued) Medicinal Plants Reported to Be Used for Malaria

Plant Species	Common Name	Family	Part Used	Preparation	Indication	References
***Mangifera indica* L.**	Mango	Anacardiaceae	Leaf, root, stem, bark, fruit, stem bark, cortex	Infusion, decoction	Malaria, fever	Adebayo and Krettli 2011; Allabi et al. 2011; Asase, Akwetey, and Achel 2010; Asase, Hesse, and Simmonds 2012; Asase and Oppong-Mensah 2009; Asase et al. 2005; Dike, Obembe, and Adebiyi 2012; Duke 1996; Ellena, Quave, and Pieroni 2012; Elliott and Brimacombe 1987; Idowu et al. 2010; Jamir, Sharma, and Dolui 1999; Koudouvo et al. 2011; Kvist et al. 2006; Milliken 1997a; Nadembega et al. 2011; Namukobe et al. 2011; Oni 2010; Rasoanaivo et al. 1992; Ruiz et al. 2011; Titanji, Zofou, and Ngemenya 2008; Tchacondo et al. 2012; Traore et al. 2013; Zirihi et al. 2005
Mangifera minor Blume	NA	Anacardiaceae	Leaf	Not stated	Malaria	Prescott, Kiapranis, and Maciver 2012

(Continued)

Table A.1 (Continued) Medicinal Plants Reported to Be Used for Malaria

Plant Species	Common Name	Family	Part Used	Preparation	Indication	References
***Manihot esculenta* Crantz**	Cassava	Euphorbiaceae	Root	Infusion, maceration, cooked	Malaria, jaundice	Giovannini 2015; Tetik, Civelek, and Cakilcioglu 2013; Yetein et al. 2013
***Manihot utilisma* Pohl.**	NA	Euphorbiaceae	Leaf	Decoction	Malaria	Rasoanaivo et al. 1992
Manniophyton fulvum Muell. Arg.	NA	Euphorbiaceae	Leaf, root bark	Decoction, juice	Malaria	Muganza et al. 2012; Tetik, Civelek, and Cakilcioglu 2013
Mansoa alliacea (Lam.) A.H. Gentry	Garlic vine	Bignoniaceae	Leaf, root, stem, bark	Decoction, direct application	Fever, malaria	Milliken 1997a; Odonne et al. 2013; Ruiz et al. 2011
Maquira coriacea (Karst.) C.C. Berg	NA	Moraceae	Root, bark, cortex	Decoction	Malaria, fever	Kvist et al. 2006; Muñoz et al. 2000
Maranta arundinacea L.	Arrowroot	Marantaceae	Root	Not stated	Fever, malaria	Milliken 1997a
Mareya micrantha Muell. Arg.	Number one	Euphorbiaceae	Leaf, stem bark	Decoction	Malaria (also in combination)	Asase, Hesse, and Simmonds 2012; Guédé et al. 2010; Zirihi et al. 2005
Margaritaria discoidea (Baill.) Webster	Pheasant berry	Euphorbiaceae	Leaf	Decoction	Malaria	Traore et al. 2013
Markhamia gellatiana	NA	Bignoniaceae	Not stated	Not stated	Malaria	Tetik, Civelek, and Cakilcioglu 2013
Markhamia lutea (Benth.) K. Schum.	Siala	Bignoniaceae	Leaf, root	Decoction	Malaria	Hamill et al. 2000; Lacroix, Prado, Kamoga, Kasenene, Namukobe et al. 2011; Stangeland et al. 2011

(Continued)

Table A.1 (Continued) Medicinal Plants Reported to Be Used for Malaria

Plant Species	Common Name	Family	Part Used	Preparation	Indication	References
Markhamia sessilis Sprague	NA	Bignoniaceae	Bark, leaf	Not stated	Malaria	Mbatchi et al. 2006; Tetik, Civelek, and Cakilcioglu 2013
Markhamia tomentosa (Benth.) K. Schum.	NA	Bignoniaceae	Leaf	Decoction	Malaria	Traore et al. 2013
Marrubium incanum Desr.	Horehound	Labiatae	Aerial part	Decoction	Malaria	Pieroni et al. 2002
Marrubium vulgare L.	Horehound	Labiatae	Aerial part	Decoction	Fever, malaria	Milliken 1997a; Pieroni et al. 2002
Martynia annua L.	Devil's claw	Pedaliaceae	Leaf	Charred	Malaria	Nadembega et al. 2011
Mascagnia benthamiana (Griseb.) Anderson	NA	Malpighiaceae	Not stated	Not stated	(Malaria)	Milliken 1997a
Matricaria chamomilla L.	False chamomile	Compositae	Not stated	Not stated	Malaria	Duke 1996; Esmaeili et al. 2009
Matricaria recutita L.	German chamomile	Compositae	Flower, whole plant	Infusion (whole plant)	Fever, malaria	Milliken 1997a
Mauritia flexuosa L.	Muriti	Palmae	Petiole	Infusion	Malaria	Milliken 1997b
Maytenus acuminata (L.f.) Loes	NA	Celastraceae	Leaf, root bark	Not stated	Malaria	Muregi et al. 2007
Maytenus arbutifolia (A. Rich.) Wilczek	NA	Celastraceae	Root bark	Decoction	Malaria	Muthaura et al. 2007
Maytenus heterophylla (Echl. & Zeyh) Robson	NA	Celastraceae	Root bark	Decoction	Malaria	Muregi et al. 2007; Muthaura et al. 2007
Maytenus krukovii A.C. Sm.	NA	Celastraceae	Root bark	Decoction	Malaria	Odonne et al. 2013

(*Continued*)

Table A.1 (Continued) Medicinal Plants Reported to Be Used for Malaria

Plant Species	Common Name	Family	Part Used	Preparation	Indication	References
Maytenus macrocarpa (Ruiz & Pavr) Briq.	NA	Celastraceae	Not stated	Not stated	Malaria	Ruiz et al. 2011
Maytenus putterlickioides (Loes.) Excell & Mendonca	NA	Celastraceae	Root bark	Decoction	Malaria	Muthaura et al. 2007
Maytenus senagalensis (Lam.) Excell [Syn: *Gymnosporia senegalensis* (Lam.) Loes.]	Confetti tree	Celastraceae	Leaf, root, stem, root bark	Decoction, burnt powder	Malaria	Gessler et al. 1995; Muregi et al. 2007; Nadembega et al. 2011; Nanyingi et al. 2008; Rukunga et al. 2009; Tchacondo et al. 2012
Maytenus spp.	NA	Celastraceae	Cortex	Not stated	Malaria	Kvist et al. 2006
Maytenus undata (Thunb.) Blakelock	NA	Celastraceae	Leaf, root bark	Decoction	Malaria	Muthaura et al. 2007
Melaleuca leucadendron L.	Paperbark tree, white teatree	Myrtaceae	Leaf	Not stated	Fever, malaria	Duke 1996; Milliken 1997a
Melanthera scandens Schumach. & Thonn.	NA	Compositae	Whole plant, leaf	Decoction	Malaria	Guédé et al. 2010; Namukobe et al. 2011; Zirihi et al. 2005
Melia azedarach L.	China berry, white cedar, umbrella tree	Meliaceae	Bark, root, flower, leaf, root bark	Decoction	Fever, malaria, quinine substitute	Milliken 1997a; Rasoanaivo et al. 1992; Titanji, Zofou, and Ngemenya 2008
***Melia indica* Brandis**	NA	Meliaceae	Not stated	Not stated	Malaria	Duke 1996
Melicope pteleifolia (Champ. ex Benth.)	NA	Rutaceae	Not stated	Not stated	Malaria	Ghorbani et al. 2011

(Continued)

Table A.1 (Continued) Medicinal Plants Reported to Be Used for Malaria

Plant Species	Common Name	Family	Part Used	Preparation	Indication	References
Melodinus monogynus Roxb.	NA	Apocynaceae	Leaf, root, bark	Not stated	Malaria	Duke 1996; Shankar, Sharma, and Deb 2012
Mentha sp.	NA	Labiatae	Not stated	Not stated	Malaria	Milliken 1997a
Mentha spicata L.	Spearmint	Labiatae	Aerial part	Infusion (consumed and as bath)	Malaria	Mesfin et al. 2012
Mentha sylvestris	NA	Labiatae	Not stated	Not stated	Malaria	Tetik, Civelek, and Cakilcioglu 2013
Mentha viridis	NA	Labiatae	Not stated	Not stated	Malaria	Duke 1996
Mesona wallichiana Benth.	Black cincau	Lauraceae	Root	Decoction	Malaria	Shankar, Sharma, and Deb 2012
Mezoneuron benthamianum Baill.	NA	Fabaceae	Leaf	Decoction	Malaria	Traore et al. 2013
Michelia alba DC.	NA	Magnoliaceae	Not stated	Not stated	Malaria	Duke 1996
***Michelia champaca* L.**	Michelia	Magnoliaceae	Not stated	Not stated	Malaria	Duke 1996
Miconia spp.	NA	Melastomataceae	Root	Decoction	Fever, malaria	Milliken 1997a,b
Microdesmis keayana J. Leonard	NA	Pandaceae	Leaf	Not stated	Malaria*	Zirihi et al. 2005
Microglossa angolensis	NA	Compositae	Not stated	Not stated	Malaria	Tetik, Civelek, and Cakilcioglu 2013
Microglossa pyrifolia Lam. Kuntze	NA	Compositae	Leaf, root bark, stem	Decoction	Malaria	Duke 1996; Guédé et al. 2010; Jamir, Sharma, and Dolui 1999; Stangeland et al. 2011; Tetik, Civelek, and Cakilcioglu 2013; Zirihi et al. 2005

(Continued)

Table A.1 (Continued) Medicinal Plants Reported to Be Used for Malaria

Plant Species	Common Name	Family	Part Used	Preparation	Indication	References
Microglossa sp.	NA	Compositae	Root	Decoction	Malaria	Geissler et al. 2002
Mikania cordifolia (L.f.) Willd.	Florida keys hempvine	Compositae	Whole plant, stem	Infusion, decoction	Malaria	Duke 1996; Milliken 1997a
Mikania guaco Humb. & Bompl.	NA	Compositae	Leaf, whole plant	Decoction, juice	Malaria (prevention and treatment), fever	Bertani et al. 2005; Duke 1996; Milliken 1997a; Vigneron et al. 2005
Mikania hookeriana DC.	NA	Compositae	Not stated	Not stated	Malaria	Milliken 1997a
***Mikania micrantha* Kunth.**	Mile-a-minute	Compositae	Leaf	Juice	Malaria (treatment and prevention)	Sharma, Chhangte, and Dolui 2001; Vigneron et al. 2005
Mikania officinalis Mart.	NA	Compositae	Not stated	Not stated	Malaria	Milliken 1997a
Mikania scandens (L.) Willd.	Climbing hempvine	Compositae	Whole plant	Decoction	Fever, malaria	Milliken 1997a
Milicia excelsa Welw. C.C. Berg	African teak, counterwood	Moraceae	Root, stem	Not stated	Malaria	Adebayo and Krettli 2011; Tetik, Civelek, and Cakilcioglu 2013
Millettia diptera Gagn.	NA	Fabaceae	Bark, seed, pericarp	Not stated	Malaria*	Nguyen-Pouplin et al. 2007
Millettia griffoniana	NA	Euphorbiaceae	Seed	Not stated	Malaria	Titanji, Zofou, and Ngemenya 2008
Millettia laurentii	NA	Fabaceae	Not stated	Not stated	Malaria	Tetik, Civelek, and Cakilcioglu 2013
Millettia versicolor Welw. ex Bak	NA	Fabaceae	Leaf	Not stated	Malaria	Mbatchi et al. 2006

(Continued)

Table A.1 (Continued) Medicinal Plants Reported to Be Used for Malaria

Plant Species	Common Name	Family	Part Used	Preparation	Indication	References
Millettia zechiana Harms	NA	Fabaceae	Stem bark	Decoction	Malaria	Guédé et al. 2010; Zirihi et al. 2005
Mimosa flexuosa Benth.	NA	Fabaceae	Not stated	Not stated	Malaria	Milliken 1997a
Mimosa pudica **L.**	Touch-me-not	Fabaceae	Root	Decoction	Fever, malarial fever, malaria	Inta, Trisonthi, and Trisonthi 2013; Milliken 1997a; Pattanaik et al. 2008
Minquartia guianensis Aublet	NA	Olacaceae	Cortex, bark	Decoction	Malaria	Kvist et al. 2006; Odonne et al. 2013; Ruiz et al. 2011
Minthostachys mollis (Kunth.) Meisn.	NA	Labiatae	Whole plant	Infusion	Malaria	Tene et al. 2007
Mitacarpus scaber	NA	Rubiaceae	Not stated	Not stated	Malaria	Tetik, Civelek, and Cakilcioglu 2013
Mitracarpus megapotamicus (Spreng.) Kuntze	NA	Rubiaceae	Leaf, stem	Not stated	Malaria	Tabuti 2008
Mitragyna inermis (Willd.) O Ktze.	NA	Rubiaceae	Leaf, leafy stem, twig (also in combination), root	Maceration, maceration of roots with leaves of *Trichilia emetica* and *Combretum micranthum* (oral and bath), decoction (leaf: oral and as bath)	Malaria (cure and prevention), fever	Adebayo and Krettli 2011; Asase et al. 2005; Betti and Yemefa 2011; Diarra et al. 2015; Koudouvo et al. 2011; Nadembega et al. 2011; Tor-anyiin, Sha'ato, and Oluma 2003; Traore et al. 2013

(*Continued*)

Table A.1 (Continued) Medicinal Plants Reported to Be Used for Malaria

Plant Species	Common Name	Family	Part Used	Preparation	Indication	References
Mohria caffrorum (L.) Desv.	NA	Schizaeaceae	Aerial part	Decoction	Malaria	Rasoanaivo et al. 1992
Mollugo nudicaulis Lamk.	Nakedstem carpetweed	Aizoaceae	Leaf, root	Infusion	Malaria	Novy 1997
Mollugo pentaphylla	NA	Molluginaceae	Not stated	Not stated	Malaria	Duke 1996
Momordica balsamina L.	Southern balsam pear	Cucurbitaceae	Leaf, stem, bark, root	Decoction, cooked, relish (especially during rainy season)	Malaria (cure and prevention)	Adebayo and Krettli 2011; Bruschi et al. 2011; Duke 1996; Ngarivhume et al. 2015
***Momordica charantia* L.**	Bitter gourd, bitter melon	Cucurbitaceae	Leaf, fruit, whole plant, aerial part	Decoction, infusion (whole plant), juice (leaf), syrup (whole plant), bath (whole plant), mashed and added to food	Fever, malaria	Asase and Oppong-Mensah 2009; Calderón, Simithy-Williams, and Gupta 2012; Chander et al. 2014; Dike, Obembe, and Adebiyi 2012; Duke 1996; Hajdu and Hohmann 2012; Mesia et al. 2008; Milliken 1997a,b; Rasoanaivo et al. 1992; Tetik, Civelek, and Cakilcioglu 2013; Yetein et al. 2013;
Momordica cochinchinensis (Lour.) Spreng.	Balsam pear	Cucurbitaceae	Leaf	Not stated	Malaria	Duke 1996; Nguyen-Pouplin et al. 2007
Momordica condensata	NA	Cucurbitaceae	Not stated	Not stated	Malaria	Tetik, Civelek, and Cakilcioglu 2013

(Continued)

Table A.1 (Continued) Medicinal Plants Reported to Be Used for Malaria

Plant Species	Common Name	Family	Part Used	Preparation	Indication	References
Momordica foetida Scrum. Et Thonn.	NA	Cucurbitaceae	Leaf, whole plant	Infusion, Decoction, juice, relish (especially during rainy season)	Malaria (also in combination), prevention also	Gessler et al. 1995; Ngarivhume et al. 2015; Stangeland et al. 2011; Titanji, Zofou, and Ngemenya 2008
Monanthotaxis sp.	NA	Annonaceae	Leaf	Decoction	Malaria	Asase et al. 2005
Monechma subsessile	NA	Acanthaceae	Leaf	Infusion, decoction	Malaria	Stangeland et al. 2011
Monniera trifolia Rich.	NA	Rutaceae	Leaf	Decoction	Malaria	Bertani et al. 2005
***Monochoria hastata* Solms.**	Arrow-leafed pondweed, hastate-leaved pondweed, Chacha layer	Pontederiaceae	Whole plant	Decoction (in combination)	Malaria, fever	Elliott and Brimacombe 1987
Monodora myristica Dun.	Calabash nutmeg	Annonaceae	Leaf, seed, fruit	Decoction	Malaria (also in combination)	Asase, Hesse, and Simmonds 2012; Nadembega et al. 2011; Okpekon et al. 2004; Tetik, Civelek, and Cakilcioglu 2013
Monotes kerstingii Gilg	NA	Dipterocarpaceae	Leaf	Decoction	Malaria	Traore et al. 2013
Morinda citrifolia	Mengkudu	Rubiaceae	Leaf, fruit	Decoction	Malaria	Al-adhroey et al. 2010
Morinda confusa	NA	Rubiaceae	Not stated	Not stated	Malaria	Tetik, Civelek, and Cakilcioglu 2013
Morinda geminata DC.	NA	Rubiaceae	Root bark	Decoction	Malaria	Traore et al. 2013

(*Continued*)

Table A.1 (Continued) Medicinal Plants Reported to Be Used for Malaria

Plant Species	Common Name	Family	Part Used	Preparation	Indication	References
Morinda lucida Benth.	Brimstone tree	Rubiaceae	Leaf, stem, root, bark, stem bark	Decoction, maceration, infusion (gin)	Malaria, fever (bath), jaundice	Adebayo and Krettli 2011; Asase, Akwetey, and Achel 2010, Asase, Hesse, and Simmonds 2012; Asase and Oppong-Mensah 2009; de Madureira et al. 2002; Dike, Obembe, and Adebiyi 2012; Duke 1996; Idowu et al. 2010; Koudouvo et al. 2011; Okpekon et al. 2004; Oni 2010; Tetik, Civelek, and Cakilcioglu 2013; Tor-anyiin, Sha'ato, and Oluma 2003; van Andel, Myren, and van Onselen 2012; Yetein et al. 2013
Morinda morindoides Bark.	NA	Rubiaceae	Aerial part, root bark, leaf, root	Decoction	Malaria	Idowu et al. 2010; Guédé et al. 2010; Mbatchi et al. 2006; Zirihi et al. 2005
***Moringa oleifera* Lam. [Syn: *Moringa pterygosperma* Gaertn.]**	Horse radish tree	Moraceae	Leaf, stem, root, root bark, seed	Decoction, maceration, infusion	Malaria, fever (maceration/bath), splenomegaly	Adebayo and Krettli 2011; Asase, Akwetey, and Achel 2010; Koudouvo et al. 2011; Milliken 1997; Rasoanaivo et al. 1992; Ssegawa and Kasenene 2007; Suleman et al. 2009; Titanji, Zofou, and Ngemenya 2008; Yetein et al. 2013

(*Continued*)

Table A.1 (Continued) Medicinal Plants Reported to Be Used for Malaria

Plant Species	Common Name	Family	Part Used	Preparation	Indication	References
Morus alba **L.**	White mulberry	Moraceae	Leaf	Not stated	Malaria*	Nguyen-Pouplin et al. 2007
Motandra guineensis	NA	Apocynaceae	Leaf	Decoction	Malaria	Guédé et al. 2010
Moya davana-dependens	NA	Celastraceae	Leaf	Infusion	Fever, malaria	Milliken 1997a
Munronia henryi Harms	NA	Meliaceae	Not stated	Not stated	Fever, malaria	Pei 1985
Musa paradisiaca L. [Syn: *Musa sapientum*]	French plantain	Musaceae	Root, leaf, stem bark	Decoction (alone [oral and as bath] and also with *O. americanum* and *O. gratissimum*); leaf taken with other herbs; mashed and added to food	Malaria	Asase, Akwetey, and Achel 2010; Asase and Oppong-Mensah 2009; Diarra et al. 2015; Duke 1996; Ellena, Quave, and Pieroni 2012; Kaou et al. 2008; Tetik, Civelek, and Cakilcioglu 2013; Stangeland et al. 2011; Traore et al. 2013
Musa spp.	NA	Musaceae	Leaf	Not stated	Fever, malaria	Allabi et al. 2011; Milliken 1997a
Musanga cecropiodies R. Br.	Corkwood, umbrella tree	Urticaceae	Leaf, bark	Decoction (leaf)	Malaria	Dike, Obembe, and Adebiyi 2012; Traore et al. 2013
Mussaenda glabra	NA	Rubiaceae	Not stated	Not stated	Malaria	Duke 1996
Myragina stipulosa	NA	Rubiaceae	Not stated	Not stated	Malaria	Tetik, Civelek, and Cakilcioglu 2013
Myrciaria dubia (Kunth) McVaugh	Camu-camu	Myrtaceae	Not stated	Not stated	Malaria	Ruiz et al. 2011

(*Continued*)

Table A.1 (Continued) Medicinal Plants Reported to Be Used for Malaria

Plant Species	Common Name	Family	Part Used	Preparation	Indication	References
Myrica kandtiana	NA	Myricaceae	Root	Decoction	Malaria	Stangeland et al. 2011
Myrica salicifolia Hochst. ex A. Rich.	NA	Myricaceae	Root, root bark	Pounded, infusion	Malaria, fever	Hamill et al. 2000; Muthaura et al. 2007
Myriocarpa longipes	NA	Urticaceae	Not stated	Not stated	Malaria	Duke 1996
Myristica fragrans Houtt.		Myristicaceae	Seed of fruit	Crushed on coral with water	Malaria	Duke 1996; Kaou et al. 2008
Myroxylon balsamum (L.) Harms	Balsam of Tolu	Fabaceae	Bark	Decoction	Fever, malaria	Milliken 1997a; Odonne et al. 2013
Myrsine africana L.	African boxwood	Myrsinaceae	Leaf, seed	Decoction	Malaria	Nanyingi et al. 2008; Suleman et al. 2009
Myrsine latifolia (Ruiz & Pav.) Spreng.	NA	Myrsinaceae	Bark	Not stated	Malaria	Milliken 1997a
***Myrtus communis* L.**	Common myrtle	Myrtaceae	Leaf	Infusion	Malaria	Milliken 1997a
Mytragina ciliata	NA	Rubiaceae	Not stated	Not stated	Malaria	Tetik, Civelek, and Cakilcioglu 2013
Mytragina stipulosa (DC.) O. Ktze.	NA	Rubiaceae	Leaf	Decoction	Malaria	Traore et al. 2013
Napoleona vogelli [Syn: *Napoleonaea leonensis* Hutch. & Dalz.]	NA	Lecythidaceae	Leaf, stem bark	Decoction	Malaria	Tetik, Civelek, and Cakilcioglu 2013; Traore et al. 2013
Narcissus sp.	NA	Liliaceae	Not stated	Not stated	Malaria	Duke 1996
Naregamia alata	Goanese Ipecac	Meliaceae	Not stated	Not stated	Malaria	Duke 1996

(Continued)

Table A.1 (Continued) Medicinal Plants Reported to Be Used for Malaria

Plant Species	Common Name	Family	Part Used	Preparation	Indication	References
Nauclea diderrichii (De Willd. & T. Durand) Merr.	NA	Rubiaceae	Bark	Decoction	Malaria	Yetein et al. 2013
Nauclea latifolia Sm. [Syn: *Sarcocephalus latifolius* (Sm.) E.A. Bruce]	African peach tree, country fig, Sierra Leone peach	Rubiaceae	Root, stem, leaf, bark, root bark, stem bark	Decoction (oral and as bath), infusion, maceration, juice, soaked in water or add porridge	Malaria, jaundice	Adebayo and Krettli 2011; Ajibesin et al. 2007; Asase, Akwetey, and Achel 2010; Asase and Oppong-Mensah 2009; Asase et al. 2005; Diarra et al. 2015; Dike, Obembe, and Adebiyi 2012; Duke 1996; Guédé et al. 2010; Idowu et al. 2010; Koudouvo et al. 2011; Nadembega et al. 2011; Tchacondo et al. 2012; Tor-anyiin, Sha'ato, and Oluma 2003; Traore et al. 2013; Yetein et al. 2013; Zirihi et al. 2005
Nauclea officinalis (Pierre ex Pit.) Merr.	NA	Rubiaceae	Bark	Not stated	Malaria*	Nguyen-Pouplin et al. 2007
Nauclea pobeguinii (Hua ex Pobég.) Merr.	NA	Rubiaceae	Leaf, stem bark	Decoction (leaf): oral and bath; maceration (stem bark)	Malaria	Diarra et al. 2015; Traore et al. 2013
Nectandra rodiaei	NA	Lauraceae	Not stated	Not stated	Malaria	Duke 1996
Neoboutonia macrocalyx Pax.	NA	Euphorbiaceae	Stem bark	Decoction	Malaria	Muthaura et al. 2007; Namukobe et al. 2011

(*Continued*)

Table A.1 (Continued) Medicinal Plants Reported to Be Used for Malaria

Plant Species	Common Name	Family	Part Used	Preparation	Indication	References
Neoboutonia velutina	NA	Euphorbiaceae	Not stated	Not stated	Malaria	Tetik, Civelek, and Cakilcioglu 2013
Neocinnamomum hainanianum C.K. Allen	NA	Lauraceae	Bark	Decoction	Malaria	Zheng and Xing 2009
Neolamarckia cadamba (Roxb.) Bosser	Kadam, bur-flower tree	Rubiaceae	Bark, leaf	Not stated	Malaria*	Nguyen-Pouplin et al. 2007
Neonauclea formicaria	NA	Rubiaceae	Not stated	Not stated	Malaria	Duke 1996
Nephrolepis biserrata (Sw.) Schott	Giant swordfern	Davalliaceae	Leafy stem	Decoction	Malaria	Koudouvo et al. 2011
Neurolaena lobata (L.) R. Br.	Sepi	Compositae	Stem, leaf, whole plant	Decoction (whole plant), maceration (leaf)	Malaria, fever	Duke 1996; Gupta et al. 1993; Koudouvo et al. 2011; Milliken 1997a
Newbouldia laevis Seem.	NA	Bignoniaceae	Leaf, root, root bark	Maceration, decoction	Malaria, fever	Adebayo and Krettli 2011; Allabi et al. 2011; Duke 1996; Koudouvo et al. 2011; Traore et al. 2013; Yetein et al. 2013
Nicotiana tabacum L.	Cultivated tobacco	Solanaceae	Root, leaf	Decoction, juice (leaf)	Fever, malaria	Duke 1996; Giovannini 2015; Jamir, Sharma, and Dolui 1999; Milliken 1997a; Nadembega et al. 2011
Nigella sativa	Black cumin	Ranunculaceae	Seed	Paste, infusion	Malaria (curative, prophylactic)	Al-adhroey et al. 2010; Suleman et al. 2009

(*Continued*)

Table A.1 (Continued) Medicinal Plants Reported to Be Used for Malaria

Plant Species	Common Name	Family	Part Used	Preparation	Indication	References
Nyctanthes arbor-tristis	NA	Oleaceae	Leaf	Decoction	Malaria (prevention)	Nagendrappa, Naik, and Payyappallimana 2013
Nymphaea lotus L.	White Egyptian lotus	Nymphaeaceae	Leaf, bark	Decoction	Malaria	Randrianarivelojosia et al. 2003; Tetik, Civelek and Cakilcioglu 2013
Ochrosia elliptica	Elliptic yellowwood	Apocynaceae	Not stated	Not stated	Malaria	Duke 1996
***Ocimum americanum* L. [Syn: *Ocimum basilicum* L.]**	Sweet basil	Labiatae	Aerial shoots, leaf, whole plant	Infusion, burnt powder, decoction	Fever, malaria	Allabi et al. 2011; Diarra et al. 2015; Gessler et al. 1995; Milliken 1997a; Nguta et al. 2010a,b; Sivasankari, Anandharaj, and Gunasekaran 2014; Tchacondo et al. 2012; Yetein et al. 2013
Ocimum americanum L. var. *americanum*	NA	Labiatae	Whole plant	Decoction (with *O. gratissimum*, and *Pectranthus scutellarioides*)	Malaria	Kaou et al. 2008
Ocimum angustifolium Benth.	NA	Lamiaceae	Tuber	Cold infusion; may induce nausea	Malaria	Ngarivhume et al. 2015
Ocimum campechianum Mill.	Least basil	Labiatae	Fronds, leaf, whole plant	Decoction (consumed, and as bath)	Fever, malaria (treatment and prevention, taken with other herbs)	Milliken 1997a; Vigneron et al. 2005

(Continued)

Table A.1 (Continued) Medicinal Plants Reported to Be Used for Malaria

Plant Species	Common Name	Family	Part Used	Preparation	Indication	References
Ocimum canum Sims	Hoary basil	Lamiaceae	Leaf, whole plant, stem, seed	Decoction, boiled (also in combination)	Malaria, malarial fever	Asase et al. 2005; Koudouvo et al. 2011; Nadembega et al. 2011; Prabhu et al. 2014; Rasoanaivo et al. 1992
Ocimum gratissimum L. [Syn: *Ocimum viride* Willd; *Ocimum suave* Willd.]	African basil	Labiatae	Leaf, stem, bark	Decoction, burnt powder, infusion (bark)	Malaria (also in combination), fever	Adebayo and Krettli 2011; Asase, Akwetey, and Achel 2010; Asase, Hesse, and Simmonds 2012; Gakuya et al. 2013; Geissler et al. 2002; Milliken 1997a; Nguta et al. 2010a,b; Prabhu et al. 2014; Rasoanaivo et al. 1992; Tchacondo et al. 2012; Tetik, Civelek, and Cakilcioglu 2013; Titanji, Zofou, and Ngemenya 2008; Yetein et al. 2013
Ocimum gratissimum L. var. *gratissimum*	African basil	Lamiaceae	Whole plant, leaf	Decoction (with *O. americanum*, *Musa paradisiaca*)	Malaria (also in combination)	Giday et al. 2007; Kaou et al. 2008; Muganza et al. 2012
Ocimum lamiifolium Hochst. ex Benth.	NA	Lamiaceae	Leaf	Decoction	Malaria	Karunamoorthi and Tsehaye 2012; Seid and Tsegay 2011; Stangeland et al. 2011

(*Continued*)

Table A.1 (Continued) Medicinal Plants Reported to Be Used for Malaria

Plant Species	Common Name	Family	Part Used	Preparation	Indication	References
***Ocimum sanctum* L. [Syn: *Ocimum tenuiflorum* L.]**	Holy basil	Labiatae	Leaf, root, whole plant, seed	Decoction, juice	Malaria (prevention and cure), fever	Al-adhroey et al. 2010; Duke 1996; Milliken 1997a; Nagendrappa, Naik, and Payyappallimana 2013; Shankar, Sharma, and Deb 2012; Suleman et al. 2009
Ocimum spicatum Deflers	NA	Labiatae	Leaf	Decoction	Malaria	Mesfin et al. 2012
Ocotea usambarensis Engl.	NA	Lauraceae	Root bark	Infusion (hot water)	Malaria	Muthaura et al. 2007
Odyendyea gabonensis	NA	Simaroubaceae	Leaf	Not stated	Malaria	Tetik, Civelek, and Cakilcioglu 2013
Oenocarpus bataua Mart.	NA	Palmae	Root, fruit	Not stated	Malaria	Kvist et al. 2006; Ruiz et al. 2011
Oenocarpus mapora H. Karst.	NA	Palmae	Sap	Not stated	Fever, malaria	Milliken 1997a
Olax gambecola	NA	Olacaceae	Whole plant	Decoction	Malaria	Guédé et al. 2010
Oldenlandia corymbosa L. [Syn: *Hedyotis corymbosa*]	Two-flowered oldenlandia, flat-top mille graines	Rubiaceae	Whole plant	Decoction	Fever, malaria	Duke 1996; Milliken 1997a
Oldenlandia herbacea (L.) DC. [Syn: *Hedyotis herbacea* L.]	NA	Rubiaceae	Root, leaf	Not stated	Fever, malaria	Duke 1996; Milliken 1997a
Olea capensis L.	NA	Oleaceae	Stem bark	Decoction	Malaria	Muthaura et al. 2007

(*Continued*)

Table A.1 (Continued) Medicinal Plants Reported to Be Used for Malaria

Plant Species	Common Name	Family	Part Used	Preparation	Indication	References
Olea europaea L.	Olive	Oleaceae	Bark, leaf, stem bark	Decoction	Fever, malaria	Milliken 1997a; Muthaura et al. 2007
Olea europaea L. ssp. *africana* (Mill.) P.S. Green [Syn: *Olea ferruginea*]	Brown olive, wild olive	Oleaceae	Bark	Decoction, maceration	Malaria	Duke 1996; Koch et al. 2005
Olea glandulifera	NA	Oleaceae	Not stated	Not stated	Malaria	Duke 1996
Olinia macrophylla Gilg	NA	Oliniaceae	Leaf	Infusion	Fever, malaria	Hamill et al. 2000
Opilia celtidifolia Guill. & Perr. [Syn: *Opilia amentacea* Roxb.]	NA	Opiliaceae	Leaf	Decoction (oral and as bath)	Malaria	Diarra et al. 2015; Okpekon et al. 2004; Seid and Tsegay 2011
Oreosyce africana Hook.f.	NA	Cucurbitaceae	Whole plant	Decoction	Malaria	Chifundera 2001
Orixa japonica	Japanese orixa	Rutaceae	Not stated	Not stated	Malaria	Duke 1996
Oroxylum indicum (L.) Vent	Midnight horror	Bignoniaceae	Bark	Decoction	Malaria (prevention and cure)	Nagendrappa, Naik, and Payyappallimana 2013; Wiart 2000
Oryza sativa L.	Rice	Poaceae	Corn	Maceration	Malaria	Nadembega et al. 2011
Otholobium mexicanum (L.f.) Grimes	NA	Fabaceae	Leaf	Decoction	Fever, malaria	Milliken 1997a
Otiophora pauciflora Baker	NA	Rubiaceae	Leaf	Maceration	Malaria	Chifundera 2001
Otostegia integrifolia Benth.	NA	Labiatae	Leaf	Not stated	Malaria	Giday et al. 2007
Otostegia michauxii Briq.	NA	Labiatae	Not stated	Not stated	Malaria	Esmaeili et al. 2009

(*Continued*)

Table A.1 (Continued) Medicinal Plants Reported to Be Used for Malaria

Plant Species	Common Name	Family	Part Used	Preparation	Indication	References
Otostegia persica (Burm.) Boiss.	NA	Labiatae	Not stated	Not stated	Malaria	Esmaeili et al. 2009
***Oxalis corniculata* L.**	Indian sorrel	Oxalidaceae	Whole plant	Infusion, decoction, juice	Malaria	Cai 2004; Kantamreddi and Wright 2012; Pascaline et al. 2011
Oxystelma secamone	NA	Asclepiadaceae	Not stated	Not stated	Malaria	Duke 1996
Ozoroa insignis Del	NA	Anacardiaceae	Whole plant, leaf, twig	Decoction	Malaria	Asase et al. 2005; Nadembega et al. 2011
Pachypodanthium confine	NA	Annonaceae	Not stated	Not stated	Malaria	Tetik, Civelek, and Cakilcioglu 2013
Palicourea rigida Kunth	NA	Rubiaceae	Leaf	Infusion	Malaria	Milliken 1997b
Palisota hirsuta	NA	Commelinaceae	Not stated	Not stated	Malaria	Tetik, Civelek, and Cakilcioglu 2013
Panax ginseng	Chinese ginseng	Araliaceae	Not stated	Not stated	Malaria	Duke 1996
***Panicum maximum* Jacq.**	Panic grass	Gramineae	Leaf	Decoction	Fever, malaria	Milliken 1997a
Panicum subalbidum Kunth.	Elbow buffalo grass	Poaceae	Whole plant	Maceration	Malaria	Nadembega et al. 2011
Papaver somniferum	Opium poppy	Papaveraceae	Not stated	Not stated	Malaria	Duke 1996
Pappea capensis (Eckl.) Zeyh.	NA	Sapindaceae	Bulblets	Infusion	Malaria	Koch et al. 2005
Parinari benna Scott-Elliot	NA	Chrysobalanaceae	Leaf, stem bark	Decoction, maceration	Malaria	Traore et al. 2013
Parinari excelsa Sabine	Guinea plum	Chrysobalanaceae	Stem	Decoction	Malaria	Gessler et al. 1995
Parinari polyandra	NA	Rosaceae	Leaf	Not stated	Malaria	Asase et al. 2005

(*Continued*)

Table A.1 (Continued) Medicinal Plants Reported to Be Used for Malaria

Plant Species	Common Name	Family	Part Used	Preparation	Indication	References
Parkia biglobosa (Jacq.) Benth	African locust bean	Fabaceae	Stem, fruit, leaf, stem bark, bark, oldest fruit	Decoction, maceration, decoction with leaves of *Trichilia emetica* (bark, oldest fruit: oral and bath with indigenous soap)	Malaria	Adebayo and Krettli 2011; Asase et al. 2005; Diarra et al. 2015; Nadembega et al. 2011
Parkinsonia aculeata L.	Jerusalem thorn	Fabaceae	Leaf, flower, seed	Infusion, decoction	Fever, malaria	Agra et al. 2007; Albuquerque et al. 2007; Milliken 1997a
Parquetina nigrescens (Afzel.) Bullock.	NA	Asclepiadaceae	Leaf	Decoction	Malaria	Guédé et al. 2010; Zirihi et al. 2005
Parthenium hysterophorus L.	Santa Maria feverfew	Compositae	Root, aerial part	Decoction, bath as prophylactic	Fever, malaria	Duke 1996; Milliken 1997a; Rasoanaivo et al. 1992
Passiflora edulis Sims	Purple granadilla	Passifloraceae	Leaf	Infusion, juice	Malaria	Milliken 1997a; Stangeland et al. 2011
***Passiflora foetida* L.**	Fetid passionflower	Passifloraceae	Flower, vine	Infusion, decoction	Fever, malaria	Milliken 1997a,b
***Passiflora lauriflora* L.**	Golden bellapple	Passifloraceae	Flower	Decoction	Malaria	Milliken 1997a
Passiflora nepalensis Walp.	NA	Passifloraceae	Root	Decoction	Malaria, fever	Shankar, Sharma, and Deb 2012; Sharma, Chhangte, and Dolui 2001

(Continued)

Table A.1 (Continued) Medicinal Plants Reported to Be Used for Malaria

Plant Species	Common Name	Family	Part Used	Preparation	Indication	References
Paullinia pinnata L.	Bread and cheese	Sapindaceae	Leafy stem, root, leaf	Decoction	Malaria, jaundice	Asase and Oppong-Mensah 2009; Asase et al. 2005; Koudouvo et al. 2011; Okpekon et al. 2004; Yetein et al. 2013
Paullinia splendida R.E. Schult.	NA	Sapindaceae	Leaf	Decoction	(Malaria)	Milliken 1997a
Paullinia yoco R.E. Schult. & Killip	Yoco	Sapindaceae	Stem	Infusion	Fever, malaria	Milliken 1997a
Pavetta corymbosa (DC.) F.N. Williams	NA	Rubiaceae	Leaf	Decoction	Malaria	Yetein et al. 2013
Pavetta crassipes K. Schum.	NA	Rubiaceae	Leaf	Decoction (consumed and as bath)	Malaria	Nadembega et al. 2011; Sanon et al. 2003
Pavetta schumanniana F.Hoff. ex K. Schum.	NA	Rubiaceae	Root	Cold infusion	Malaria	Ngarivhume et al. 2015.
Pavetta ternifolia	NA	Rubiaceae	Stem	Not stated	Malaria	Mahyar et al. 1991
Pectis capillaris	NA	Compositae	Not stated	Not stated	Malaria	Duke 1996
Peddia involucrata Bak.	NA	Thymelaeaceae	Leaf, bark	Decoction	Malaria	Randrianarivelojosia et al. 2003
Peganum harmala L.	Harmal peganum	Zygophyllaceae	Fruit, seed	Raw fruit, powder (seed)	Malaria, malarial fever	Duke 1996; Khajoei Nasab and Khosravi 2014; Ullah et al. 2013
Penianthus longifolius Miers	NA	Menispermaceae	Stem bark, root	Decoction, maceration (Root)	Malaria, fever	Muganza et al. 2012; Tetik, Civelek, and Cakilcioglu 2013

(*Continued*)

Table A.1 (Continued) Medicinal Plants Reported to Be Used for Malaria

Plant Species	Common Name	Family	Part Used	Preparation	Indication	References
Pennisetum pedicellatum Trin.	Annual kyasuwa grass, hairy fountain grass	Poaceae	Leaf	Decoction: oral and bath	Malaria	Diarra et al. 2015; Traore et al. 2013
***Pennisetum purpureum* Schumach.**	Elephant grass, napier grass	Poaceae	Twig, leaf	Decoction	Malaria	Allabi et al. 2011; Tetik, Civelek, and Cakilcioglu 2013; Yetein et al. 2013
Pentadiplandra brazzeana	NA	Pentadiplandraceae	Not stated	Not stated	Malaria	Tetik, Civelek, and Cakilcioglu 2013
Pentanisia ouranogyne S. Moore	NA	Rubiaceae	Root	Not stated	Malaria	Nguta et al. 2010b
***Pentas lanceolata* (Forssk.) Defleurs**	Egyptian starcluster	Rubiaceae	Root bark	Decoction	Malaria	Koch et al. 2005
Pentas longiflora	NA	Rubiaceae	Leaf	Juice	Malaria	Stangeland et al. 2011
Peperomia nigropunctata Miq.	NA	Piperaceae	Leaf, stem	Infusion	Fever, malaria	Milliken 1997a
Peperomia spp.	NA	Piperaceae	Leaf, stem	Decoction	Fever, malaria	Milliken 1997a,b
Peperomia trifolia (L.) A. Dietr.	NA	Piperaceae	Whole plant	Decoction	Fever, malaria	Milliken 1997a
Peperomia vulcanica	NA	Piperaceae	Not stated	Not stated	Malaria	Tetik, Civelek, and Cakilcioglu 2013
Pergularia daemia (Forssk.) Chiov.	Pergularia	Asclepiadaceae	Leafy stem	Decoction	Malaria	Koudouvo et al. 2011
Pericopsis elata (Harms) Van Meeuwen	NA	Fabaceae	Leaf	Not stated	Malaria*	Adebayo and Krettli 2011

(Continued)

Table A.1 (Continued) Medicinal Plants Reported to Be Used for Malaria

Plant Species	Common Name	Family	Part Used	Preparation	Indication	References
Pericopsis laxiflora (Benth.)	Kolo-kolo	Fabaceae	Leaf (alone and in combination), stem bark	Decoction (leaf: oral and as bath)	Malaria	Asase et al. 2005; Diarra et al. 2015; Tchacondo et al. 2012; Traore et al. 2013
Perilla frutescens	Beefsteak plant	Labiatae	Not stated	Not stated	Malaria	Duke 1996
Periploca linearifolia Dill. & A. Rich.	NA	Asclepiadaceae	Root bark	Decoction	Malaria	Muthaura et al. 2007; Rukunga et al. 2009
***Persea americana* Mill.**	Avocado tree	Lauraceae	Leaf, seed	Infusion, decoction	Malaria, fever	Dike, Obembe, and Adebiyi 2012; Duke 1996; Milliken 1997a,b; Traore et al. 2013
Petersianthus macrocarpus (P. Beauv.) Liben	Stinkwood tree	Lecythidaceae	Stem	Decoction	Malaria	Asase and Oppong-Mensah 2009; Ruiz et al. 2011
Petiveria alliacea L.	Guinea henweed	Phytolaccaceae	Leaf, root, stem bark, aerial part, whole plant	Decoction (also as bath), infusion, also inhalation of vapors after boiling in water	Fever, malaria (treatment with other herbs, and prevention)	Bourdy, Chavez de Michel, and Roca-Coulthard 2004; Idowu et al. 2010; Milliken 1997a,b; Odonne et al. 2013; Ruiz et al. 2011; Scarpa 2004; Vigneron et al. 2005
Peucedanum japonicum	NA	Apiaceae	Not stated	Not stated	Malaria	Duke 1996
Philodendron cf. *uleanum* Engl.	NA	Araceae	Whole plant	Infusion	Fever, malaria	Milliken 1997a
Philodendron sp.	NA	Amaryllidaceae	Whole plant, aerial roots	Decoction, infusion	Fever, malaria	Calderón, Simithy-Williams, and Gupta 2012; Milliken 1997b

(*Continued*)

Table A.1 (Continued) Medicinal Plants Reported to Be Used for Malaria

Plant Species	Common Name	Family	Part Used	Preparation	Indication	References
Phlogacanthus thyrsiflorus (Roxb.) Nees.	NA	Acanthaceae	Leaf	Not stated	Malaria	Tangjang et al. 2011
Phragmites mauritianus Kunth	NA	Gramineae	Aerial part	Decoction	Malaria	Rasoanaivo et al. 1992
Phyllagathis griffithii King	NA	Melastomataceae	Not stated	Not stated	Malaria	Duke 1996
***Phyllagathis rotundifolia* Blume**	NA	Melastomataceae	Root, leaf	Decoction	Malaria	Duke 1996; Ong and Nordiana 1999
***Phyllanthus amarus* Schum & Thonn.**	Carry me seed	Euphorbiaceae	Whole plant, aerial part, leaf, root	Decoction, infusion (aerial part)	Malaria (treatment and prevention), fever, shiver	Adebayo and Krettli 2011; Ajibesin et al. 2007; Asase, Akwetey, and Achel 2010; Asase and Oppong-Mensah 2009; Duke 1996; Idowu et al. 2010; Milliken 1997a,b; Nguyen-Pouplin et al. 2007; Tchacondo et al. 2012; Upadhyay et al. 2010; Vigneron et al. 2005; Yetein et al. 2013
Phyllanthus corcovadensis Mull. Arg.	NA	Euphorbiaceae	Whole plant	Not stated	Malaria	Milliken 1997a
***Phyllanthus emblica* L.**	Indian gooseberry	Euphorbiaceae	Not stated	Juice	Malaria, fever	Ghimire and Bastakoti 2009

(Continued)

Table A.1 (Continued) Medicinal Plants Reported to Be Used for Malaria

Plant Species	Common Name	Family	Part Used	Preparation	Indication	References
Phyllanthus muellerianus (Kuntze) Exell.	NA	Euphorbiaceae	Leaf	Decoction	Malaria	Guédé et al. 2010; Tetik, Civelek, and Cakilcioglu 2013; Zirihi et al. 2005
***Phyllanthus niruri* L.**	Gale of the wind	Euphorbiaceae	Whole plant	Infusion (bath), decoction	Fever, malaria (treatment and prevention), also in combination with *Eurycoma longifolia*	Al-adhroey et al. 2010; Duke 1996; Elliott and Brimacombe 1987; Milliken 1997a; Ruiz et al. 2011; Vigneron et al. 2005
Phyllanthus reticulatus Poir, *Phyllanthus multiflorus* Roxb.	Potato plant, seaside laurel	Euphorbiaceae	Leaf, bark, flower, root	Juice	Malaria	Kadir et al. 2014
Phyllanthus sp.	NA	Euphorbiaceae	Aerial part	Decoction	Malaria	Rasoanaivo et al. 1992
Phyllanthus stipulatus	Stipulate leaf-flower	Euphorbiaceae	Not stated	Not stated	Malaria	Duke 1996
Physalis alkekengi L.	Strawberry groundcherry	Solanaceae	Fruit	Not stated	Fever, malaria	Milliken 1997a
Physalis angulata L.	Cutleaf groundcherry	Solanaceae	Root, whole plant, shoot, leaf	Infusion, decoction, maceration	Fever, malaria	Guédé et al. 2010; Kvist et al. 2006; Milliken 1997a,b; Odonne et al. 2013; Ruiz et al. 2011; Zirihi et al. 2005
Phytolacca dodecandra L'Herit	Pokeweed	Phytolaccaceae	Leaf	Not stated	Malaria	Mesfin, Demissew, and Teklehaymanot 2009

(*Continued*)

Table A.1 (Continued) Medicinal Plants Reported to Be Used for Malaria

Plant Species	Common Name	Family	Part Used	Preparation	Indication	References
Picralima nitida (Staph) Th. & H. Dur.	NA	Apocynaceae	Stem, seed, root, fruit, stem bark	Decoction, consumed (seed)	Malaria, fever	Adebayo and Krettli 2011; Asase and Oppong-Mensah 2009; Duke 1996; Muganza et al. 2012; Titanji, Zofou, and Ngemenya 2008
Picramnia antidesma Sw.	NA	Simaroubaceae	Bark	Not stated	Fever, malaria	Duke 1996; Milliken 1997a
Picramnia pentandra Sw.	Florida bitterbush	Simaroubaceae	Leaf, bark, root	Infusion	Malaria, quinine substitute	Milliken 1997a
Picrasma excelsa Sw.	Bitterwood	Simaroubaceae	Wood	Infusion, decoction	Fever, malaria	Milliken 1997a
Picrasma javanica Bl.	NA	Simaroubaceae	Bark	Infusion	Malaria	Shankar, Sharma, and Deb 2012
Picris repens	NA	Compositae	Not stated	Not stated	Malaria	Duke 1996
Picrolemma pseudocoffea Ducke	NA	Simaroubaceae	Leaf, bark, stem, root	Infusion, decoction	Fever, malaria (used both as prevention and cure in French Guiana)	Bertani et al. 2005; Milliken 1997a; Vigneron et al. 2005
Picrorhiza kurrooa Royle ex Benth.	NA	Scrophulariaceae	Root, rhizome	Decoction	Malaria	Duke 1996; Shankar, Sharma, and Deb 2012; Willcox 2011
Picrorhiza scrophulariaefolia	NA	Scrophulariaceae	Not stated	Not stated	Malaria	Duke 1996

(*Continued*)

Table A.1 (Continued) Medicinal Plants Reported to Be Used for Malaria

Plant Species	Common Name	Family	Part Used	Preparation	Indication	References
Piliostigma reticulatum (DC.) Hochst. [Syn: *Bauhinia reticulata* DC.]	NA	Fabaceae	Leaf	Decoction: oral and bath	Malaria	Betti and Yemefa 2011; Diarra et al. 2015
Piliostigma thonningii (Schumach.) Milne-Redh. [Syn: *Bauhinia thonningii* Schum.]	Monkey bread	Fabaceae	Leaf, bark, stem bark	Infusion, maceration, powder, decoction	Malaria, malaria (in children)	Adebayo and Krettli 2011; Diarra et al. 2015; Nadembega et al. 2011; Seid and Tsegay 2011; Traore et al. 2013
Pilocarpus jaborandi Holmes	Jaborandi	Rutaceae	Leaf, bark	Infusion	Fever, malaria	Milliken 1997a
Pinellia tuberifera	Crow dipper	Araceae	Not stated	Not stated	Malaria	Duke 1996
***Piper aduncum* L.**	Higuillo de hoja menuda	Piperaceae	Seed	Infusion, suspension of seeds	Malaria, fever	Milliken 1997a
Piper betel	Betel pepper	Piperaceae	Leaf	Chewed, decoction, maceration	Malaria, fever (children)	Al-adhroey et al. 2010; Chander et al. 2014; Duke 1996; Leaman et al. 1995
Piper capense L.f.	NA	Piperaceae	Root	Decoction	Malaria	Koch et al. 2005
Piper guineense Schumach. & Thonn.	Pepper	Piperaceae	Leaf	Decoction	Malaria (in combination)	Asase, Hesse, and Simmonds 2012
Piper hispidum Sw.	Jamaican pepper	Piperaceae	Leaf, stem	Decoction	Malaria, fever	Duke 1996; Gupta et al. 1993; Mesfin, Demissew, and Teklehaymanot 2009; Milliken 1997a

(*Continued*)

Table A.1 (Continued) Medicinal Plants Reported to Be Used for Malaria

Plant Species	Common Name	Family	Part Used	Preparation	Indication	References
Piper longum L.	Indian long pepper	Piperaceae	Fruit, root	Decoction, juice (fruit)	Malaria (prevention and cure), malarial fever	Nagendrappa, Naik, and Payyappallimana 2013; Namsa, Mandal, and Tangjang 2011; Prabhu et al. 2014; Sharma, Chhangte, and Dolui 2001
Piper marginatum Jacq.	Marigold pepper	Piperaceae	Leaf, root, whole plant	Decoction	Fever, malaria	Bertani et al. 2005; Milliken 1997a; Vigneron et al. 2005
Piper mullesua Buch. Ham.	NA	Piperaceae	Leaf, fruit	Not stated	Malaria	Shankar, Sharma, and Deb 2012
***Piper nigrum* L.**	Black pepper	Piperaceae	Fruit, seed	Direct application (fruit), decoction	Malaria (prevention and cure), (malaria)	Duke 1996; Kamaraj et al. 2012; Milliken 1997a; Nadembega et al. 2011; Nagendrappa, Naik, and Payyappallimana 2013; Tetik, Civelek, and Cakilcioglu 2013; Upadhyay et al. 2010
Piper peltatum L. [Syn: *Piper umbellatum; Pothomorphe peltata* (L.) Miq.]	Monkey's hand	Piperaceae	Leaf	Nasal drops, infusion (consumed, and as bath)	Malaria (including cerebral), fever	Akendengue 1992; Milliken 1997a,b; Ruiz et al. 2011; Tetik, Civelek, and Cakilcioglu 2013
Piper pyrifolium Vahl	NA	Piperaceae	Aerial part, fruit	Decoction, crushed on a coral with water	Malaria	Kaou et al. 2008; Rasoanaivo et al. 1992

(Continued)

Table A.1 (Continued) Medicinal Plants Reported to Be Used for Malaria

Plant Species	Common Name	Family	Part Used	Preparation	Indication	References
***Piper sarmentosum* Roxb.**	Wild pepper, kadok	Piperaceae	Leaf	Decoction	Malaria, malarial fever	Ong and Nordiana 1999; Rahman et al. 1999
Pistia stratiotes L.	NA	Araceae	Whole plant	Decoction	Malaria	Koudouvo et al. 2011
Pithecellobium (Albizia) panurense Spruce ex Benth	NA	Fabaceae	Root	Infusion	(Malaria)	Milliken 1997a
Pithecellobium laetum Benth.	NA	Fabaceae	Aerial part	Infusion	Malaria	Roumy et al. 2007
Pithecellobium sp.	NA	Fabaceae	Bark	Decoction	Fever, malaria	Milliken 1997a,b
***Pittosporum ferrugineum* Ait.**	Belalang puak	Pittosporaceae	Not stated	Not stated	Malaria	Duke 1996
Pittosporum viridiflorum Sims	Cape cheesewood	Pittosporaceae	Stem bark	Decoction	Malaria	Muthaura et al. 2007
Plantago amplexicaulis	NA	Plantaginaceae	Not stated	Not stated	Malaria	Duke 1996
Plantago intermedia L.	NA	Plantaginaceae	Aerial part	Infusion	Malaria	Uzun et al. 2004
***Plantago major* L.**	Common plantain	Plantaginaceae	Whole plant, leaf, root	Tincture, infusion, decoction, juice (leaf)	Fever, malaria	Milliken 1997a; Sharma, Chhangte, and Dolui 2001
Plectranthus amboinicus (Lour.) Spreng.	Mexican mint	Labiatae	Whole plant	Decoction (with *O. americanum*, and *O. gratissimum*)	Malaria	Kaou et al. 2008

(*Continued*)

Table A.1 (Continued) Medicinal Plants Reported to Be Used for Malaria

Plant Species	Common Name	Family	Part Used	Preparation	Indication	References
Plectranthus barbatus Andr.	Forskohlii	Labiatae	Leaf, whole plant	Decoction, infusion	Malaria	Githinji and Kokwaro 1993; Milliken 1997b; Namukobe et al. 2011; Nguta et al. 2010a,b; Vigneron et al. 2005
Plectranthus cf. *forskohlii*	NA	Labiatae	Leaf	Decoction	Malaria	Jamir, Sharma, and Dolui 1999; Stangeland et al. 2011
Plectranthus sp.	NA	Labiatae	Leaf	Juice	Malaria	Stangeland et al. 2011
Pleonotoma melioides (S. Moore) A. H. Gentry	NA	Bignoniaceae	Not stated	Not stated	Malaria	Milliken 1997a
Pluchea ovalis	NA	Compositae	Leaf	Decoction, maceration	Malaria	Stangeland et al. 2011
Plumbago zeylanica Linn.	Wild leadwort	Plumbaginaceae	Root, leaf	Infusion (root or leaf), cold infusion (root)	Malaria (prevention and cure)	Duke 1996; Giday et al. 2007; Ngarivhume et al. 2015; Norscia and Borgognini-Tarli 2006; Poonam and Singh 2009
***Plumeria rubra* L.**	Temple tree	Apocynaceae	Bark	Not stated	Malaria*	Nguyen-Pouplin et al. 2007
Podocarpus latifolius (Thunb.) R. Br. ex Mirb.	East African yellowwood	Podocarpaceae	Root	Decoction	Malaria	Koch et al. 2005
Podocarpus sprucei Parl.	NA	Podocarpaceae	Leaf	Infusion	Malaria	Tene et al. 2007
Podophyllum versipelle	Chinese mayapple	Berberidaceae	Not stated	Not stated	Malaria	Duke 1996

(*Continued*)

Table A.1 (Continued) Medicinal Plants Reported to Be Used for Malaria

Plant Species	Common Name	Family	Part Used	Preparation	Indication	References
Pogonopus speciosus (Jacq.) K. Schum.	NA	Rubiaceae	Bark	Decoction	Malaria	Milliken 1997a
Pogostemon glaber Benth.	NA	Labiatae	Not stated	Decoction	Malaria	Ghorbani et al. 2011
Pollia condensata	NA	Commelinaceae	Not stated	Not stated	Malaria	Tetik, Civelek, and Cakilcioglu 2013
Polyalthia suaveolens Engl. & Diels	NA	Annonaceae	Stem, fruit, root bark, stem bark	Decoction	Malaria, fever	Mesia et al. 2008; Muganza et al. 2012; Tsabang et al. 2012
Polycarpa glabrifolia	NA	Caricaceae	Not stated	Not stated	Malaria	Tetik, Civelek, and Cakilcioglu 2013
Polyceratocarpus sp.	NA	Annonaceae	Stem	Decoction	Malaria	Tsabang et al. 2012
Polydora serratuloides (DC.) H. Rob. [Syn: *Vernonia perrottetii* Sch.Bip. ex Walp.]	NA	Compositae	Leaf	Decoction: oral and bath	Malaria	Diarra et al. 2015
Polygala micrantha	NA	Polygalaceae	Not stated	Not stated	Malaria	Milliken 1997a
***Polygala paniculata* L.**	Orosne	Polygalaceae	Whole plant, root, leaf, flower	Decoction, cooked with meat in bamboo	Fever, malaria	Calderón, Simithy-Williams, and Gupta 2012; Jorim et al. 2012; Milliken 1997a
Polygala persicariaefolia D.C.	NA	Poaceae	Whole plant	Decoction	Malaria	Shankar, Sharma, and Deb 2012
Polygonatum officinale	Angular Solomon's seal	Liliaceae	Not stated	Not stated	Malaria	Duke 1996

(*Continued*)

Table A.1 (Continued) Medicinal Plants Reported to Be Used for Malaria

Plant Species	Common Name	Family	Part Used	Preparation	Indication	References
Polygonum aviculare	Prostrate knotweed	Polygonaceae	Not stated	Not stated	Malaria	Duke 1996
Polygonum bistorta	Meadow bistort	Polygonaceae	Not stated	Not stated	Malaria	Duke 1996
Polygonum punctatum Elliot	Dotted smartweed	Polygonaceae	Not stated	Not stated	Fever, malaria	Duke 1996; Milliken 1997a
Polymnia uvedalia	NA	Compositae	Not stated	Not stated	Malaria	Duke 1996
Polyouratea hexasperma (A. St-Hil.) Tiegh. [Syn: *Gomphia hexasperma* A. St.-Hil.]	NA	Ochnaceae	Not stated	Not stated	Quinine substitute	Cosenza et al. 2013
Polypodium decumanum (Willd.) J. Sm. [Syn: *Phlebodium decumanum* (Willd.) J. Sm.]	Creeping golden polypody	Polypodiaceae	Rhizome	Maceration	Malaria, fever	Kvist et al. 2006; Milliken 1997a
***Polyscias filicifolia* (C. Moore ex E. Fourn.) L.H. Bailey**	Fern-leaf aralia	Araliaceae	Leaf	Fresh	Malaria	Jorim et al. 2012
***Pongamia pinnata* L. (Pierre)**	Mempari	Fabaceae	Leaf	Paste, decoction	Malaria (prevention)	Mallik, Panda, and Padhy 2012; Nagendrappa, Naik, and Payyappallimana 2013
Populus alba	White poplar	Salicaceae	Not stated	Not stated	Malaria	Duke 1996
Poraqueiba sericea Tulasne	NA	Icacinaceae	Cortex	Not stated	Malaria	Kvist et al. 2006

(*Continued*)

Table A.1 (Continued) Medicinal Plants Reported to Be Used for Malaria

Plant Species	Common Name	Family	Part Used	Preparation	Indication	References
Porophyllum gracile Benth.	Slender poreleaf	Compositae	Not stated	Decoction	Malaria	Dimayuga and Agundez 1986
Porophyllum seemannii	NA	Compositae	Not stated	Not stated	Malaria	Duke 1996
Porterandia cladantha (K. Schum.) Keay	NA	Rubiaceae	Leaf	Not stated	Malaria	Mbatchi et al. 2006
***Portulaca oleracea* L.**	Little hogweed	Portulacaceae	Whole plant	Not stated	Fever, malaria	Milliken 1997a
Portulaca sp.	NA	Portulacaceae	Stem	Infusion	Malaria	Milliken 1997b
Potalia amara Aubl.	Sacha mangua	Loganiaceae	Aerial part, leaf, whole plant	Decoction	Fever, malaria	Kvist et al. 2006; Milliken 1997a
Potalia resinifera Mart.	NA	Loganiaceae	Not stated	Not stated	Malaria	Ruiz et al. 2011
Potamogeton javanicus Hass Karl	NA	Potamogetonaceae	Aerial part	Decoction, infusion	Malaria	Rasoanaivo et al. 1992
Potentilla discolor	NA	Rosaceae	Not stated	Not stated	Malaria	Duke 1996
Potentilla reptans L.	NA	Rosaceae	Aerial part	Decoction	Malaria	Pieroni, Quave, and Santoro 2004
Potomorphe sp. Miq.	NA	Piperaceae	Not stated	Not stated	Malaria	Elisabetsky and Shanley 1994
Pouteria caimito (Ruiz & Pav.) Radlk.	Abiu	Sapotaceae	Bark, leaf	Not stated	Fever, malaria	Kvist et al. 2006; Milliken 1997a
Premna angolensis Curke	NA	Verbenaceae	Bark	Not stated	Malaria*	de Madureira et al. 2002

(*Continued*)

Table A.1 (Continued) Medicinal Plants Reported to Be Used for Malaria

Plant Species	Common Name	Family	Part Used	Preparation	Indication	References
Premna chrysoclada (Bojer) Gurke	NA	Verbenaceae	Not stated	Not stated	Malaria	Gathirwa et al. 2011
Premna latifolia	NA	Verbenaceae	Bark	Decoction	Malaria (prevention)	Nagendrappa, Naik, and Payyappallimana 2013
***Premna serratifolia* L.**	Coastal premna, creek premna	Verbenaceae	Root, leaf	Decoction	Malaria, fever	Rasoanaivo et al. 1992; Waruruai et al. 2011
Prosopis africana (Guill. & Perr.) Taub.	African mesquite	Fabaceae	Leaf, stem, bark	Maceration (oral and as bath), decoction	Malaria	Adebayo and Krettli 2011; Diarra et al. 2015; Traore et al. 2013
Prunus africana Hook.f.	Red stinkwood	Rosaceae	Stem bark	Decoction	Malaria	Muthaura et al. 2007; Namukobe et al. 2011
Prunus cerasoides D. Don.	Sour cherry, wild Himalayan cherry	Rosaceae	Bark	Decoction	Malaria	Sharma, Chhangte, and Dolui 2001
Prunus domestica L.	European plum	Rosaceae	Leaf	Decoction	Malaria	Milliken 1997a
Prunus mume	Japanese apricot	Rosaceae	Not stated	Not stated	Malaria	Duke 1996
Prunus persica (L.) Batsch	Peach	Rosaceae	Seed, leaf, root	Infusion	Malaria (in combination), prevention	Giday et al. 2007; Ngarivhume et al. 2015
Prunus sphaerocarpa Sw.	Myrtle laurel cherry, West Indies cherry	Rosaceae	Bark	Infusion	Malaria (prevention)	Milliken 1997a
Prunus virginiana	Black cherry	Rosaceae	Not stated	Not stated	Malaria	Duke 1996
Pseudarthria hookeri	NA	Fabaceae	Leaf, root	Decoction	Malaria	Jamir, Sharma, and Dolui 1999; Stangeland et al. 2011
Pseudobombax munguba (Mart. and Zucc.) Dugand	NA	Bombacaceae	Cortex	Not stated	Malaria	Kvist et al. 2006

(*Continued*)

Table A.1 (Continued) Medicinal Plants Reported to Be Used for Malaria

Plant Species	Common Name	Family	Part Used	Preparation	Indication	References
Pseudocedrela kotschyi (Schweinf.) Harms.	Dry-zone cedar	Meliaceae	Twig, leaf, root	Powder, decoction	Malaria	Allabi et al. 2011; Asase et al. 2005; Tchacondo et al. 2012
Pseudolmedia spuria	False breadnut	Moraceae	Not stated	Not stated	Malaria	Duke 1996
Pseudoprotorhus longifolius (A. Rich.) Engl.	NA	Anacardiaceae	Leaf	Juice	Malaria	Gessler et al. 1995; Mbatchi et al. 2006
Pseudoprotorhus longifolius H. Perr.	NA	Anacardiaceae	Leaf	Decoction	Malaria	Rasoanaivo et al. 1992
Pseudospondias microcarpa (A. Rich.) Engl.	NA	Anacardiaceae	Stem bark	Decoction	Malaria	Traore et al. 2013
Pseudoxandra cuspidata Maas.	NA	Annonaceae	Bark	Decoction	Malaria	Bertani et al. 2005
Psiadia punctulata (D.C.) Vatke	NA	Apocynaceae	Root	Not stated	Malaria	Koch et al. 2005
***Psidium guajava* L.**	Guava tree	Myrtaceae	Leaf, root, stem, bark (resin)	Decoction, infusion	Fever, malaria (also in combination)	Adebayo and Krettli 2011; Asase, Hesse, and Simmonds 2012; Asase and Oppong-Mensah 2009; Dike, Obembe, and Adebiyi 2012; Gessler et al. 1995; Milliken 1997a; Nadembega et al. 2011; Ruiz et al. 2011; Tetik, Civelek, and Cakilcioglu 2013

(*Continued*)

Table A.1 (Continued) Medicinal Plants Reported to Be Used for Malaria

Plant Species	Common Name	Family	Part Used	Preparation	Indication	References
Psidium personii McVaugh	NA	Myrtaceae	Leaf	Decoction	Malaria	Asase, Akwetey, and Achel 2010
Psidium sp.	NA	Myrtaceae	Leaf	Decoction	Malaria	Milliken 1997b
Psittacanthus cordatus (Hoffmanns.) Blume	NA	Loranthaceae	Not stated	Not stated	Malaria	Duke 1996; Milliken 1997a
Psoralea pentaphylla	NA	Fabaceae	Not stated	Not stated	Malaria	Duke 1996
Psorospermum febrifugum Hochr.	NA	Hypericaceae	Leaf, stem bark	Decoction	Malaria	Tetik, Civelek, and Cakilcioglu 2013; Traore et al. 2013
Psorospermum senegalense Spach.	NA	Clusiaceae	Leaf	Decoction	Malaria (sometimes in combination)	Jansen et al. 2010
Psychotria sp.	NA	Rubiaceae	Leaf	Not stated	Malaria	Milliken 1997b
Psychotria ipecacuanha Stokes	Ipecac	Rubiaceae	Root	Powdered, aqueous extract	Fever, malaria	Milliken 1997a
Psychotria kirkii Hiern	NA	Rubiaceae	Root	Decoction	Malaria	Gessler et al. 1995
Ptelea trifoliata	Common hoptree	Rutaceae	Not stated	Not stated	Malaria	Duke 1996
Pteleopsis suberosa Engl. & Diels.	NA	Combretaceae	Leaf	Decoction: oral and bath	Malaria	Diarra et al. 2015
Pterocarpus erinaceus Poir.	Barwood	Fabaceae	Bark, leaf (in combination), stem bark	Decoction (oral and bath), maceration	Malaria	Asase et al. 2005; Diarra et al. 2015; Traore et al. 2013; Yetein et al. 2013

(Continued)

Table A.1 (Continued) Medicinal Plants Reported to Be Used for Malaria

Plant Species	Common Name	Family	Part Used	Preparation	Indication	References
***Pterocarpus macrocarpus* Kurz.**	Burma padauk	Fabaceae	Leaf	Not stated	Malaria*	Nguyen-Pouplin et al. 2007
***Pterocarpus rohrii* Vahl.**	NA	Fabaceae	Bark, leaf	Decoction, juice, infusion	Malaria, fever	Bertani et al. 2005; Milliken 1997a
Pterocarpus santalinoides L'Hérit. ex DC.	NA	Fabaceae	Leaf	Decoction	Malaria	Traore et al. 2013
Pterocarpus soyauxii	Barwood	Fabaceae	Not stated	Not stated	Malaria	Tetik, Civelek, and Cakilcioglu 2013
Punica granatum	Pomegranate	Lythraceae	Bark	Decoction	Malaria	Duke 1996
Pupalia lappacea (L.) Juss	NA	Amaranthaceae	Whole plant	Decoction	Malaria, jaundice	Nadembega et al. 2011; Yetein et al. 2013
Pupalia micrantha Hauman	NA	Amaranthaceae	Root, root bark	Decoction	Malaria	Mesfin et al. 2012
Pycananthus angolensis	NA	Myrtaceae	Not stated	Not stated	Malaria	Tetik, Civelek, and Cakilcioglu 2013
Pycnanthus angolensis (Welw. Ward)	Boxboard, African nutmeg	Myristicaceae	Bark, stem, stem bark	Decoction	Malaria	Asase and Oppong-Mensah 2009; de Madureira et al. 2002; Zirihi et al. 2005
Pycnostachys eminii Gurke	NA	Labiatae	Leaf	Decoction (per os and bath)	Malaria	Ssegawa and Kasenene 2007
Quassia africana Baill.	NA	Simaroubaceae	Leaf, root, root bark	Decoction, maceration	Malaria, fever	Mbatchi et al. 2006; Muganza et al. 2012; Tetik, Civelek, and Cakilcioglu 2013

(*Continued*)

Table A.1 (Continued) Medicinal Plants Reported to Be Used for Malaria

Plant Species	Common Name	Family	Part Used	Preparation	Indication	References
***Quassia amara* L.**	Quassia wood	Simaroubaceae	Leaf, stem, bark, root, wood	Decoction, infusion	Fever, malaria (used both as prevention and cure in French Guiana)	Adebayo and Krettli 2011; Bertani et al. 2005; Duke 1996; Milliken 1997a; Vigneron et al. 2005
Quassia undulata (Guill. & Perr.) D Dietr.	NA	Simaroubaceae	Leaf, stem	Not stated	Malaria*	Adebayo and Krettli 2011
Quiina guianensis Aubl.	NA	Ochnaceae	Not stated	Not stated	Quinine substitute	Cosenza et al. 2013
Randia echinocarpa	NA	Rubiaceae	Not stated	Not stated	Malaria	Duke 1996
Randia spinosa (Jacq.) Karst.	Emetic nut	Rubiaceae	Fruit	Decoction	Malaria	Muñoz, Sauvain, Bourdy, Arrázola et al. 2000
Randia talangnigna DC.	NA	Rubiaceae	Aerial part	Decoction	Malaria	Rasoanaivo et al. 1992
Randia fasciculata (Roxb.) DC.	NA	Rosaceae	Leaf	Decoction	Malaria	Shankar, Sharma, and Deb 2012
Ranunculus japonicus	Japanese buttercup	Ranunculaceae	Not stated	Not stated	Malaria	Duke 1996
Ranunculus muricatus	Spinyfruit buttercup	Ranunculaceae	Not stated	Not stated	Malaria	Duke 1996
Ranunculus trichophyllus	Threadleaf crowfoot	Ranunculaceae	Not stated	Not stated	Malaria	Duke 1996
Ranunculus zuccarinii	NA	Ranunculaceae	Not stated	Not stated	Malaria	Duke 1996
Rauvolfia duckei Marcgr.	NA	Apocynaceae	Not stated	Not stated	Malaria	Milliken 1997a

(Continued)

Table A.1 (Continued) Medicinal Plants Reported to Be Used for Malaria

Plant Species	Common Name	Family	Part Used	Preparation	Indication	References
Rauvolfia macrophylla	NA	Apocynaceae	Stem	Not stated	Malaria	Tetik, Civelek, and Cakilcioglu 2013
Rauvolfia obscura	NA	Apocynaceae	Stem	Not stated	Malaria	Tetik, Civelek, and Cakilcioglu 2013
Rauvolfia samarensis	NA	Apocynaceae	Not stated	Not stated	Malaria	Duke 1996
Rauvolfia serpentina (L.) Benth. ex Kurz	Serpentine wood	Apocynaceae	Root, stem	Tablet, juice	Malaria (prevention [root], treatment [juice])	Nagendrappa, Naik, and Payyappallimana 2013
Rauvolfia tetraphylla **L.**	Be-still tree	Apocynaceae	Root	Decoction	Fever, malaria	Milliken 1997a
Rauvolfia vomitoria Afzel.	Poison devil's pepper	Apocynaceae	Root, bark, leaf, root bark	Decoction	Malaria	Adebayo and Krettli 2011; Guédé et al. 2010; Idowu et al. 2010; Tetik, Civelek, and Cakilcioglu 2013; Zirihi et al. 2005
Rauwolfia sp.	NA	Apocynaceae	Root	Not stated	Malaria	Kvist et al. 2006
Remijia amphithrix Standl.	NA	Rubiaceae	Bark	Infusion	Malaria	Milliken 1997a
Remijia ferruginea (A. St.-Hil.) DC.; *Cinchona ferruginea* A. St.-Hil.	NA	Rubiaceae	Not stated	Not stated	Quinine substitute	Cosenza et al. 2013

(*Continued*)

Table A.1 (Continued) Medicinal Plants Reported to Be Used for Malaria

Plant Species	Common Name	Family	Part Used	Preparation	Indication	References
Remijia firmula (Mart.) Wedd. [Syn: *Cinchona bergeniana* Mart.; *Cinchona firmula* Mart.]	NA	Rubiaceae	Not stated	Not stated	Quinine substitute	Cosenza et al. 2013
Remijia macrophylla (H. Karst.) Benth & Hook. F. ex Flueck	NA	Rubiaceae	Bark	Infusion	Malaria	Milliken 1997a
Remijia macrocnemia (Mart.) Wedd. [Syn: *Cinchona macrocnemia* Mart.]	NA	Rubiaceae	Not stated	Not stated	Quinine substitute	Cosenza et al. 2013
Remijia pedunculata (H. Karst.) Flueck	NA	Rubiaceae	Bark	Infusion	Malaria	Milliken 1997a
Remijia peruviana Standl.	NA	Rubiaceae	Cortex	Not stated	Fever, malaria	Kvist et al. 2006; Milliken 1997a
Remijia purdieana Wedd.	NA	Rubiaceae	Bark	Infusion	Malaria	Milliken 1997a
Remijia sp.	NA	Rubiaceae	Not stated	Not stated	Malaria	Duke 1996
Remijia trianae Wernham	NA	Rubiaceae	Bark	Infusion	Malaria	Milliken 1997a
Remijia ulei Krause	NA	Rubiaceae	Bark	Infusion	Malaria	Milliken 1997a
Remijia vellozii (A. St.-Hil.) DC. [Syn: *Cinchona vellosii* A. St.-Hil.]	NA	Rubiaceae	Not stated	Not stated	Quinine substitute	Cosenza et al. 2013

(*Continued*)

Table A.1 (Continued) Medicinal Plants Reported to Be Used for Malaria

Plant Species	Common Name	Family	Part Used	Preparation	Indication	References
Renealmia guianensis Maas	NA	Zingiberaceae	Root, leaf	Decoction (consumed, and as bath)	Fever, malaria	Milliken 1997a,b
Rhamnus prinoides L.	NA	Rhamnaceae	Root bark, leaf	Decoction	Malaria	Koch et al. 2005; Muregi et al. 2007; Muthaura et al. 2007
Rhamnus staddo L.	NA	Rhamnaceae	Leaf, fruit, root, root bark	Decoction	Malaria, fever	Koch et al. 2005; Muregi et al. 2007; Muthaura et al. 2007; Nanyingi et al. 2008
Rhaphiostylis beninensis (Hook.f.) Planch.	NA	Icacinaceae	Leaf	Decoction	Malaria	Traore et al. 2013
Rheum officinale	Chinese rhubarb	Polygonaceae	Not stated	Not stated	Malaria	Duke 1996
Rhigiocarya racemifera Miers	NA	Menispermaceae	Leaf	Not stated	Malaria*	Zirihi et al. 2005
Rhizophora mangle L.	Red mangrove	Rhizophoraceae	Resin from trunk	Not stated	Malaria	Milliken 1997a
Rhododendron molle	Chinese azalea, Japanese azalea	Ericaceae	Not stated	Not stated	Malaria	Duke 1996
Rhoicissus tridentata (L.f.)	NA	Vitaceae	Leaf	Not stated	Malaria	Nanyingi et al. 2008
Rhus natalensis Bernh. ex Kraus	NA	Anacardiaceae	Not stated	Decoction	Malaria	Gathirwa et al. 2011; Nanyingi et al. 2008
Rhus semialata	Chinese gall	Anacardiaceae	Not stated	Not stated	Malaria	Duke 1996
Rhus taratana (Bak.) H. Perr.	NA	Anacardiaceae	Leaf	Decoction	Malaria	Rasoanaivo et al. 1992

(*Continued*)

Table A.1 (Continued) Medicinal Plants Reported to Be Used for Malaria

Plant Species	Common Name	Family	Part Used	Preparation	Indication	References
Rhynchosia sp.	NA	Fabaceae	Leaf, bark	Decoction	Fever, malaria	Ssegawa and Kasenene 2007
Rhynchosia viscosa	NA	Fabaceae	Not stated	Maceration	Malaria	Stangeland et al. 2011
Ribes nigrum L.	European blackcurrant	Grossulariaceae	Leaf	Infusion	Malaria	Milliken 1997a
***Ricinus communis* L.**	Castor oil plant	Euphorbiaceae	Leaf, whole plant, root	Infusion, direct application, decoction (also with *O. americanum,* and *O. gratissimum*)	Fever, malaria	Kaou et al. 2008; Milliken 1997a; Nanyingi et al. 2008; Nguta et al. 2010a,b
Rosa canina L.	Dog rose	Rosaceae	Not stated	Decoction	Malaria	Qureshi and Bhatti 2008
Rothmannia hispido K. Schum.	NA	Rubiaceae	Leaf, root	Decoction	Malaria (in combination)	Asase, Hesse, and Simmonds 2012
Rottboelia exaltata L.F.	Guinea fowlgrass, itch grass	Gramineae	Leaf	Decoction	Malaria	Nguta et al. 2010b
Roucheria calophylla Planch.	NA	Linaceae	Bark	Decoction	Malaria	Milliken 1997a
Roucheria punctata (Ducke) Ducke	NA	Linaceae	Bark	Decoction	Malaria	Roumy et al. 2007
Rourea coccinea (Thonn. ex Schmach.) Benth.	NA	Connaraceae	Leaf	Decoction	Malaria	Yetein et al. 2013
Roylea calycina (Roxb.) Briq.	NA	Labiatae	Not stated	Juice	Malaria, fever	Jain and Puri 1984
Rubus ellipticus Sm.	Yellow Himalayan raspberry	Rosaceae	Root	Decoction	Malaria	Shankar, Sharma, and Deb 2012

(Continued)

Table A.1 (Continued) Medicinal Plants Reported to Be Used for Malaria

Plant Species	Common Name	Family	Part Used	Preparation	Indication	References
Rumex abyssinicus	NA	Polygonaceae	Leaf, stem	Juice	Malaria	Karunamoorthi and Tsehaye 2012; Tetik, Civelek, and Cakilcioglu 2013
Rumex crispus L.	Curly dock	Polygonaceae	Leaf, root	Decoction	Fever, malaria	Milliken 1997a
Rumex obtusifolius L.	Bitter dock	Polygonaceae	Leaf, root	Decoction	Fever, malaria	Milliken 1997a
Rumex usambarensis (Dammer) Dammer.	NA	Polygonaceae	Leaf	Decoction	Malaria	Ssegawa and Kasenene 2007
Rustia formosa (Cham. & Schtdl.) Klotzsch [Syn: *Exostema formosum* Cham. & Schltdl.]	NA	Rubiaceae	Not stated	Not stated	Quinine substitute	Cosenza et al. 2013
Ruta graveolens L.	Common rue	Rutaceae	Flower	Not stated	Malaria	Duke 1996
Rytigynia umbellulata (Hiern) Robyns	NA	Rubiaceae	Leaf	Decoction	Malaria	Yetein et al. 2013
Saba senegalensis (A.DC.) Pichon	Liane saba	Apocynaceae	Leaf	Decoction: oral and bath	Malaria	Diarra et al. 2015
Sabatia elliottii	NA	Gentianaceae	Not stated	Not stated	Malaria	Duke 1996
Sabdariffa rubra Kostel.	Florida cranberry, rozelle	Malvaceae	Leaf	Juice	Malaria	Odonne et al. 2013

(*Continued*)

Table A.1 (Continued) Medicinal Plants Reported to Be Used for Malaria

Plant Species	Common Name	Family	Part Used	Preparation	Indication	References
Sabicea villosa Willd. ex Roem. & Schult.	Woolly woodvine	Rubiaceae	Leaf	Infusion	Malaria	Milliken 1997a; Roumy et al. 2007
Saccharum officinarum L.	Sugarcane	Gramineae	Stem, twig	Decoction, juice, infusion	Malaria, fever	Adebayo and Krettli 2011; Ajibesin et al. 2007; Asase, Akwetey, and Achel 2010; Milliken 1997a; Ruiz et al. 2011; Yetein et al. 2013
Saldinia sp.	NA	Rubiaceae	Aerial part	Decoction	Malaria	Rasoanaivo et al. 1992
Salix alba	White willow	Salicaceae	Bark	Decoction, powder	Fever, malaria	Duke 1996; Milliken 1997a
Salix babylonica	Weeping willow	Salicaceae	Not stated	Not stated	Malaria	Duke 1996
Salix fragilis	Crack willow	Salicaceae	Not stated	Not stated	Malaria	Duke 1996
Salix humboldtiana Willd.	Humboldt's willow	Salicaceae	Leaf, bark, cortex	Decoction	Fever, malaria	Kvist et al. 2006; Milliken 1997a
Salix purpurea	Purpleosier willow	Salicaceae	Not stated	Not stated	Malaria	Duke 1996
Salix taxifolia	Yewleaf willow	Salicaceae	Not stated	Not stated	Malaria	Duke 1996
Salsola somalensis N.E. Br.	NA	Chenopodiaceae	Leaf	Not stated	Malaria	Suleman et al. 2009
Salvadora persica L.	Toothbrush tree	Salvadoraceae	Leaf, root	Decoction	Malaria	Mesfin et al. 2012; Muthee et al. 2011; Norscia and Borgognini-Tarli 2006
Salvia lavanduloides Kunth	NA	Labiatae	Leaf	Infusion	Malaria	Duke 1996; Milliken 1997a
Salvia shannonii J. D. Sm.	NA	Labiatae	Whole plant	Decoction	Malaria	Milliken 1997a

(Continued)

Table A.1 (Continued) Medicinal Plants Reported to Be Used for Malaria

Plant Species	Common Name	Family	Part Used	Preparation	Indication	References
Sambucus australis Cham. & Schltr.	Southern elder	Caprifoliaceae	Leaf	Not stated	Fever, malaria	Milliken 1997a
Sanicula marilandica	Maryland sanicle	Apiaceae	Not stated	Not stated	Malaria	Duke 1996
Sansevieria liberica Hort. ex Gerome & Labroy	NA	Dracaenaceae	Rhizome, leaf, root	Decoction	Malaria	Koudouvo et al. 2011; Yetein et al. 2013
Sansevieria trifasciata	Mother-in-law's tongue, snake plant, viper's bowstring hemp	Agavaceae	Not stated	Not stated	Malaria	Duke 1996
Sarcocephalus officinalis Pierre ex Pit.	NA	Rubiaceae	Bark	Not stated	Malaria	Hout et al. 2006
Satyrium nepalense D. Don	NA	Orchidaceae	Tuber	Powder	Malaria	Bano et al. 2014; Shankar, Sharma, and Deb 2012
Saururus loureiri	NA	Saururaceae	Not stated	Not stated	Malaria	Duke 1996
Schinus molle L.	Peruvian peppertree	Anacardiaceae	Seed	Not stated	Malaria	Giday et al. 2007
Schismatoclada concinna Bak.	NA	Rubiaceae	Root bark	Decoction	Malaria	Rasoanaivo et al. 1992
Schismatoclada farahimpensis Bak.	NA	Rubiaceae	Root bark	Decoction	Malaria	Rasoanaivo et al. 1992
Schismatoclada viburnoides Bak.	NA	Rubiaceae	Root bark	Decoction	Malaria	Rasoanaivo et al. 1992
Schkuhria pinnata (Lam.) Kuntze	Pinnate false threadleaf	Compositae	Seed, stem, whole plant	Infusion (hot or cold water)	Malaria	Duke 1996; Milliken 1997a; Muthaura et al. 2007

(Continued)

Table A.1 (Continued) Medicinal Plants Reported to Be Used for Malaria

Plant Species	Common Name	Family	Part Used	Preparation	Indication	References
***Schleichera oleosa* (Lour.) Oken**	Lac tree, Ceylon oak	Sapindaceae	Bark	Not stated	Malaria	Duke 1996; Hout et al. 2006
Schultesia guianensis (Aubl.) Malme	NA	Gentianaceae	Whole plant	Decoction	Fever, malaria	Milliken 1997a
Schultesia lisianthoides (Griseb.) Benth. & Hook.f. ex. Hemsl.	NA	Gentianaceae	Not stated	Not stated	Malaria	Duke 1996; Milliken 1997a
Schumanniophyton magnificum	NA	Rubiaceae	Not stated	Not stated	Malaria	Tetik, Civelek, and Cakilcioglu 2013
Scleria barteri Boeck.	NA	Cyperaceae	Whole plant	Not stated	Malaria	Mbatchi et al. 2006
Scleria hirtella Sw.	Riverswamp nutrush	Cyperaceae	Not stated	Not stated	Malaria	Milliken 1997a
Sclerocarya caffra Sond.	NA	Anacardiaceae	Leaf	Decoction, inhalation	Malaria	Rasoanaivo et al. 1992
***Scoparia dulcis* L.**	Licorice weed	Scrophulariaceae	Stem, leaf, whole plant, aerial part	Decoction, infusion	Fever, malaria	Adebayo and Krettli 2011; Calderón, Simithy-Williams, and Gupta 2012; Milliken 1997a; Rasoanaivo et al. 1992; Ruiz et al. 2011; Tetik, Civelek, and Cakilcioglu 2013; Traore et al. 2013
Scopolia japonica	NA	Solanaceae	Not stated	Not stated	Malaria	Duke 1996
Scrophularia dulcis	NA	Scrophulariaceae	Leaf	Decoction (with sugar)	Malaria	Tangjang et al. 2011

(*Continued*)

Table A.1 (Continued) Medicinal Plants Reported to Be Used for Malaria

Plant Species	Common Name	Family	Part Used	Preparation	Indication	References
Scrophularia oldhami	NA	Scrophulariaceae	Not stated	Not stated	Malaria	Duke 1996
Scutellaria galericulata	Marsh skullcap	Labiatae	Not stated	Not stated	Malaria	Duke 1996
Secamone africana (Oliv.) Bullock	NA	Asclepiadaceae	Leaf	Not stated	Malaria	Namukobe et al. 2011
Securidaca longipedunculata Fres.	NA	Polygalaceae	Root, stem bark, leaf	Decoction (leaf: oral and as bath), powder (root for both: oral and as bath)	Malaria	Diarra et al. 2015; Duke 1996; Nadembega et al. 2011; Nguta et al. 2010a,b; Tchacondo et al. 2012
Selaginella pallescens Sping	NA	Selaginellaceae	Not stated	Not stated	Malaria	Milliken 1997a
Senecio hoffmannii Klatt.	NA	Compositae	Not stated	Not stated	Malaria	Duke 1996; Milliken 1997a
Senecio ompricaefolius (ex DC.) H. Humb.	NA	Compositae	Aerial part	Decoction	Malaria	Rasoanaivo et al. 1992
Senecio salignus	NA	Compositae	Not stated	Not stated	Malaria	Duke 1996
Senecio scandens	NA	Compositae	Not stated	Not stated	Malaria	Duke 1996
Senecio syringitolius O. Hofmann.	NA	Compositae	Leaf	Decoction	Malaria	Nguta et al. 2010a,b
Senencio nandensis S. Moore	NA	Compositae	Leaf	Decoction, infusion (bath)	Malaria	Ssegawa and Kasenene 2007
Senencio petitianus A. Rich.	NA	Compositae	Stem	Infusion	Fever, malaria	Ssegawa and Kasenene 2007

(*Continued*)

Table A.1 (Continued) Medicinal Plants Reported to Be Used for Malaria

Plant Species	Common Name	Family	Part Used	Preparation	Indication	References
***Senna alata* (L.) Roxb.**	Emperor's candlestick	Fabaceae	Leaf, root	Decoction (curative and preventive), infusion	Malaria (also in combination), fever	Asase, Akwetey, and Achel 2010, 2012; Elliott and Brimacombe 1987; Guédé et al. 2010; Hajdu and Hohmann 2012; Milliken 1997a; Tetik, Civelek, and Cakilcioglu 2013; Vigneron et al. 2005; Zirihi et al. 2005
Senna didymobotrya	Popcorn senna	Fabaceae	Leaf, root, root bark	Infusion, juice, decoction	Malaria	Geissler et al. 2002; Gessler et al. 1995; Jamir, Sharma, and Dolui 1999; Jeruto et al. 2008; Muthaura et al. 2007; Stangeland et al. 2011
***Senna hirsuta* (L.) Irwin & Barneby [Syn: *Cassia hirsuta*]**	Woolly senna	Fabaceae	Leaf, seed	Decoction, infusion	Fever, malaria	Milliken 1997a; Tetik, Civelek, and Cakilcioglu 2013
Senna italica Mill.	Port Royal senna	Fabaceae	Leaf	Decoction (oral and as bath)	Malaria	Diarra et al. 2015; Mesfin et al. 2012
Senna longiracemosa (Vatke) Lock	NA	Fabaceae	Root	Decoction	Malaria	Samuelsson et al. 1991
***Senna obtusifolia* (L.) Irwin & Barneby**	Sicklepod	Fabaceae	Root, whole plant, leaf	Decoction (oral and as bath), infusion	Fever, malaria	Diarra et al. 2015; Gessler et al. 1995; Milliken 1997a,b

(*Continued*)

Table A.1 (Continued) Medicinal Plants Reported to Be Used for Malaria

Plant Species	Common Name	Family	Part Used	Preparation	Indication	References
***Senna occidentalis* (L.) Link**	Coffee senna	Fabaceae	Seed, leaf (also in combination), root, bark, whole plant, aerial part, stem bark	Decoction (oral and as bath), infusion, juice (leaf), burnt powder	Fever, malaria, quinine substitute, shiver	Adebayo and Krettli 2011; Asase, Akwetey, and Achel 2010; Asase et al. 2005; Brandão et al. 1992; Diarra et al. 2015; Duke 1996; Gessler et al. 1995; Jamir, Sharma, and Dolui 1999; Kaou et al. 2008; Milliken 1997a,b; Nadembega et al. 2011; Nguta et al. 2010a,b; Novy 1997; Rasoanaivo et al. 1992; Ssegawa and Kasenene 2007; Tchacondo et al. 2012; Tetik, Civelek, and Cakilcioglu 2013; Willcox 2011; Yetein et al. 2013; Zirihi et al. 2005
Senna reticulata (Willd.) Irwin & Barneby	NA	Fabaceae	Root, flower	Decoction	Fever, malaria	Kvist et al. 2006; Milliken 1997a; Ruiz et al. 2011; Vigneron et al. 2005
Senna septemtrionalis (Viv.) H.S. Irwin & Barneby	Buttercup bush	Fabaceae	Root	Decoction	Malaria	Ngarivhume et al. 2015

(*Continued*)

Table A.1 (Continued) Medicinal Plants Reported to Be Used for Malaria

Plant Species	Common Name	Family	Part Used	Preparation	Indication	References
***Senna siamea* (Lam.) H.S. Irwin & Barneby [Syn: *Cassia siamea* Lam.]**	Siamese cassia	Fabaceae	Stem, leaf, bark, root, flower, bulblet	Decoction (for leaf: consumed and as bath), infusion (flower), maceration (root; oral and bath)	Malaria, fever (flower)	Adebayo and Krettli 2011; Al-adhroey et al. 2010; Asase, Akwetey, and Achel 2010, Asase, Hesse, and Simmonds 2012; Diarra et al. 2015; Mbatchi et al. 2006; Nadembega et al. 2011; Sanon et al. 2003; Traore et al. 2013; Yetein et al. 2013
Senna singueana Del [Syn: *Cassia singueana*]	NA	Fabaceae	Leaf, root	Decoction	Malaria, fever	Adebayo and Krettli 2011; Nadembega et al. 2011; Nanyingi et al. 2008
Senna spp.	NA	Fabaceae	Not stated	Not stated	Malaria	Milliken 1997b
Senna tora (L.) Roxb.	NA	Fabaceae	Leaf	Not stated	Malaria	Chander et al. 2014
Sericocomposis hildebrandtii Schinz.	NA	Amaranthaceae	Root	Not stated	Malarial complications	Muthee et al. 2011
Serjania racemosa	NA	Sapindaceae	Not stated	Not stated	Malaria	Duke 1996
Sesamum indicum	Sesame	Pedaliaceae	Not stated	Not stated	Malaria	Duke 1996
Sesbania pachycarpa de Candole	NA	Fabaceae	Not stated	Not stated	Malaria	Nadembega et al. 2011
Sesbania sesban (L.) Merr.	NA	Fabaceae	Leaf	Decoction	Malaria	Namukobe et al. 2011

(Continued)

Table A.1 (Continued) Medicinal Plants Reported to Be Used for Malaria

Plant Species	Common Name	Family	Part Used	Preparation	Indication	References
Setaria megaphylla (Steud.) Dur. & Schinz	Bigleaf bristlegrass	Poaceae	Leaf	Not stated	Malaria*	Adebayo and Krettli 2011
Setaria sp.	NA	Poaceae	Whole plant	Cooked with meat in bamboo	Malaria	Jorim et al. 2012
Sida acuta **Burm. F.**	Common wireweed	Malvaceae	Whole plant, stem, leaf	Decoction, maceration, infusion, juice	Fever, malaria, quinine substitute	Adebayo and Krettli 2011; Cosenza et al. 2013; Duke 1996; Milliken 1997a; Nadembega et al. 2011; Nguyen-Pouplin et al. 2007
Sida rhombifolia **L.**	Cuban jute	Malvaceae	Root, leaf, whole plant	Decoction, maceration, infusion	Malaria, fever, splenomegaly	Milliken 1997a; Rasoanaivo et al. 1992; Shankar, Sharma, and Deb 2012; Tetik, Civelek, and Cakilcioglu 2013
Sida urens	Tropical fanpetals	Malvaceae	Not stated	Not stated	Malaria	Tetik, Civelek, and Cakilcioglu 2013
Siegesbeckia octoaristata	NA	Compositae	Not stated	Not stated	Malaria	Milliken 1997a
Siegesbeckia orientalis	St. Paul's wort	Compositae	Root	Decoction	Malaria	Duke 1996; Stangeland et al. 2011
Silene flos-cuculi L. Gretuer & Burdet	Ragged robin	Caryophyllaceae	Flower	Decoction (added to wine)	Malaria	Leto et al. 2013
Simaba cedron Planch.	Cedron	Simaroubaceae	Fruit, flower, bark, seed, root, wood	Infusion	Malaria, fever	Duke 1996; Joly et al. 1987; Milliken 1997a

(*Continued*)

Table A.1 (Continued) Medicinal Plants Reported to Be Used for Malaria

Plant Species	Common Name	Family	Part Used	Preparation	Indication	References
Simaba ferruginea A. St.-Hil.	NA	Simaroubaceae	Root, bark	Decoction	Fever, malaria	Milliken 1997a
Simaba morettii Feuillet	NA	Simaroubaceae	Bark	Decoction	Malaria	Milliken 1997a
Simarouba amara Aubl.	Bitterwood	Simaroubaceae	Wood, root	Not stated	Fever, malaria, (malaria)	Milliken 1997a
Simarouba glauca	Paradise tree	Simaroubaceae	Not stated	Not stated	Malaria	Duke 1996
Sinapsis arvensis L.	Charlock mustard	Cruciferae	Root	Decoction	Fever, malaria	Milliken 1997a
Siparuna guianensis (Aubl.)	NA	Monimiaceae	Leaf, bark, whole plant	Decoction (consumed, and as bath), juice	Fever, malaria	Bertani et al. 2005; Milliken 1997a,b; Vigneron et al. 2005
Siparuna pauciflora (Beurl.) A. DC.	NA	Monimiaceae	Whole plant	Decoction (consumed, and as bath)	Fever, malaria	Duke 1996; Milliken 1997a
Siparuna poeppigii (Tul.) A. DC.	NA	Monimiaceae	Leaf	Not stated	Malaria	Vigneron et al. 2005
Siparuna sp.	NA	Siparunaceae	Leaf	Infusion, decoction	Fever, malaria	Calderón, Simithy-Williams, and Gupta 2012; Odonne et al. 2013
Siphonochilus aethiopicus (Schweinf.) B.L. Burtt	Natal ginger, wild ginger	Zingiberaceae	Whole plant	Decoction	Malaria	Nadembega et al. 2011
Smilax china	China root	Smilacaceae	Not stated	Not stated	Malaria	Duke 1996

(Continued)

Table A.1 (Continued) Medicinal Plants Reported to Be Used for Malaria

Plant Species	Common Name	Family	Part Used	Preparation	Indication	References
Smilax spinosa	NA	Smilacaceae	Not stated	Not stated	Malaria	Duke 1996
Smilax spp.	NA	Smilacaceae	Leaf, root	Decoction	Fever, malaria	Kvist et al. 2006; Milliken 1997b
Solanecio manii (Hook.f.) C. Jeffrey	NA	Compositae	Leaf, stem	Juice, infusion (also as bath)	Malaria	Ssegawa and Kasenene 2007
Solanum dulcamara	Climbing nightshade	Solanaceae	Not stated	Not stated	Malaria	Duke 1996
Solanum erianthum D. Don	Potatotree	Solanaceae	Root, leaf	Decoction	Malaria*	Adebayo and Krettli 2011; Milliken 1997a
***Solanum grandiflorum* Ruiz & Pav.**	NA	Solanaceae	Fruit	Infusion	Malaria	Milliken 1997a
Solanum incanum L.	Nightshade	Solanaceae	Root, leaf, root bark	Decoction	Malaria	Muthaura et al. 2007; Nguta et al. 2010a,b; Rukunga et al. 2009
Solanum indicum L.	Indian nightshade	Solanaceae	Aerial part, fruit	Decoction	Fever, malaria*	Rasoanaivo et al. 1992; Zirihi et al. 2005
Solanum leucocarpon Dunal	NA	Solanaceae	Leaf	Not stated	Malaria	Vigneron et al. 2005
Solanum micranthum Willd. ex Roem. & Schult.	NA	Solanaceae	Leaf	Not stated	Malaria	Milliken 1997a
Solanum nigrum L.	Black nightshade	Solanaceae	Root, fruit	Direct application, consumed	Fever, malaria	Pushpangadan and Atal 1984; Zirihi et al. 2005
Solanum nitidum Ruiz & Pav.	NA	Solanaceae	Root	Decoction	Malaria	Fernandez, Sandi, and Kokoska 2003
Solanum paniculatum L.	NA	Solanaceae	Bark, root, fruit	Maceration (alcoholic)	Fever, malaria	Milliken 1997a

(Continued)

Table A.1 (Continued) Medicinal Plants Reported to Be Used for Malaria

Plant Species	Common Name	Family	Part Used	Preparation	Indication	References
Solanum pseudoquina A. St.-Hil.	NA	Solanaceae	Not stated	Not stated	Quinine substitute	Cosenza et al. 2013
Solanum sp.	NA	Solanaceae	Root, root bark, leaf	Decoction	Malaria	Mesfin et al. 2012; Milliken 1997b
Solanum syzymbrifolium L.	NA	Solanaceae	Fruit	Infusion	Malaria	Chifundera 2001
***Solanum torvum* Sw.**	Turkey berry	Solanaceae	Fruit, leaf	Decoction, juice	Malaria, fever	Asase, Akwetey, and Achel 2010; Elliott and Brimacombe 1987
Solanum vairum Cl.	NA	Solanaceae	Root	Decoction	Malaria	Shankar, Sharma, and Deb 2012
Solenostemon latifolius (Benth.) J.K. Morton	NA	Labiatae	Leaf	Infusion, decoction	Malaria	Ssegawa and Kasenene 2007
***Solenostemon monostachyus* (P. Beauv.) Briq.**	Monkey's potato	Labiatae	Leaf	Not stated	Malaria	Adebayo and Krettli 2011; Ajibesin et al. 2007
Solidago virgoaurea	NA	Compositae	Not stated	Not stated	Malaria	Duke 1996
Sonchus oleraceus L.	Common sowthistle	Compositae	Leaf	Infusion	Malaria	Stangeland et al. 2011
Sorghum guineense Stapf.	Sweet sorghum, broomcorn	Poaceae	Not stated	Not stated	Malaria	Nadembega et al. 2011
Sorindeia juglandifolia (A. Rich) Planch. ex Oliv.	NA	Anacardiaceae	Leaf	Decoction	Malaria	Traore et al. 2013
Soymida febrifuga	Indian redwood	Meliaceae	Not stated	Not stated	Malaria	Duke 1996

(Continued)

Table A.1 (Continued) Medicinal Plants Reported to Be Used for Malaria

Plant Species	Common Name	Family	Part Used	Preparation	Indication	References
***Spathodea campanulata* P. Beauv.**	African tuliptree	Bignoniaceae	Stem, root	Decoction	Malaria	Adebayo and Krettli 2011; Asase and Oppong-Mensah 2009; Tetik, Civelek, and Cakilcioglu 2013
Sphaeranthus suaveolens (Forsk.) DC.	NA	Compositae	Whole plant	Infusion (hot/cold water)	Malaria	Muthaura et al. 2007
Sphenocentrum jollyanum Pierre	NA	Menispermaceae	Root	Not stated	Malaria*	Adebayo and Krettli 2011
Spigelia marilandica	Woodland pinkroot	Loganiaceae	Not stated	Not stated	Malaria	Duke 1996
Spilanthes oleraceae L.	Spilanthes	Compositae	Flower	Decoction	Malaria	Willcox 2011
Spirospermum penduliflorum Thou.	NA	Menispermaceae	Root, stem bark	Decoction	Malaria, adjunct to chloroquine and quinine	Rasoanaivo et al. 1992
Spondias mombin Jacq.	Yellow mombin	Anacardiaceae	Leaf, bark	Decoction	Malaria, fever	Allabi et al. 2011; Koudouvo et al. 2011; Milliken 1997b; Ruiz et al. 2011; Tetik, Civelek, and Cakilcioglu 2013; Traore et al. 2013; Yetein et al. 2013
***Spondias pinnata* (Koenig et L.f.) Kurtz**	NA	Anacardiaceae	Bark	Not stated	Malaria*	Nguyen-Pouplin et al. 2007
Spondias purpurea	Purple mombin	Anacardiaceae	Not stated	Not stated	Malaria	Duke 1996
Sporobolus helveolus	NA	Poaceae	Not stated	Not stated	Malaria	Duke 1996

(*Continued*)

Table A.1 (Continued) Medicinal Plants Reported to Be Used for Malaria

Plant Species	Common Name	Family	Part Used	Preparation	Indication	References
Stachytarpheta angustifolia Vahl	NA	Verbenaceae	Whole plant	Maceration	Malaria	Nadembega et al. 2011
Stachytarpheta cayennensis (Rich.) Vahl	Cayenne porterweed	Verbenaceae	Whole plant, leaf, branch, shoot	Decoction, juice, infusion	Fever, malaria	Adebayo and Krettli 2011; Kvist et al. 2006; Milliken 1997a,b; Odonne et al. 2013; Tetik, Civelek, and Cakilcioglu 2013
***Stachytarpheta jamaicensis* (L.) Vahl**	Lightblue snakeweed	Verbenaceae	Whole plant	Infusion, decoction	Fever, malaria	Duke 1996; Milliken 1997a
Stachytarpheta straminea Moldenke.	NA	Verbenaceae	Not stated	Not stated	Malaria	Ruiz et al. 2011
Steganotaenia araliacea Hochst.	NA	Umbelliferae	Stem	Infusion	Malaria	Gessler et al. 1995
Stenocline inuloides DC.	NA	Compositae	Leaf	Decoction	Malaria, fever	Rasoanaivo et al. 1992
Stephania hermandifolia (Willd) Walp.	NA	Menispermaceae	Root	Decoction	Malaria	Namsa, Mandal, and Tangjang 2011
Stephania japonica Miers.	Ivyweed, tapevine	Menispermaceae	Tuber	Powder	Malaria, malarial fever	Pattanaik et al. 2008; Shankar, Sharma, and Deb 2012
***Sterculia apetala* (Jacq.) H. Karst.**	Panama tree	Malvaceae	Bark	Decoction	Malaria	Duke 1996; Milliken 1997a

(Continued)

Table A.1 (Continued) Medicinal Plants Reported to Be Used for Malaria

Plant Species	Common Name	Family	Part Used	Preparation	Indication	References
Sterculia setigera Del.	NA	Malvaceae	Bark, stem, leaf	Decoction: oral and as bath	Malaria	Adebayo and Krettli 2011; Asase et al. 2005; Diarra et al. 2015; Tor-anyiin, Sha'ato, and Oluma 2003
Stereospermum suaveolens	Fragrant padri tree	Bignoniaceae	Not stated	Not stated	Malaria	Duke 1996
Sterospermum kunthiamum	NA	Bignoniaceae	Root, leaf, stem	Decoction	Malaria	Adebayo and Krettli 2011; Tor-anyiin, Sha'ato, and Oluma 2003
Stevia eupatoria	NA	Compositae	Not stated	Not stated	Malaria	Duke 1996
Stigmaphyllon convolvulifolium (Cav.) A. Juss.	NA	Malpighiaceae	Leaf	Decoction	(Malaria)	Milliken 1997a
Stigmaphyllon sinuatum (DC.) A. Juss.	NA	Malpighiaceae	Leaf	Decoction	Fever, (malaria)	Milliken 1997a
Striga hermonthica (Del.) Benth.	Purple witchweed	Scrophulariaceae	Whole plant	Not stated	Malaria*	Adebayo and Krettli 2011
Strophanthus hispidus DC.	Brown strophanthus, poison arrowvine	Apocynaceae	Leaf, root, root bark, fruit	Decoction	Malaria	Traore et al. 2013; Yetein et al. 2013
Struchium sparganophorum (L.) Kuntze	Yerba de faja	Compositae	Leaf	Not stated	Malaria*	de Madureira et al. 2002
Strychnopsis thouarsii Baill.	NA	Menispermaceae	Root bark, leaf	Decoction	Malaria	Rasoanaivo et al. 1992
Strychnos colubrina	NA	Loganiaceae	Not stated	Not stated	Malaria	Duke 1996

(Continued)

Table A.1 (Continued) Medicinal Plants Reported to Be Used for Malaria

Plant Species	Common Name	Family	Part Used	Preparation	Indication	References
Strychnos fendleri Sprague & Sandwith	NA	Loganiaceae	Bark	Not stated	Malaria	Milliken 1997a
Strychnos henningsii Gilg	NA	Loganiaceae	Stem bark	Decoction	Malaria	Muthaura et al. 2007
Strychnos icaja	Curare	Loganiaceae	Not stated	Not stated	Malaria	Tetik, Civelek, and Cakilcioglu 2013
Strychnos innocua Del.	NA	Loganiaceae	Leaf	Not stated	Malaria	Asase et al. 2005; Nadembega et al. 2011
Strychnos mostuoides Leeuwenberg	NA	Loganiaceae	Aerial part	Decoction	Malaria	Rasoanaivo et al. 1992
Strychnos potatorum L.f.	Clearing nut tree	Loganiaceae	Root, stem	Decoction	Malaria	Ngarivhume et al. 2015
Strychnos pseudoquina A. St.-Hil.	NA	Loganiaceae	Bark	Infusion, decoction	Fever, malaria, quinine substitute	Cosenza et al. 2013; Milliken 1997a
Strychnos spinosa Lam.	Monkey orange, natal orange	Loganiaceae	Leaf, bark	Not stated	Malaria	Asase et al. 2005; Nadembega et al. 2011; Zirihi et al. 2005
Strychnos spp.	NA	Loganiaceae	Bark	Decoction	Malaria	Milliken 1997b
Strychnos wallichiana Steud. ex DC.	NA	Loganiaceae	Bark	Decoction	Malaria	Chander et al. 2014
Stylosanthes erecta P. Beauv.	Nigerian stylo	Fabaceae	Whole plant	Decoction	Malaria	Nadembega et al. 2011
Sweetia panamensis	NA	Fabaceae	Not stated	Not stated	Malaria	Duke 1996

(Continued)

Table A.1 (Continued) Medicinal Plants Reported to Be Used for Malaria

Plant Species	Common Name	Family	Part Used	Preparation	Indication	References
Swertia angustifolia Buch.-Ham. ex D. Don	NA	Gentianaceae	Whole plant	Not stated	Malaria	Bahadur, Münzbergová, and Timsina 2010
Swertia chirata Buch. Hamilt.	NA	Gentianaceae	Stem, whole plant	Juice, decoction	Malaria	Milliken 1997a; Willcox 2011
Swertia chirayata Ham.	NA	Gentianaceae	Stem	Infusion	Malaria	Poonam and Singh 2009
Swertia chirayita (Roxb.) Karsten.	NA	Gentianaceae	Not stated	Juice	Malaria	Jain and Puri 1984
Swertia nervosa Wall.	NA	Gentianaceae	Whole plant	Decoction	Malaria, malarial fever	Malla, Gauchan, and Chhetri 2015; Shankar, Sharma, and Deb 2012
Swertia racemosa C.B. Clarke	NA	Gentianaceae	Whole plant	Not stated	Malaria	Bahadur, Münzbergová, and Timsina 2010
***Swietenia macrophylla* King**	Honduras mahogany	Meliaceae	Bark	Decoction	Fever, malaria	Duke 1996; Milliken 1997a
***Swietenia mahagoni* (L.) Jacq.**	West Indian mahogany	Meliaceae	Leaf, bark	Decoction	Fever, malaria	Duke 1996; Milliken 1997a
Synedrela nodiflora Th.	Nodeweed	Compositae	Root	Not stated	Malaria	Nadembega et al. 2011
Syringa vulgaris L.	Common lilac	Oleaceae	Fruit, bark	Decoction	Fever, malaria	Duke 1996; Milliken 1997a
***Syzygium aromaticum* (L.) Merr. et Perry**	Clove	Myrtaceae	Flower	Decoction	Malaria	Chander et al. 2014
Syzygium cordatum Hochst. ex Krauss	Water tree, water berry	Myrtaceae	Root	Decoction	Malaria	Gessler et al. 1995

(*Continued*)

Table A.1 (Continued) Medicinal Plants Reported to Be Used for Malaria

Plant Species	Common Name	Family	Part Used	Preparation	Indication	References
Syzygium guineense (Willd.) DC.	NA	Myrtaceae	Leaf	Decoction	Malaria	Traore et al. 2013
***Tabebuia aurea* (Silva Manso) Benth. & Hook.f. ex S. Moore**	Caribbean trumpet tree	Bignoniaceae	Stem bark	Decoction (with *Tabebuia impetiginosa*)	Malaria	Hajdu and Hohmann 2012; Milliken 1997a
Tabebuia impetiginosa (Mart. ex DC.) Standl.	NA	Bignoniaceae	Stem bark	Decoction (with *Tabebuia aurea*)	Malaria	Hajdu and Hohmann 2012
Tabebuia ochracea (Cham.) Standl. ssp. *neochrysantha* (A. H. Gentry)	NA	Bignoniaceae	Bark	Decoction	Malaria	Milliken 1997a
***Tabebuia rosea* (Bertol.) DC.**	Pink trumpet tree	Bignoniaceae	Bark, leaf	Infusion, decoction	Malaria	Milliken 1997a
Tabernaemontana crassa	NA	Apocynaceae	Stem	Not stated	Malaria	Tetik, Civelek, and Cakilcioglu 2013
***Tabernaemontana elegans* Stapf.**	Toad tree	Apocynaceae	Root	Infusion	Malaria	Ngarivhume et al. 2015.
Tabernaemontana pachysiphon Stapf.	Giant pinwheel flower	Apocynaceae	Root, stem bark	Not stated	Malaria	Gakuya et al. 2013
Tabernaemontana penduliflora	NA	Apocynaceae	Root	Not stated	Malaria	Tetik, Civelek, and Cakilcioglu 2013
Tabernaemontana sp.	NA	Apocynaceae	Not stated	Not stated	Malaria	Milliken 1997a
***Tacca chantrieri* Andre**	Batflower, cat's whiskers, devil flower	Taccaceae	Root	Not stated	Malaria*	Nguyen-Pouplin et al. 2007

(*Continued*)

Table A.1 (Continued) Medicinal Plants Reported to Be Used for Malaria

Plant Species	Common Name	Family	Part Used	Preparation	Indication	References
Tachia guianensis Aubl.	NA	Gentianaceae	Root	Infusion, decoction	Fever, malaria	Milliken 1997a
Tacillus umbellifer (Schult.) Danser	NA	Loranthaceae	Leaf	Juice and ground powder applied	Malaria	Malla, Gauchan, and Chhetri 2015
Tagetes erecta L.	Aztec marigold	Compositae	Leaf	Infusion (bath), decoction	Fever, malaria	Duke 1996; Milliken 1997a; Rasoanaivo et al. 1992; Ruiz et al. 2011
Tagetes filifolia Lagaska	Irish lace	Compositae	Not stated	Not stated	Malaria	Milliken 1997a
Tagetes florida	NA	Compositae	Not stated	Not stated	Malaria	Duke 1996
Tagetes lucida Cav.	Sweet-scented marigold	Compositae	Whole plant	Decoction	Fever, malaria	Duke 1996; Milliken 1997a
***Tagetes patula* L.**	French marigold	Compositae	Leaf	Juice (external), decoction	Fever, malaria	Milliken 1997a; Rasoanaivo et al. 1992
Talauma mexicana (DC.) G. Don	NA	Magnoliaceae	Bark	Decoction	Malaria	Duke 1996; Milliken 1997a
Talinum portulacifolium (Forssk.) Asch. ex Schweinf.	Flameflower	Portulacaceae	Leaf	Not stated	Malaria	Titanji, Zofou, and Ngemenya 2008
***Tamarindus indica* L.**	Tamarind	Fabaceae	Fruit, leaf, bark, stem bark	Decoction (alone and with *Senna obtusifolia*, juice (leaf), infusion (leaf, fruit), maceration (fruit)	Fever, malaria (leaf, fruit)	Asase et al. 2005; Betti and Yemefa 2011; Diarra et al. 2015; El-Kamali and El-Khalifa 1999; Musa et al. 2011; Nadembega et al. 2011; Nguta et al. 2010b; Titanji, Zofou, and Ngemenya 2008; Traore et al. 2013

(*Continued*)

Table A.1 (Continued) Medicinal Plants Reported to Be Used for Malaria

Plant Species	Common Name	Family	Part Used	Preparation	Indication	References
Tamarix chinensis	Five-stamen tamarisk	Tamaricaceae	Not stated	Not stated	Malaria	Duke 1996
Tambourissa leptophylla (Tul.) A. DC.	NA	Monimiaceae	Fruit	Crushed on coral with water	Malaria	Kaou et al. 2008
Tanacetum parthenium (L.) Schultz-Bip.	Feverfew	Compositae	Flower	Not stated	Malaria	Milliken 1997a
Tapinanthus dodoneifolius (D.C.) Danser	NA	Loranthaceae	Leaf	Decoction	Malaria	Asase, Hesse, and Simmonds 2012
Tapinanthus paradoxa	NA	Loranthaceae	Whole plant	Decoction	Malaria	Nadembega et al. 2011
Tapinanthus sessilifolius (P. Beau.) van Tiegh.	NA	Loranthaceae	Leaf	Not stated	Malaria*	Adebayo and Krettli 2011
Taraxacum officinale Wiggers	Common dandelion	Compositae	Whole plant	Juice	Malaria, fever	Milliken 1997a; Monigatti, Bussmann, and C.S. Weckerle 2013; Shankar, Sharma, and Deb 2012
Taraxacum panalptnum van Soest	NA	Compositae	Not stated	Not stated	Malaria	Neves et al. 2009
Tasmannia piperita (Hook.f.) Miers.	NA	Winteraceae	Leaf	Fresh/dry	Malaria	Jorim et al. 2012
Teclea nobilis Delile	NA	Rutaceae	Aerial part	Decoction	Malaria	Lacroix, Prado, Kamoga, Kasenene, Namukobe et al. 2011

(*Continued*)

Table A.1 (Continued) Medicinal Plants Reported to Be Used for Malaria

Plant Species	Common Name	Family	Part Used	Preparation	Indication	References
Teclea simplicifolia (Eng) Verdoon	NA	Rutaceae	Root, flower, stem bark	Decoction	Malaria (including cerebral), fever	Muthaura et al. 2007; Nanyingi et al. 2008; Nguta et al. 2010a,b; Rukunga et al. 2009
***Tectona grandis* Lour.**	Teak	Verbenaceae	Leaf	Decoction	Malaria	Asase, Akwetey, and Achel 2010; Duke 1996; Koudouvo et al. 2011; Tchacondo et al. 2012; Yetein et al. 2013
Tephrosia bracteolata Guill & Perr	NA	Fabaceae	Whole plant	Decoction	Malaria	Nadembega et al. 2011
Tephrosia purpurea (L.) Pers.	Fishpoison	Fabaceae	Root	Not stated	Malaria	Gakuya et al. 2013
Terminalia albida	NA	Combretaceae	Leaf, stem bark	Decoction	Malaria	Madge 1998; Traore et al. 2013
Terminalia avicennoides Guill. & Perr.	NA	Combretaceae	Stem	Not stated	Malaria*	Adebayo and Krettli 2011; Nadembega et al. 2011
Terminalia boivinii Tul.	NA	Combretaceae	Leaf	Decoction	Malaria	Norscia and Borgognini-Tarli 2006
Terminalia brevipes Pampan.	NA	Combretaceae	Bark	Decoction	Malaria	Samuelsson et al. 1992
Terminalia catappa L.	Sea almond	Combretaceae	Leaf, bark, fruit, stem	Decoction	Malaria, fever	Asase and Oppong-Mensah 2009; Milliken 1997a; Tor-anyiin, Sha'ato, and Oluma 2003
Terminalia chebula L.	Myrobalan	Combretaceae	Fruit	Not stated	Malaria	Tangjang et al. 2011

(*Continued*)

Table A.1 (Continued) Medicinal Plants Reported to Be Used for Malaria

Plant Species	Common Name	Family	Part Used	Preparation	Indication	References
Terminalia glaucescens Planch. ex Benth.	NA	Combretaceae	Leaf, root, trunk bark	Not stated	Malaria	Okpekon et al. 2004; Magassouba et al. 2007
Terminalia ivorensis	Ivory Coast almond	Combretaceae	Stem	Decoction	Malaria	Asase and Oppong-Mensah 2009; Tetik, Civelek, and Cakilcioglu 2013
Terminalia latifolia Blanco	NA	Combretaceae	Leaf	Not stated	Malaria*	Adebayo and Krettli 2011
Terminalia macroptera	NA	Combretaceae	Root bark, leaf, stem bark	Decoction (consumed and as bath)	Malaria	Nadembega et al. 2011; Sanon et al. 2003; Tetik, Civelek, and Cakilcioglu 2013; Traore et al. 2013
Terminalia superba	Superb terminalia	Combretaceae	Leaf, root	Maceration (oral and as bath)	Malaria	Diarra et al. 2015; Tetik, Civelek, and Cakilcioglu 2013
Tessaria integrifolia Ruiz & Pav.	NA	Compositae	Leaf	Decoction	Malaria, fever	Milliken 1997a; Muñoz, Sauvain, Bourdy, Callapa, Rojas et al. 2000
Tetracera alnifolia Willd.	NA	Dilleniaceae	Leaf	Decoction	Malaria	Mbatchi et al. 2006; Traore et al. 2013
Tetracera spp.	NA	Dilleniaceae	Not stated	Not stated	Malaria	Milliken 1997a

(Continued)

Table A.1 (Continued) Medicinal Plants Reported to Be Used for Malaria

Plant Species	Common Name	Family	Part Used	Preparation	Indication	References
Tetracera potatoria Afzel.	NA	Dilleniaceae	Leaf	Decoction	Malaria	Traore et al. 2013
Tetracera volubilis L.	NA	Dilleniaceae	Root, leaf	Decoction infusion, infusion (leaf)	Fever, malaria	Milliken 1997a
Tetradenia sp.	NA	Labiatae	Root	Decoction	Malaria	Stangeland et al. 2011
***Tetrapleura tetraptera* (Schumm. & Thonn.) Taub.**	NA	Fabaceae	Fruit, stem	Decoction	Malaria	Adebayo and Krettli 2011; Asase and Oppong-Mensah 2009; Muganza et al. 2012; Tetik, Civelek, and Cakilcioglu 2013
Tetrapterys discolor (G. Mey.) DC.	NA	Malpighiaceae	Leaf	Decoction	Malaria	Milliken 1997a
Tetrorchidium didymostemon	NA	Euphorbiaceae	Leaf	Juice, decoction	Malaria	Stangeland et al. 2011
Teucrium chamaedrys L.	Wall germander	Labiatae	Aerial part	Decoction	Malaria	di Tizio et al. 2012; Pieroni, Quave, and Santoro 2004
Teucrium cubense Jacq.	Small coastal germander	Labiatae	Whole plant	Decoction	Fever, malaria	Milliken 1997a
***Thalia geniculata* L.**	Bent alligator-flag	Marantaceae	Leaf	Decoction	Malaria	Yetein et al. 2013
Thalictrum foliolosum L.	Asian meadow rue	Ranunculaceae	Root, rhizome	Not stated	Malaria	Duke 1996; Shankar, Sharma, and Deb 2012
Thelypteris sp.	NA	Thelypteridaceae	Leaf	Juice	Malaria	Odonne et al. 2013
Theobroma bicolor Humbl. & Bonpl.	Peruvian cacao, tiger cacao	Malvaceae	Not stated	Not stated	Malaria	Ruiz et al. 2011

(*Continued*)

Table A.1 (Continued) Medicinal Plants Reported to Be Used for Malaria

Plant Species	Common Name	Family	Part Used	Preparation	Indication	References
***Theobroma cacao* L.**	Cacao	Malvaceae	Leaf, stem bark	Decoction	Malaria	Asase and Oppong-Mensah 2009; Asase, Akwetey, and Achel 2010; Idowu et al. 2010
***Thespesia populnea* (L.) Sol. ex Correa**	Portia tree, bendytree, Pacific rosewood	Malvaceae	Leaf	Infusion	Fever, malaria	Duke 1996; Milliken 1997a
***Thevetia peruviana* (Pers.) K. Schum.**	Luckynut	Apocynaceae	Latex, seed, bark, leaf	Decoction (leaf)	Fever, malaria	Duke 1996; Milliken 1997a; Nguyen-Pouplin et al. 2007
Thomandersia hensii De Wild. & T. Durand	NA	Acanthaceae	Leaf, stem	Decoction	Malaria	Muganza et al. 2012; Tetik, Civelek, and Cakilcioglu 2013
Thujia occidentalis	Arborvitae	Cupressaceae	Not stated	Not stated	Malaria	Duke 1996
***Thunbergia alata* Boj.**	Blackeyed Susan vine	Acanthaceae	Leaf, stem	Infusion	Fever, malaria	Hamill et al. 2000
Timonius timon	NA	Rubiaceae	Leaf	Not stated	Malaria	Mahyar et al. 1991
Tinospora cordifolia Miers.	NA	Menispermaceae	Stem	Juice, decoction (prevention)	Malaria (prevention and cure)	Duke 1996; Poonam and Singh 2009; Upadhyay et al. 2010
***Tinospora crispa* (L.) Miers**	NA	Menispermaceae	Stem, leaf, root	Decoction, maceration in rum/white wine	Malaria (treatment and prevention)	Al-adhroey et al. 2010; Bertani et al. 2005; Duke 1996; Milliken 1997a; Nguyen-Pouplin et al. 2007; Ong and Nordiana 1999; Rahman et al. 1999; Vigneron et al. 2005
Tinospora sinensis	Chinese tinospora	Menispermaceae	Not stated	Not stated	Malaria	Duke 1996

(Continued)

Table A.1 (Continued) Medicinal Plants Reported to Be Used for Malaria

Plant Species	Common Name	Family	Part Used	Preparation	Indication	References
***Tithonia diversifolia* (Hemsl.) A. Gray**	Tree marigold	Compositae	Leaf, flower, aerial part	Decoction, maceration (children), infusion (leaf)	Malaria	Adebayo and Krettli 2011; de Madureira et al. 2002; Kichu et al. 2015; Milliken 1997a; Stangeland et al. 2011; Tetik, Civelek, and Cakilcioglu 2013
Tithonia rotundifolia (Mill.) S.F. Blake	Clavel de muerto	Compositae	Leaf	Decoction	Malaria, fever	Duke 1996; Milliken 1997a
Toddalia aculeata	NA	Rutaceae	Not stated	Not stated	Malaria	Duke 1996
Toddalia asiatica Lamk.	NA	Rutaceae	Root, leaf, aerial part, root bark, Tuber	Decoction, infusion (per os and bath)	Malaria	Duke 1996; Jamir, Sharma, and Dolui 1999; Muregi et al. 2007; Muthaura et al. 2007; Ngarivhume et al. 2015; Novy 1997; Rasoanaivo et al. 1992; Stangeland et al. 2011
Trachelospermum lucidum	NA	Apocynaceae	Not stated	Not stated	Malaria	Duke 1996
Trema commersonii Boj.	NA	Ulmaceae	Aerial part	Decoction	Malaria	Rasoanaivo et al. 1992
Trema orientalis Blume [Syn: *Trema guineensis*]	Charcoal tree, poison peach, peach cedar	Ulmaceae	Aerial part, bark	Decoction (aerial part and bark), bath (bark), eaten as food (bark)	Malaria	Rasoanaivo et al. 1992; Tetik, Civelek, and Cakilcioglu 2013; Traore et al. 2013
***Triadica cochinchinensis* Lour.**	Mouse deer's delight	Euphorbiaceae	Bark	Decoction, bath, eaten as food	Malarial fever	Junsongduang et al. 2014

(Continued)

Table A.1 (Continued) Medicinal Plants Reported to Be Used for Malaria

Plant Species	Common Name	Family	Part Used	Preparation	Indication	References
Tribulus cistoides L.	Jamaican feverplant	Zygophyllaceae	Whole plant	Decoction	Malaria	Milliken 1997a
Tricalysia cryptocalyx Baker	NA	Rubiaceae	Leaf	Decoction	Malaria	Norscia and Borgognini-Tarli 2006
Trichanthera gigantea (Humb. & Bonpl.) Nees	NA	Acanthaceae	Leafy branch tips	Decoction	Malaria	Milliken 1997a
Trichilia cipo (A. Juss.) C. DC.	NA	Meliaceae	Bark	Infusion	Fever, malaria	Milliken 1997a
***Trichilia emetica* Vahl.**	Christmas bells, Ethiopian mahogany	Meliaceae	Leaf, root	Powder; oral and bath: decoction, maceration (root)	Malaria	Diarra et al. 2015; Seid and Tsegay 2011; Tchacondo et al. 2012
Trichilia gilletii	NA	Meliaceae	Not stated	Not stated	Malaria	Tetik, Civelek, and Cakilcioglu 2013
Trichilia havanensis Jacq.	NA	Meliaceae	Bark, fruit	Infusion (fruit), maceration in alcohol (fruit), bath	Malaria	Alfaro 1984; Duke 1996; Milliken 1997a
Trichilia monadelpha (Thonn.) J.de Wilde	NA	Meliaceae	Stem	Decoction	Malaria	Asase and Oppong-Mensah 2009
Trichilia prieureana A. Juss.	NA	Meliaceae	Leaf	Decoction	Fever, malaria, shiver	Yetein et al. 2013
Tricholepis furcata Candolle	NA	Compositae	Not stated	Not stated	Malaria	Yaseen et al. 2015
Trichosanthes cucumerina	Snakegourd	Cucurbitaceae	Not stated	Not stated	Malaria	Duke 1996
Triclisia dictyophylla Diels	NA	Menispermaceae	Leaf	Decoction	Malaria, fever	Muganza et al. 2012

(Continued)

Table A.1 (Continued) Medicinal Plants Reported to Be Used for Malaria

Plant Species	Common Name	Family	Part Used	Preparation	Indication	References
Triclisia gelletii	NA	Menispermaceae	Not stated	Not stated	Malaria	Duke 1996
Triclisia macrocarpa (Baill.) Diel	NA	Menispermaceae	Root bark, stem bark	Decoction	Malaria	Rasoanaivo et al. 1992
Tricoscypha ferruginea	NA	Anacardiaceae	Leaf, stem	Not stated	Malaria	Tetik, Civelek, and Cakilcioglu 2013
***Tridax procumbens* L.**	Coatbuttons	Compositae	Whole plant, leaf	Decoction, consumed	Fever, malaria	Hamill et al. 2000; Mesia et al. 2008; Milliken 1997a
Trimelia bakeri Gilg.	NA	Samydaceae	Not stated	Not stated	Malaria	Muthee et al. 2011
Trimeria bakeri Gilg.	NA	Flacourtiaceae	Root	Grinding	Malaria	Jamir, Sharma, and Dolui 1999
Trimeria grandifolia ssp. *tropica*	NA	Flacourtiaceae	Leaf	Decoction	Malaria	Stangeland et al. 2011
***Triplaris americana* L.**	Ant tree	Polygonaceae	Stem	Decoction	Malaria	Sanz-Biset et al. 2009
Triplaris cf. *poeppigiana* Wedd.	NA	Polygonaceae	Bark	Decoction	Malaria	Odonne et al. 2013
***Triplaris weigeltiana* Kuntze**	NA	Polygonaceae	Bark, cortex	Infusion	Malaria	Kvist et al. 2006; Milliken 1997a
Triplotaxis stellulifera	NA	Compositae	Not stated	Not stated	Malaria	Tetik, Civelek, and Cakilcioglu 2013
Tristellateia madagascariensis Poir	NA	Malpighiaceae	Leaf	Decoction	Malaria, fever	Randrianarivelojosia et al. 2003
Tristiropsis sp.	NA	Sapindaceae	Leaf	Fresh	Malaria	Jorim et al. 2012

(Continued)

Table A.1 (Continued) Medicinal Plants Reported to Be Used for Malaria

Plant Species	Common Name	Family	Part Used	Preparation	Indication	References
Tropidia cucurligioides Lindl.	NA	Orchidaceae	Not stated	Not stated	Malaria	Duke 1996
Turnera diffusa Willd. ex Schult.	Mexican holly	Turneraceae	Leaf	Infusion	Fever, malaria	Duke 1996; Milliken 1997a
***Turnera ulmifolia* L.**	Ramgoat dashalong	Turneraceae	Whole plant	Decoction, tincture, powder	Fever, malaria	Milliken 1997a
Turraea mombassana Hiern ex DC.	NA	Meliaceae	Root	Not stated	Malaria	Muthee et al. 2011
Turreanthus africanus	NA	Meliaceae	Bark	Decoction	Malaria (also in combination)	Asase, Hesse, and Simmonds 2012; Tetik, Civelek, and Cakilcioglu 2013
Tussilago farfara L.	Coltsfoot	Compositae	Leaf	Decoction	Malaria	Leto et al. 2013
Uapaca paludosa Aubrev. et Leandri	NA	Euphorbiaceae	Bark, leaf	Not stated	Malaria	Mbatchi et al. 2006
Uapaca togoensis Pax	NA	Euphorbiaceae	Leaf, root, stem bark	Decoction: oral, and bath (leaf and root)	Malaria	Diarra et al. 2015; Traore et al. 2013
Uncaria guianensis (Aubl.) Gmel.	NA	Rubiaceae	Cortex, bark, stem	Decoction	Malaria	Kvist et al. 2006; Odonne et al. 2013; Ruiz et al. 2011
Unonopsis floribunda Diels	NA	Annonaceae	Cortex	Not stated	Malaria	Kvist et al. 2006
Unonopsis spectabilis Diels	NA	Annonaceae	Bark	Not stated	Malaria	Milliken 1997a
Uraria chamae P. Beauv	NA	Annonaceae	Leaf, root	Decoction	Malaria	Yetein et al. 2013

(Continued)

Table A.1 (Continued) Medicinal Plants Reported to Be Used for Malaria

Plant Species	Common Name	Family	Part Used	Preparation	Indication	References
Uraria lagopodioides	NA	Fabaceae	Not stated	Not stated	Malaria	Duke 1996
Uraria picta (Jacq.) Desv.	NA	Fabaceae	Leaf	Decoction	Malaria	Koudouvo et al. 2011
Uraria prunellaefolia	NA	Fabaceae	Not stated	Not stated	Malaria	Duke 1996
***Urena lobata* L.**	Caesarweed	Malvaceae	Whole plant	Not stated	Malaria*	Nguyen-Pouplin et al. 2007
Urera baccifera (L.) Gaudich. ex Wedd.	Scratchbush	Urticaceae	Root	Decoction is drunk and then vomited	Malaria	Giovannini 2015
Urera laciniata Goudot ex Wedd.	NA	Urticaceae	Leaf	Infusion (per os and bath)	Malaria	Valadeau et al. 2010
Urophyllum lyallii (Bak.) Bremek.	NA	Rubiaceae	Root bark, leaf	Decoction	Adjunct to chloroquine and quinine, splenomegaly	Rasoanaivo et al. 1992
Urtica dioica L.	Stinging nettle	Urticaceae	Whole plant	Raw, infusion, decoction	Malaria	Milliken 1997a
Urtica massaica Mildbr.	Forest nettle, Maasai stinging nettle	Urticaceae	Root	Not stated	Malaria	Rukunga et al. 2009
Usnea spp.	NA	Parmeliaceae	Whole plant	Not stated	Malaria	Muthee et al. 2011
Uvaria acuminata Oliv.	NA	Annonaceae	Not stated	Not stated	Malaria	Gathirwa et al. 2011
Uvaria afzelii Sc. Elliot	NA	Annonaceae	Not stated	Not stated	Malaria	Okpekon et al. 2004
Uvaria ambongoensis (Baill.) Diels	NA	Annonaceae	Leaf	Infusion	Malaria	Norscia and Borgognini-Tarli 2006

(*Continued*)

Table A.1 (Continued) Medicinal Plants Reported to Be Used for Malaria

Plant Species	Common Name	Family	Part Used	Preparation	Indication	References
Uvaria chamae P. Beauv.	Finger root	Annonaceae	Leaf, root	Decoction, maceration (leaf)	Malaria, jaundice (maceration)	Adebayo and Krettli 2011; Allabi et al. 2011; Koudouvo et al. 2011; Tetik, Civelek, and Cakilcioglu 2013; Traore et al. 2013
Uvaria leptocladon Oliv.	NA	Annonaceae	Root	Not stated	Malaria	Norscia and Borgognini-Tarli 2006
Uvaria sp.	NA	Annonaceae	Stem	Decoction	Malaria	Tsabang et al. 2012
Uvariodendron spp.	NA	Annonaceae	Leaf	Not stated	Malaria	Tetik, Civelek, and Cakilcioglu 2013
Vandellia sessiliflora Benth.	NA	Scrophulariaceae	Whole plant	Decoction	Malaria	Shankar, Sharma, and Deb 2012
Vangueria madagascariensis Gmel.	Voa vanga	Rubiaceae	Stem bark	Decoction	Malaria	Muthaura et al. 2007
Vepris ampody H. Perr.	NA	Rutaceae	Leaf, bark	Decoction	Malaria	Randrianarivelojosia et al. 2003
Vepris lanceolata (Lam.) G. Don	White ironwood	Rutaceae	Root, leaf	Decoction	Malaria	Gessler et al. 1995
Verbascum thapsus	Common mullein	Scrophulariaceae	Not stated	Not stated	Malaria	Duke 1996
Verbena litoralis Kunth	Seashore vervain	Verbenaceae	Whole plant, leaf, shoot	Juice, decoction, infusion	Fever, malaria	Duke 1996; Kvist et al. 2006; Milliken 1997a; Roumy et al. 2007; Ruiz et al. 2011; Sanz-Biset et al. 2009; Teklehaymanot and Giday 2010
Verbena officinalis L.	Herb-of-the-cross	Verbenaceae	Whole plant	Maceration	Malaria, fever	Valadeau et al. 2009

(*Continued*)

Table A.1 (Continued) Medicinal Plants Reported to Be Used for Malaria

Plant Species	Common Name	Family	Part Used	Preparation	Indication	References
Vernonia adoensis [Syn: *Baccharoides adoensis* (Sch. Bip. ex Walp.) H. Rob.]	NA	Compositae	Leaf, flower, root	Decoction, juice, infusion	Malaria	Ngarivhume et al. 2015; Stangeland et al. 2011
Vernonia ampandrandavensis Bak.	NA	Compositae	Aerial part	Decoction	Malaria	Rasoanaivo et al. 1992
Vernonia amygdalina Delile. [Syn: *Gymnanthemum amygdalinum* A. Chev.]	Bitterleaf	Compositae	Leaf, root, stem	Decoction, juice, infusion, maceration	Malaria (also in combination with *Citrus limon*), also used for prevention	Adebayo and Krettli 2011; Asase, Akwetey, and Achel 2010; Asase and Oppong-Mensah 2009; Asase et al. 2005; de Madureira et al. 2002; Geissler et al. 2002; Hamill et al. 2000; Lacroix, Prado, Kamoga, Kasenene, Namukobe et al. 2011; Idowu et al. 2010; Jamir, Sharma, and Dolui 1999; Karunamoorthi and Tsehaye 2012; Namukobe et al. 2011; Ssegawa and Kasenene 2007; Stangeland et al. 2011; Tetik, Civelek, and Cakilcioglu 2013; Titanji, Zofou, and Ngemenya 2008; Tor-anyiin, Sha'ato, and Oluma 2003; Vigneron et al. 2005; Yetein et al. 2013

(*Continued*)

Table A.1 (Continued) Medicinal Plants Reported to Be Used for Malaria

Plant Species	Common Name	Family	Part Used	Preparation	Indication	References
Vernonia auriculifera Hiern	NA	Compositae	Root, leaf	Decoction, infusion (hot/cold water)	Malaria	Hamill et al. 2000; Muthaura et al. 2007
Vernonia brachycalyx O. Hoffm.	NA	Compositae	Leaf	Infusion (hot/cold water)	Malaria	Muthaura et al. 2007
Vernonia brachiata Benth. ex Oerst.	NA	Compositae	Leaf	Decoction	Malaria	Milliken 1997a
Vernonia brazzavillensis Aubrev. ex Compere	NA	Compositae	Bark, leaf, root	Not stated	Malaria	Mbatchi et al. 2006
Vernonia campanea S. Moore	NA	Compositae	Flower, leaf	Decoction	Malaria, fever	Hamill et al. 2000
Vernonia chapelieri Drak.	NA	Compositae	Aerial part	Decoction	Malaria	Rasoanaivo et al. 1992
***Vernonia cinerea* (L.) Less.**	Little ironweed	Compositae	Leaf, whole plant	Decoction	Malaria	Adebayo and Krettli 2011; Allabi et al. 2011; Duke 1996; Jain and Puri 1984
Vernonia colorata (Willd.) Drake.	NA	Compositae	Leaf	Decoction (oral and as bath)	Malaria	Diarra et al. 2015; Nadembega et al. 2011; Traore et al. 2013
Vernonia condensata Baker	NA	Compositae	Leaf	Infusion	Malaria	Milliken 1997b
Vernonia conferta	NA	Compositae	Not stated	Not stated	Malaria	Tetik, Civelek, and Cakilcioglu 2013
Vernonia jugalis Oliver et Hiern	NA	Compositae	Leaf	Decoction	Malaria, fever	Hamill et al. 2000

(*Continued*)

Table A.1 (Continued) Medicinal Plants Reported to Be Used for Malaria

Plant Species	Common Name	Family	Part Used	Preparation	Indication	References
Vernonia lasiopus O. Hoffm.	NA	Compositae	Leaf, root, root bark, stem bark	Decoction, juice, infusion	Malaria	Hamill et al. 2000; Jamir, Sharma, and Dolui 1999; Muregi et al. 2007; Muthaura et al. 2007; Stangeland et al. 2011
Vernonia pectoralis Bak.	NA	Compositae	Aerial part	Decoction	Malaria	Rasoanaivo et al. 1992
Vernonia sp.	NA	Compositae	Aerial part	Decoction	Malaria	Rasoanaivo et al. 1992
Vernonia trichodesma Bak.	NA	Compositae	Leaf	Decoction	Malaria	Rasoanaivo et al. 1992
Veronica cymbalaria Bodard	Glandular speedwell	Scrophulariaceae	Aerial part	Decoction	Malaria	De Natale and Pollio 2007
Vetiveria zizanoides Nosk.	NA	Poaceae	Root	Powder	Malarial fever	Prabhu et al. 2014
Viburnum nudum	Possumhaw	Caprifoliaceae	Not stated	Not stated	Malaria	Duke 1996
Viburnum obovatum	Small-leaf arrowwood	Caprifoliaceae	Not stated	Not stated	Malaria	Duke 1996
Vicia hirsuta	Tiny vetch	Fabaceae	Not stated	Not stated	Malaria	Duke 1996
Vigna subterranea (L.) Verdc.	Bambarra groundnut	Fabaceae	Not stated	Not stated	Malaria	Nadembega et al. 2011
***Vigna unguiculata* (L.) Walp.**	Cowpea	Fabaceae	Not stated	Not stated	Malaria	Nadembega et al. 2011
Vincetoxicum atratum	Blackend swallowwort	Asclepiadaceae	Not stated	Not stated	Malaria	Duke 1996
Viola odorata L.	Sweet violet	Violaceae	Not stated	Infusion	Malarial fever	Katuura et al. 2007
Virola calophylla Warb.	Virola	Myristicaceae	Bark	Infusion, decoction	Malaria	Milliken 1997a; Roumy et al. 2007
Virola carinata (Benth.) Warb.	White ucuba	Myristicaceae	Bark	Infusion	Malaria	Milliken 1997a

(*Continued*)

Table A.1 (Continued) Medicinal Plants Reported to Be Used for Malaria

Plant Species	Common Name	Family	Part Used	Preparation	Indication	References
Virola sp.	NA	Myristicaceae	Leaf, bark	Infusion	Fever, malaria	Milliken 1997a,b
Viscum album	European mistletoe	Viscaceae	Not stated	Not stated	Malaria	Duke 1996
Vismia guianensis (Aubl.) Choisy	NA	Guttiferae	Latex, leaf	Decoction	Fever, malaria	Milliken 1997a
Vismia guineensis (L.) Choisy	NA	Hypericaceae	Leaf, stem bark	Decoction	Malaria	Traore et al. 2013
Vismia sessilifolia (Aubl.) DC.	NA	Guttiferae	Leaf	Decoction	Fever, malaria	Milliken 1997a
Vitellaria paradoxa C.F. Gaertn. [Syn: *Butyrospermum parkii* (G. Don) Kotschy]	Shea butter tree	Sapotaceae	Leaf, stem bark	Decoction, infusion	Malaria, fever	Diarra et al. 2015; Jansen et al. 2010; Nadembega et al. 2011; Traore et al. 2013
Vitex cannabifolia	NA	Labiatae	Not stated	Not stated	Malaria	Duke 1996
Vitex doniana Sweet	NA	Verbenaceae	Leaf	Decoction	Malaria	Nadembega et al. 2011; Traore et al. 2013
Vitex madiensis Oliv.	NA	Lamiaceae	Leaf	Decoction: oral and bath	Malaria	Diarra et al. 2015
***Vitex negundo* L.**	Chinese chastetree	Verbenaceae	Leaf	Not stated	Malaria*	Nguyen-Pouplin et al. 2007
Vitex peduncularis Wall.	NA	Verbenaceae	Bark, leaf, stem, root	Decoction, infusion	Fever, malaria	Duke 1996; Namsa, Mandal, and Tangjang 2011; Shankar, Sharma, and Deb 2012; Sharma, Chhangte, and Dolui 2001
***Vitex trifolia* L.**	Simpleleaf chastetree	Verbenaceae	Leaf	Decoction	Malaria	Chander et al. 2014
Vitis japonica Thunb.	Bushkiller	Vitaceae	Not stated	Not stated	Malaria	Duke 1996

(Continued)

Table A.1 (Continued) Medicinal Plants Reported to Be Used for Malaria

Plant Species	Common Name	Family	Part Used	Preparation	Indication	References
Voacanga africana	Small-fruit wild frangipani	Apocynaceae	Stem	Not stated	Malaria	Tetik, Civelek, and Cakilcioglu 2013
***Waltheria indica* L.**	Uhaloa	Malvaceae	Aerial part, leaf	Decoction (oral and as bath)	Malaria	Diarra et al. 2015; Jansen et al. 2010; Nadembega et al. 2011
Warburgia ugandensis Sprague	East African greenbark, pepperbark tree	Canellaceae	Bark, leaf	Consumed	Malaria	Gakuya et al. 2013; Ssegawa and Kasenene 2007
Watsonia sp.	NA	Iridaceae	Not stated	Not stated	Malaria	Milliken 1997a
***Wedelia biflora* DC.**	Sea oxeye	Compositae	Not stated	Not stated	Malaria	Duke 1996
Wedelia mossambicensis Oliver	NA	Compositae	Leaf, root	Decoction	Malaria, fever	Hamill et al. 2000
Willughbeia sp.	NA	Apocynaceae	Not stated	Not stated	Malaria	Duke 1996
Wissadula amplissima var. *rostrata*	Big yellow velvetleaf	Malvaceae	Whole plant	Decoction	Malaria	Nadembega et al. 2011
Withania somnifera (L.) Dunal	Withania	Solanaceae	Whole plant, root, root bark	Decoction	Malaria	Gakuya et al. 2013; Mesfin et al. 2012; Muregi et al. 2007; Muthaura et al. 2007; Namsa, Mandal, and Tangjang 2011
***Wrightia dubia* (Sims.) Spreng.**	NA	Apocynaceae	Leaf	Not stated	Malaria*	Nguyen-Pouplin et al. 2007

(Continued)

Table A.1 (Continued) Medicinal Plants Reported to Be Used for Malaria

Plant Species	Common Name	Family	Part Used	Preparation	Indication	References
Xanthium spinosum L.	Spiny cocklebur	Compositae	Whole plant, leaf, root	Infusion (leaf), decoction (root)	Malaria, fever	Duke 1996; Milliken 1997a
Xanthium strumarium L.	Rough cocklebur	Compositae	Leaf	Not stated	Malaria	Duke 1996; Shankar, Sharma, and Deb 2012
Xeroderris stuhlmannii	NA	Fabaceae	Leaf (alone and in combination), root	Maceration (root), decoction (leaf): both orally and as bath	Malaria	Asase et al. 2005; Diarra et al. 2015
***Ximenia americana* L.**	Tallow wood	Olacaceae	Leaf	Decoction: oral and bath	Malaria	Diarra et al. 2015; Grønhaug et al. 2008; Nadembega et al. 2011; Traore et al. 2013
Ximenia caffra Sond.	NA	Olacaceae	Not stated	Decoction	Malaria, fever	Nanyingi et al. 2008
Xylopia spp.	NA	Annonaceae	Not stated	Not stated	Malaria	Milliken 1997a
Xylopia aethiopica A. Rich.	Ethiopian pepper	Annonaceae	Leaf, fruit	Decoction, maceration (alone; fruits also with roots of *Senna siamea*)	Malaria (also in combination), jaundice (maceration)	Asase, Hesse, and Simmonds 2012; Diarra et al. 2015; Koudouvo et al. 2011; Yetein et al. 2013
Xylopia parviflora	NA	Annonaceae	Seed	Not stated	Malaria	Tetik, Civelek, and Cakilcioglu 2013
Xylopia phloiodora	NA	Annonaceae	Seed	Not stated	Malaria	Tetik, Civelek, and Cakilcioglu 2013
Xylopia vielana Pierre ex Fin. et Gagnep.	NA	Annonaceae	Bark	Not stated	Malaria*	Nguyen-Pouplin et al. 2007
Xymolox monosperma	NA	Annonaceae	Seed, leaf, stem	Not stated	Malaria	Tetik, Civelek, and Cakilcioglu 2013

(*Continued*)

Table A.1 (Continued) Medicinal Plants Reported to Be Used for Malaria

Plant Species	Common Name	Family	Part Used	Preparation	Indication	References
Zanha golungensis Hiern	NA	Sapindaceae	Not stated	Maceration	Malaria	Bruschi et al. 2011
Zanthoxylum avicennae DC.	NA	Rutaceae	Leaf	Not stated	Malaria	Zheng and Xing 2009
Zanthoxylum chalybeum Engl.	NA	Rutaceae	Root, stem, leaf, root bark, stem bark	Decoction	Malaria	Gakuya et al. 2013; Geissler et al. 2002; Gessler et al. 1995; Muthaura et al. 2007; Nguta et al. 2010a,b; Rukunga et al. 2009; Ssegawa and Kasenene 2007; Titanji, Zofou, and Ngemenya 2008
Zanthoxylum gilletti	NA	Rutaceae	Stem bark	Decoction	Malaria	Guédé et al. 2010
Zanthoxylum hamiltonianum Wall.	NA	Rutaceae	Root	Decoction	Malaria	Namsa, Mandal, and Tangjang 2011
Zanthoxylum hermaphroditum Willd.	NA	Rutaceae	Bark	Infusion (in rum/ white wine)	Fever, malaria	Milliken 1997a
Zanthoxylum lemarei	NA	Rutaceae	Not stated	Not stated	Malaria	Tetik, Civelek, and Cakilcioglu 2013
Zanthoxylum leprieurii	NA	Rutaceae	Not stated	Not stated	Malaria	Tetik, Civelek, and Cakilcioglu 2013
Zanthoxylum pentandrum (Aubl.) R. Howard	NA	Rutaceae	Bark	Maceration (in wine/rum)	Malaria	Milliken 1997a
Zanthoxylum perrotteti DC.	NA	Rutaceae	Bark	Infusion (in rum/ white wine)	Fever, malaria	Milliken 1997a

(*Continued*)

Table A.1 (Continued) **Medicinal Plants Reported to Be Used for Malaria**

Plant Species	Common Name	Family	Part Used	Preparation	Indication	References
Zanthoxylum piperitum	Japanese pepper	Rutaceae	Not stated	Not stated	Malaria	Duke 1996
Zanthoxylum rhoifolium Lam.	NA	Rutaceae	Bark	Decoction, maceration (in wine/rum)	Malaria	Bertani et al. 2005; Milliken 1997a; Vigneron et al. 2005
Zanthoxylum tsihanimpotsa H.Perr.	NA	Rutaceae	Stem bark; leaf, bark	Decoction	Malaria	Randrianarivelojosia et al. 2003; Rasoanaivo et al. 1992
Zanthoxylum usambarense (Engl.) Kokwaro	NA	Rutaceae	Root, fruit, seed, stem bark	Decoction	Malaria	Nanyingi et al. 2008; Muthaura et al. 2007; Muthee et al. 2011
Zanthoxylum xanthoxyloides (Lam.) Waterman	NA	Rutaceae	Root, stem bark	Decoction	Malaria	Koudouvo et al. 2011; Nadembega et al. 2011
Zanthoxylum zanthoxyloides (Lam.) Zepernick & Timler [Syn: *Fagara zanthxyloides* Lam.]	Senegal prickly-ash	Rutaceae	Stem, stem bark, root	Decoction (stem, stem bark), intranasally as snuff (root), maceration (root)	Malaria, fever	Adebayo and Krettli 2011; Allabi et al. 2011; Asase and Oppong-Mensah 2009; Diarra et al. 2015; Hamill et al. 2000; Traore et al. 2013
***Zea mays* L.**	Corn	Poaceae	Fruit, leaf, root	Maceration, decoction (also in combination)	Malaria	Koudouvo et al. 2011; Nadembega et al. 2011; Semenya et al. 2012
Zehneria scabra L.f. Sond.	NA	Cucurbitaceae	Fruit, root, leaf	Juice (root)	Malaria	Giday et al. 2007, 2010; Karunamoorthi and Tsehaye 2012; Muthee et al. 2011
Zingiber cassumunar Roxb.	Bengal ginger, cassumar ginger	Zingiberaceae	Not stated	Not stated	Malaria	Duke 1996

(*Continued*)

Table A.1 (Continued) Medicinal Plants Reported to Be Used for Malaria

Plant Species	Common Name	Family	Part Used	Preparation	Indication	References
Zingiber mioga	Japanese ginger	Zingiberaceae	Not stated	Not stated	Malaria	Duke 1996
***Zingiber officinale* Roscoe**	Garden ginger	Zingiberaceae	Fruit, rhizome, root	Decoction, infusion, maceration; burnt on charcoal and inhaled nasally	Malaria	Adebayo and Krettli 2011; Asase, Hesse, and Simmonds 2012; Duke 1996; Idowu et al. 2010; Karunamoorthi and Tsehaye 2012; Kvist et al. 2006; Milliken 1997a; Nagendrappa, Naik, and Payyappallimana 2013; Tetik, Civelek, and Cakilcioglu 2013; Tor-anyiin, Sha'ato, and Oluma 2003; Yetein et al. 2013
Zinnia peruviana (L.) L.	Peruvia zinnia	Compositae	Leaf	Not stated	Malaria	Tabuti 2008
***Ziziphus jujuba* Mill.**	Chinese date, jujube	Rhamnaceae	Leaf	Decoction: oral and bath	Malaria	Diarra et al. 2015
Ziziphus mucronata Willd.	Buffalo thorn	Rhamnaceae	Leaf	Decoction: oral and bath	Malaria	Diarra et al. 2015
***Ziziphus mauritania* Lam.**	Indian jujube	Rhamnaceae	Leaf, root	Decoction	Malaria	Betti and Yemefa 2011; Nadembega et al. 2011
Ziziphus spina-christi (L.) Desf.	Christ's thorn	Rhamnaceae	Root, fruit	Infusion, consumed (fruit)	Malaria	El-Kamali and El-Khalifa 1999; Musa et al. 2011
Zygia longifolia (Humb. & Bonpl. ex Willd.) Britton & Rose	NA	Fabaceae	Bark, bark resin, leaf	Decoction, eaten, infusion	(Malaria), malaria, fever	Milliken 1997a; Roumy et al. 2007

Note: Plants in bold are those that can be found in Singapore. (Malaria) refers to symptoms of malaria.
* Indication was not distinguished between malaria or fever.

Table A.2 Medicinal Plants Used for Malaria from Ethnobotanical Surveys That Provided Details on Preparation

Species	Part Used	Preparation	Remark	Country	Reference
			Acanthaceae		
Adhatoda vasica Nees	Leaf, root	Decoction	Decoction taken PO thrice; also repels insects	Assam, Northeast India	Namsa, Mandal, and Tangjang 2011
	Leaf	Powder	1–2 teaspoons BD	India	Poonam and Singh 2009
Adhatoda zeylanica Medicus	Leaf	Decoction	Water from decoction used for bathing, leaf paste applied to whole body to cure chronic fever/malaria	Northeast India	Shankar, Sharma, and Deb 2012
Andrographis paniculata Wall. ex Nees	Leaf	Not stated	Crushed raw leaves taken PO for 2 days twice daily with half a glass of milk	Northeast India	Shankar, Sharma, and Deb 2012
	Whole plant	Decoction	Decoction taken PO BD; also used as an insect repellent	Assam, Northeast India	Namsa, Mandal, and Tangjang 2011
Justicia flava Vahl	Leaf	Decoction	Decocted with leaves of *Ocimum viride*, *Cymbopogon citratus*, and *Hoslundia opposita;* taken TDS	Ghana	Asase, Hesse, and Simmonds 2012
			Amaranthaceae		
Amaranthus hybridus	Leaf	Decoction	1 cup TDS for 4–5 days	Msambweni, Kenya	Nguta et al. 2010a
Pupalia micrantha Hauman	Root, root bark	Decoction	Boiled and consumed	Ethiopia	Mesfin et al. 2012
			Amaryllidaceae		
Allium sativum	Rhizome	Pulverized paste	Taken for 5 days, dosed by teaspoons	Ethiopia	Suleman et al. 2009

(Continued)

Table A.2 (Continued) Medicinal Plants Used for Malaria from Ethnobotanical Surveys That Provided Details on Preparation

Species	Part Used	Preparation	Remark	Country	Reference
			Anacardiaceae		
Mangifera indica L.	Leaf, stem, root	Decoction	BD for 2–3 days	Ghana	Asase, Hesse, and Simmonds 2012
	Leaf	Decoction	Decoct with cut fruits of *Citrus aurantiifolia* for around 1 hour Dosage regimen: one cup TDS until recovery	West Ghana	Asase, Akwetey, and Achel 2010
			Annonaceae		
Annickia chlorantha (Oliv.)	Stem	Decoction	Stem bark removed by scraping with machete; decoct 500 g in 3 liters of water for 20 min Dosage regimen: PO 250 ml TDS for 15 days	Cameroon	Tsabang et al. 2012
	Stem	Decoction	Decoction of 300 g of stem bark from this plant and *Rauvolfia vomitoria* and *Fagara macrophylla* in 4 liters of water for 20 min (one recipe also included stem bark of *Nauclea latifolia*) Dosage regimen: PO 250 ml TDS for 10 days	Cameroon	Tsabang et al. 2012
Annona squamosa L.	Stem	Decoction	PO thrice for larvicidal activity	Assam, Northeast India	Namsa, Mandal, and Tangjang 2011

(*Continued*)

Table A.2 (Continued) Medicinal Plants Used for Malaria from Ethnobotanical Surveys That Provided Details on Preparation

Species	Part Used	Preparation	Remark	Country	Reference
Annona muricata L.	Fruit, leaf	Decoction	Boiled with stem bark of *Erythrophleum ivorensis, Anopyxis klaineana, Cocos nucifera, Turraeanthus africanus, Alstonia boonei,* fruit of *Citrus aurantiifolia, C. sinesis*, root of *Annona muricata* and *Thaumatococcus* sp. into a mixture and used in a bath where females sit	Ghana	Asase, Hesse, and Simmonds 2012
	Leaf	Decoction	Decoct handful of leaves in 3 liters of water for 20 min Dosage regimen: PO 250 ml OD for 7 days	Cameroon	Tsabang et al. 2012
Annona senegalensis Pers.	Leaf	Decoction	About 300 ml of decoction QDS for 10 days; symptoms expected to resolve during period of therapy, interpreted as cure	Tiv, Nigeria	Tor-anyiin, Sha'ato, and Oluma 2003
	Root	Decoction	500 g of root stem bark decocted in 3 liters of water for 20 min Dosage regimen: 250 ml OD for 15 days	Cameroon	Tsabang et al. 2012
	Leaf	Decoction	100 g of young leaves each of this plant and *Piptostigma thonningii*, 100 g of leaves of *Senna alata*, 100 g of *Chrysanthellum americana*, *Lippia multiflora*, 300 g of *Terminalia glaucescens*, 300 g root stem of *Nauclea latifolia*, and 100 g of *Ocimum gratissimum* are decocted in 5 liters of water until it is reduced to 3 liters; roots of *N. latifolia* should be harvested at sunrise or sunset	Cameroon	Tsabang et al. 2012

(*Continued*)

Table A.2 (Continued) Medicinal Plants Used for Malaria from Ethnobotanical Surveys That Provided Details on Preparation

Species	Part Used	Preparation	Remark	Country	Reference
Annona squamosa Linn.	Leaf	Decoction	150 g decocted in 3 liters of water for 20 min Dosage regimen: 250 ml BD for 10 days	Cameroon	Tsabang et al. 2012
Greenwayodendron sp.	Leaf	Decoction	Adult dose: 1 cup TDS until recovery Pediatric dose: half cup TDS until recovery	West Ghana	Asase, Akwetey, and Achel 2010
Monodora myristica Dun.	Leaf	Decoction	Together with stem bark of *Erythrophleum ivorensis, Anopyxis klaineana, Cocos nucifera, Turraeanthus africanus* and *Alstonia boonei*, and root of *Annona muricata, Thaumatococcus* sp., *Zingiber officinale, Piper guineense* and *Xylopia aethiopica* are pounded, decocted, and used as enema	Ghana	Asase, Hesse, and Simmonds 2012
Polyalthia suaveolens Engl. & Diels	Stem	Decoction	500 g decocted in 3 liters of water for 20 min Dosage regimen: 250 ml TDS for 15 days	Cameroon	Tsabang et al. 2012
	Fruit	Decoction	One serrated fruit decocted in 2 liters of water for 15 min Dosage regimen: 250 ml TDS for 7 days	Cameroon	Tsabang et al. 2012
Polyceratocarpus sp.	Stem	Decoction	500 g decocted in 3 liters of water until half the volume Dosage regimen: 250 ml TDS	Cameroon	Tsabang et al. 2012
Pseudoxandra cuspidata Maas.	Bark	Decoction	200 g inner bark decocted for 15 min in 500 ml of water, and left to cool	French Guiana	Bertani et al. 2005
Uvaria sp.	Stem	Decoction	500 g decocted in 3 liters of water until half the volume Dosage regimen: 250 ml TDS	Cameroon	Tsabang et al. 2012

(*Continued*)

Table A.2 (Continued) Medicinal Plants Used for Malaria from Ethnobotanical Surveys That Provided Details on Preparation

Species	Part Used	Preparation	Remark	Country	Reference
			Apiaceae		
Anethum graveolens L.	Leaf, root	Decoction	Decoction of pounded leaves and root is consumed	Ethiopia	Mesfin et al. 2012
Foeniculum vulgare Miller	Root	Infusion	Boiled and consumed as tea	Ethiopia	Mesfin et al. 2012
			Apocynaceae		
Acokanthera schimperi (A.D.C.) Schweeinf.	Leaf	Juice	Leaves grounded and mixed in water, sieved through thin cloth, filtrate drunk	Ethiopia	Mesfin et al. 2012
Alstonia boonei De Wild.	Leaf	Decoction	1 teacup of decoction TDS	Ghana	Asase, Hesse, and Simmonds 2012
Alstonia scholaris R. Br.	Bark	Pill	1 pill of dried and grinded bark and honey is taken 2–3 times daily until recovery	Bangladesh	Kadir et al. 2014
	Root, bark	Decoction	Decoction taken PO twice	Assam, Northeast India	Namsa, Mandal, and Tangjang 2011
Geissospermum laeve (Vell.) Miers	Bark	Decoction	500 g, grated, decocted in 2 liters of water for 15 min Dosage regimen: 1 cup BD until symptoms resolve	Bolivia	Muñoz, Sauvain, Bourdy, Callapa, Bergeron et al. 2000
	Bark	Decoction	40 g decocted in 1 liter of water for 15 min and left to cool	French Guiana	Bertani et al. 2005
	Bark	Maceration	40 g macerated in 250 ml of rum for 1 month for prevention	French Guiana	Bertani et al. 2005
Geissospermum argenteum Woodson	Bark	Maceration	40 g macerated in 250 ml of rum for 1 month for prevention	French Guiana	Bertani et al. 2005
Laudolphia buchananii (Hall.f) Stapf.	Leaf	Decoction	1 cup TDS for 3–4 days	Msambweni, Kenya	Nguta et al. 2010a

(Continued)

Table A.2 (Continued) Medicinal Plants Used for Malaria from Ethnobotanical Surveys That Provided Details on Preparation

Species	Part Used	Preparation	Remark	Country	Reference
Melodinus monogynus Roxb.	Leaf, root, bark	Not stated	Contains narcotic poison	Northeast India	Shankar, Sharma, and Deb 2012
			Araceae		
Amorphophallus angolensis N.E. Br.	Tuber	Maceration	1 glass BD	Congo	Chifundera 2001
			Arecaceae		
Cocos nucifera L.	Root	Decoction	60 g decocted in 1 liter of water for about 45 min; taken TDS Pediatric dose: half the aforementioned dose	West Ghana	Asase, Akwetey, and Achel 2010
Elaeis guineensis Jacq.	Root	Decoction	Consume until recovery	West Ghana	Asase, Akwetey, and Achel 2010
Euterpe precatoria	Root	Decoction	300 g decocted in 800 ml water for 15 min, and left to cool	French Guiana	Bertani et al. 2005
			Aristolochiaceae		
Aristolochia indica f.	Root	Decoction	Decoction in hot water taken PO once	Assam, Northeast India	Namsa, Mandal, and Tangjang 2011
			Asclepiadaceae		
Leptadenia hastata (Pers.) Decne.	Root, root bark	Decoction	Fresh/dried root grounded and boiled	Ethiopia	Mesfin et al. 2012
			Asparagaceae		
Asparagus africanus Lam.	Leaf	Juice	Fresh or dried leaves pounded and mixed with that of *Aloe* sp. and consumed	Ethiopia	Mesfin et al. 2012

(*Continued*)

Table A.2 (Continued) Medicinal Plants Used for Malaria from Ethnobotanical Surveys That Provided Details on Preparation

Species	Part Used	Preparation	Remark	Country	Reference
			Balanitaceae		
Balanites rotundifolia (van Tieghem) Blatter	Leaf	Decoction	Fresh leaves boiled and consumed	Ethiopia	Mesfin et al. 2012
			Berberidaceae		
Berberis aristata DC.	Root	Not stated	Root bark used as tonic	Northeast India	Shankar, Sharma, and Deb 2012
			Bignoniaceae		
Sterospermum kunthiamum	Root	Decoction	About 300 ml taken TDS for 6 days	Tiv, Nigeria	Tor-anyiin, Sha'ato, and Oluma 2003
			Bixaceae		
Bixa orellana L.	Root	Decoction	300 g boiled in 500 ml of water for 20 min and left to cool	French Guiana	Bertani et al. 2005
			Bombacaceae		
Adansonia digitata Linn.	Leaf	Decoction	1 cup taken TDS for 3 to 4 days until recovery	Msambweni, Kenya	Nguta et al. 2010a
Cynoglossum glochidion Wall.	Root	Not stated	Root pounded/powdered and mixed with water, 10 g taken BD	Northeast India	Shankar, Sharma, and Deb 2012

(Continued)

Table A.2 (Continued) Medicinal Plants Used for Malaria from Ethnobotanical Surveys That Provided Details on Preparation

Species	Part Used	Preparation	Remark	Country	Reference
			Capparidaceae		
Maerua oblongifolia (Forssk.) A. Rich.	Leaf	Decoction	Leaves pounded and boiled with goat milk; alternatively, it is consumed as a mixture with leaves of *Withania somnifera*	Ethiopia	Mesfin et al. 2012
			Caricaceae		
Carica papaya L.	Leaf	Decoction	Dried leaves decocted with leaves of *Ocimum viride* and *Hoslundia opposita*, fruit of *Citrus aurantiifolia* and *C. sinesis*, root of *Cassia siamea*, dried leaves of *Capsicum annuum* L., and drunk	Ghana	Asase, Hesse, and Simmonds 2012
	Leaf	Decoction	Dried leaves decocted; 1 cup BD taken	West Ghana	Asase, Akwetey, and Achel 2010
	Leaf, twig	Not stated	Concoction of fresh leaves and twigs from *C. papaya*, *M. indica*, *P. guajava*, and *C. citratus* taken TDS for 5 days; can also be boiled and steam inhaled BD during treatment	Tiv, Nigeria	Tor-anyiin, Sha'ato, and Oluma 2003
	Root	Decoction	Decoct 300 g of freshly cut roots in 3 liters of water for 1 hour, and leave to cool	French Guiana	Bertani et al. 2005
	Leaf	Decoction	4 dry yellow leaves placed in 2.5 liters of water, decocted for 20 min, and left to cool	French Guiana	Bertani et al. 2005
	Seed	Decoction	Taken for 3 days, dosed by cups	Ethiopia	Suleman et al. 2009
	Leaf	Consumed	Add the following to water: dry leaves of *Musa paradisiaca*, little unripe fruits and leaves of *Citrus limon*, leaves of *Cassia occidentalis,* leaves of *Mangifera indica*, leaves of *Ziziphus mauritiana*, and drink	Italy and Senegal	Ellena, Quave, and Pieroni 2012

(*Continued*)

Table A.2 (Continued) Medicinal Plants Used for Malaria from Ethnobotanical Surveys That Provided Details on Preparation

Species	Part Used	Preparation	Remark	Country	Reference
			Chenopodiaceae		
Halothamnus somalensis (N.E. Br.) Botsch.	Root, root bark	Decoction	Pounded and boiled with sugar and goat milk	Ethiopia	Mesfin et al. 2012
			Combretaceae		
Combretum padoides Engl and Diels.	Leaf	Decoction	1 cup TDS for 3–5 days	Msambweni, Kenya	Nguta et al. 2010a
Terminalia catappa	Leaf	Decoction	About 300 ml of aqueoua decoction of yellowing or red fresh leaves taken every 6 hours until resolution of symptoms	Tiv, Nigeria	Tor-anyiin, Sha'ato, and Oluma 2003
Terminalia chebula L.	Fruit	Not stated	Fruit is dried and powdered and mixed with honey and *Piper mullesua* L.	Arunchal Pradesh, India	Milliken 1997
			Commelinaceae		
Commelina erecta L.	Leaf, stem	Decoction	100 g of fresh leaves and stems, decocted for 5 min in 200 ml water, left to cool	French Guiana	Bertani et al. 2005
	Leaf, stem	Juice	100 g of fresh leaves and stems pounded in cold water and filtered	French Guiana	Bertani et al. 2005
			Compositae		
Acanthospermum hispidum L.	Whole plant	Decoction	Dosage regimen: 1 cup TDS for 3–5 days Pediatric dose: half cup TDS 3–5 days	West Ghana	Asase, Akwetey, and Achel 2010

(*Continued*)

Table A.2 (Continued) Medicinal Plants Used for Malaria from Ethnobotanical Surveys That Provided Details on Preparation

Species	Part Used	Preparation	Remark	Country	Reference
Artemisia vulgaris L.	Leaf	Decoction	Decoction taken PO thrice with honey; also used as insect repellent	Assam, Northeast India	Namsa, Mandal, and Tangjang 2011
Chromolaena odorata	Leaf, twig	Not stated	Aqueous concoction can be taken with that of *Ocimum gratissimum* if available; 300 ml taken TDS–QDS for 5 days; use for bathing BD	Tiv, Nigeria	Tor-anyiin, Sha'ato, and Oluma 2003
Conyza pyrrhopappa Sch.Bip. ex A. Rich.	Leaf, stem	Direct application	Fresh leaves and stem are crushed and used for bathing	Ethiopia	Mesfin et al. 2012
Helianthus annuus L.	Flower, leaf	Decoction	Decoction taken with honey	Northeast India	Shankar, Sharma, and Deb 2012
Mikania guaco Humb. & Bompl.	Leaf	Decoction	50 g fresh leaves decocted in 300 ml water for 5 min and left to cool	French Guiana	Bertani et al. 2005
	Leaf	Juice	50 g fresh leaves pounded with some cold water and filtered		
Senecio syringitolius O. Hoffman.	Leaf	Decoction	1 cup TDS for 3–4 days	Mambweni, Kenya	Nguta et al. 2010a
Taraxacum officinale Wigg.	Whole plant	Not stated	Powder used	Northeast India	Shankar, Sharma, and Deb 2012
Tessaria integrifolia Ruiz & Pav.	Leaf	Decoction	Decocted with *Hymenachne donacifolia* and some termite eggs, to be taken while hot	Bolivia	Muñoz, Sauvain, Bourdy, Callapa, Rojas et al. 2000

(*Continued*)

Table A.2 (Continued) Medicinal Plants Used for Malaria from Ethnobotanical Surveys That Provided Details on Preparation

Species	Part Used	Preparation	Remark	Country	Reference
Vernonia amygdalina Delile.	Leaf	Juice	Fresh leaves squeezed with clean water and filtered; about 300 ml of bitter filtrate is taken BD–TDS for 5 days	Tiv, Nigeria	Tor-anyiin, Sha'ato, and Oluma 2003
		Decoction	50 g of fresh leaves decocted in 1 litre of water, drink 1 cup TDS	West Ghana	Asase, Akwetey, and Achel 2010
			Pediatric dose: half a cup TDS		
		Infusion	Alternative: leaves can be mashed and infusion taken as required		
			Cucurbitaceae		
Cucumis ficifolius A. Rich.	Whole plant	Decoction	Whole plant crushed, boiled, and consumed	Ethiopia	Mesfin et al. 2012
Gerranthus lobatus (Cogn.) Jeffrey	Root	Decoction	1 cup TDS for 3–4 days	Msambweni, Kenya	Nguta et al. 2010a
Gymnopetalum cochinchinensis Kurz	Root	Decoction	BD	Assam, Northeast India	Namsa, Mandal, and Tangjang 2011
Oreosyce africana Hook.f.	Whole plant	Decoction	A crushed handful is mixed with a handful of stem barks from *Persea americana* and is decocted with 1.5 liters of tap water; full glass taken BD for 2 days	Congo	Chifundera 2001
			Euphorbiaceae		
Acalypha indica L.	Aerial part	Decoction	Pounded and boiled with water, then drunk	Ethiopia	Mesfin et al. 2012
Alchornea cordifolia (Schumach. & Thonn.) Müll. Arg	Root	Decoction	Boil roots of this herb with that of *Cassia siamea*; decoction taken TDS until cured	Ghana	Asase, Hesse, and Simmonds 2012
	Leaf	Decoction	Decocted with rhizomes of *Imperata cylindrica*; about 300 ml taken TDS for 7 days	Tiv, Nigeria	Tor-anyiin, Sha'ato, and Oluma 2003

(Continued)

Table A.2 (Continued) Medicinal Plants Used for Malaria from Ethnobotanical Surveys That Provided Details on Preparation

Species	Part Used	Preparation	Remark	Country	Reference
Bridelia ferruginea Benth.	Leaf	Decoction	Fresh/dried leaves decocted and about 300 ml taken TDS for 7 days	Tiv, Nigeria	Tor-anyiin, Sha'ato, and Oluma 2003
Croton macrostachis	Leaf	Decoction	Taken for 5 days, dosed by cups	Ethiopia	Suleman et al. 2009
Croton tiglium L.	Leaf, flower	Not stated	Powder consumed with a glass of water BD until cured	Northeast India	Shankar, Sharma, and Deb 2012
Jatropha curcas L.	Seed	Consumed	Outer cover of seed removed; inside part swallowed with camel milk or chewed	Ethiopia	Mesfin et al. 2012
Manihot esculenta Crantz	Root	Cooked	Peeled and cooked roots are fermented to make chichi, and juice from leaves of *Nicotiana tabacum* are added and used orally	Amazonian Ecuador	Giovannini 2015
Mareya micrantha Muell. Arg.	Leaf	Decoction	Decocted with cut fruits of *C. sinensis*; taken TDS; lime decreases potency of decoction but prevents emesis and gastric discomfort	Ghana	Asase, Hesse, and Simmonds 2012
Phyllanthus amarus Schum & Thonn.	Whole plant	Decoction	50 g decocted in 1 liter of water Adult dosage regimen: 1 cup TDS after food until recovery; may cause dizziness Pediatric dosage: half the adult dose Sweeten with honey or sugar if required	West Ghana	Asase, Akwetey, and Achel 2010
Phyllanthus reticulatus Poir	Leaf, bark, flower, root	Juice	1 spoonful for juice of leaf, bark, flower, and root given 3 or 4 times daily in polydipsia for 2–3 days	Bangladesh	Kadir et al. 2014
Ricinus communis L.	Root, leaf	Decoction	1 cup TDS for 3–4 days	Msambweni, Kenya	Nguta et al. 2010a
Securinega virosa (Roxb. Ex Willd) Baill.	Whole plant	Decoction	Decoct with leaves of *Morinda lucida* and consume decoction as required until recovery	West Ghana	Asase, Akwetey, and Achel 2010

(*Continued*)

Table A.2 (Continued) Medicinal Plants Used for Malaria from Ethnobotanical Surveys That Provided Details on Preparation

Species	Part Used	Preparation	Remark	Country	Reference
			Fabaceae		
Crotolaria occulta Grab	Whole plant	Juice	Taken with warm water	Northeast India	Shankar, Sharma, and Deb 2012
Desmodium gangeticum DC.	Whole plant	Juice	2 teaspoons BD	India	Poonam and Singh 2009
Hymenaea courbaril L.	Bark	Decoction	15 × 10 cm piece cut into smaller pieces, decocted in 800 ml of water for 15 min, and left to cool	French Guiana	Bertani et al. 2005
Indigofera articulata Gouan	Root	Infusion	Root crushed and boiled as tea, then filtered using thin cloth, and consumed	Ethiopia	Mesfin et al. 2012
Indigofera coerulea Roxb.	Leaf	Decoction	Leaves pounded and boiled	Ethiopia	Mesfin et al. 2012
Pterocarpus rohrii Vahl.	Leaf	Decoction	15 fresh leaves decocted in 1.5 liters of water for 10 min, and left to cool	French Guiana	Bertani et al. 2005
	Leaf	Juice	15 leaves pounded in cold water and filtered	French Guiana	Bertani et al. 2005
	Bark	Decoction	15 × 10 cm piece cut into smaller pieces, decocted in 800 ml water for 15 min, and left to cool	French Guiana	Bertani et al. 2005
Senna alata (L.) Roxb.	Leaf	Decoction	Decoct 1 liter in 1 liter of water 1 cup taken as a single dose only; may cause frequent gastric emptying	West Ghana	Asase, Akwetey, and Achel 2010
	Leaf	Decoction	PO TDS	Ghana	Asase, Hesse, and Simmonds 2012
Senna italica Mill.		Decoction	Dried leaves powdered, boiled with water, and added with goat/camel milk	Ethiopia	Mesfin et al. 2012

(Continued)

Table A.2 (Continued) Medicinal Plants Used for Malaria from Ethnobotanical Surveys That Provided Details on Preparation

Species	Part Used	Preparation	Remark	Country	Reference
Senna occidentalis (L.) Link	Seed	Decoction	Dried seeds, ground, and about 2 teaspoonfuls decocted in 0.5 liters of water for 5 min; taken BD until recovery	West Ghana	Asase, Akwetey, and Achel 2010
	Root, leaf	Decoction	1 cup TDS for 3 to 4 days	Msambweni, Kenya	Nguta et al. 2010a
Senna siamea (Lam.) H.S. Irwin & Barneby	Stem	Decoction	Consume as required until recovery	West Ghana	Asase, Akwetey, and Achel 2010
	Root	Decoction	PO TDS; causes frequent diuresis	Ghana	Asase, Hesse, and Simmonds 2012
Tamarindus indica L.		Infusion	Infusion kept overnight, consumed after taking goat soup	Ethiopia	Mesfin et al. 2012
			Flacourtiaceae		
Flacourtia indica (Burm.f) Merr.	Root, stem bark, leaf	Decoction	1 cup TDS for 3–4 days	Msambweni, Kenya	Nguta et al. 2010a
			Gemlinaceae		
Swertia chirayata Ham.	Stem	Infusion	100–150 ml BD	India	Poonam and Singh 2009
			Gentianaceae		
Coutoubea racemosa Aubl.	Whole plant	Decoction	3 whole plants decocted in 700 ml water for 15 min, then left to cool	French Guiana	Bertani et al. 2005
Coutoubea spicata Aubl.	Whole plant	Decoction	3 whole plants including roots are decocted in 1 liter of water for 20 min, then left to cool	French Guiana	Bertani et al. 2005

(Continued)

Table A.2 (Continued) **Medicinal Plants Used for Malaria from Ethnobotanical Surveys That Provided Details on Preparation**

Species	Part Used	Preparation	Remark	Country	Reference
Irlbachia alata (Aubl.) Maas	Leaf	Decoction	200 g fresh leaves decocted in 300 ml of water for 10 min, and left to cool	French Guiana	Bertani et al. 2005
	Leaf	Juice	150 g fresh leaves pounded with 100 ml of cold water, then filtered	French Guiana	Bertani et al. 2005
	Root	Decoction	150 g decocted in 200 ml of water for 10 min and left to cool	French Guiana	Bertani et al. 2005
Swertia nervosa Wall.	Whole plant	Decoction	Whole plant boiled in water and filtered, about 5 teaspoons twice daily	Nepal	Malla, Gauchan, and Chhetri 2015
			Hydnoraceae		
Hydnora johannis Becc.	Root	Decoction	Fresh root crushed and boiled	Ethiopia	Mesfin et al. 2012
			Lamiaceae		
Ajuga remota	Leaf	Decoction	Taken for 3 days; dosed by cups	Ethiopia	Suleman et al. 2009
Clerodendrum viscosum Vent.	Leaf	Decoction	Boiled extract of leaves taken once daily in the morning for 7 days	Bangladesh	Kadir et al. 2014
Mentha spicata L.	Aerial part	Infusion	Can be consumed and also used for bathing	Ethiopia	Mesfin et al. 2012
Ocimum basilicum L.	Leaf	Decoction	1 cup TDS for 3–4 days	Msambweni, Kenya	Nguta et al. 2010a
Ocimum canum Sims	Leaf	Decoction	8–10 pinches taken for 10 days	India	Prabhu et al. 2014
Ocimum gratissimum L.	Leaf	Decoction	40–60 g decocted in 1 liter of water; taken TDS until recovery	West Ghana	Asase, Akwetey, and Achel 2010
Ocimum sanctum L.	Leaf	Juice	Juice taken PO TDS with honey; also an insect repellent	Assam, Northeast India	Namsa, Mandal, and Tangjang 2011
Ocimum spicatum Deflers	Leaf	Decoction	Leaves are powdered and boiled; sometimes taken with coffee	Ethiopia	Mesfin et al. 2012

(*Continued*)

Table A.2 (Continued) Medicinal Plants Used for Malaria from Ethnobotanical Surveys That Provided Details on Preparation

Species	Part Used	Preparation	Remark	Country	Reference
Ocimum suave Willd.	Leaf	Decoction	1 cup TDS for 3–5 days	Msambweni, Kenya	Nguta et al. 2010a
	Bark	Decoction	1 teacup 3 times daily for 3 weeks	India	Prabhu et al. 2014
Ocimum viride Willd.	Leaf	Decoction	Decoction taken as desired	Ghana	Asase, Hesse, and Simmonds 2012
Plectranthus barbatus Andr.	Leaf	Decoction	1 cup TDS for 3–4 days	Msambweni, Kenya	Nguta et al. 2010a
			Lauraceae		
Cinnamomum bejolghota (Buch. Ham)	Bark, leaf	Decoction	Bark and leaves boiled with leaves of *Anacolosa crassipes*; water used for bathing, steam inhaled, water taken internally	Northeast India	Shankar, Sharma, and Deb 2012
			Lecythidaceae		
Grias neuberthii J.F. Macbr.	Bark	Decoction	Scraped bark mixed with water and boiled; decoction is drunk and vomited	Amazonian Ecuador	Giovannini 2015
			Loranthaceae		
Tapinanthus dodoneifolius (D.C.) Danser	Leaf	Decoction	Leaves dried at room temperature, and decocted; taken before food	Ghana	Asase, Hesse, and Simmonds 2012
			Lythraceae		
Lawsonia inermis L.	Root, root bark	Juice	Grounded root bark squeezed by cloth; juice added to soup of camel or goat meat and boiled for 4 hours and consumed	Ethiopia	Mesfin et al. 2012

(*Continued*)

Table A.2 (Continued) Medicinal Plants Used for Malaria from Ethnobotanical Surveys That Provided Details on Preparation

Species	Part Used	Preparation	Remark	Country	Reference
			Malvaceae		
Sterculia setigera	Bark	Not stated	About 300 ml of aqueous concoction with fresh or dry bark from this herb and that of *Parkia biglobosa* is taken TDS until symptoms resolve	Tiv, Nigeria	Tor-anyiin, Sha'ato, and Oluma 2003
Theobroma cacao L.	Leaf	Decoction	Decoct for 1 hour, consume as required until recovery	West Ghana	Asase, Akwetey, and Achel 2010
			Meliaceae		
Azadirachta indica (A. Juss) L.	Leaf	Decoction	Decoction taken PO once with honey; also used as insect repellent	Assam, Northeast India	Namsa, Mandal, and Tangjang 2011
	Leaf	Decoction	1 cup TDS until recovery; alternatives: decoct with fruits of *A. cosmosus* or mash leaves in water	West Ghana	Asase, Akwetey, and Achel 2010
	Leaf, root	Decoction	PO TDS	Ghana	Asase, Hesse, and Simmonds 2012
	Leaf, twig	Decoction, Maceration	About 300 ml of macerated extract or decoction in clean water taken TDS–QDS for 10 days; symptoms expected to resolve by end of therapy; interpreted as cure	Tiv, Nigeria	Tor-anyiin, Sha'ato, and Oluma 2003
	Leaf	Maceration	Fresh apical leaves (buds) pounded and soaked in water; filtrate consumed	Ethiopia	Mesfin et al. 2012
	Bark	Infusion	1 glass twice daily for 2–3 weeks	India	Prabhu et al. 2014
	Root bark, stem bark, leaf	Decoction	1/4 glass TDS for 2–3 days	Msambweni, Kenya	Nguta et al. 2010a

(*Continued*)

Table A.2 (Continued) Medicinal Plants Used for Malaria from Ethnobotanical Surveys That Provided Details on Preparation

Species	Part Used	Preparation	Remark	Country	Reference
	Stem bark, root, leaf	Decoction	Decocted and given as a drink when required with whole plant of *Solanum lycopersicum* L., *Zingiber officinale* Boehmer., and *Capsicum frutescens* L. Decocted and drank as required with fruits or leaves of *Citrus limon* L. Decocted and drank as required with leaves, stem bark, or roots of *Khaya anthotheca* C.D.C, whole plant of *Sarcophrynium brachystachys* Schumann., and fruit of *Citrus aurantiifolia* (Christm.) Swingle	Ghana	Asase and Kadera 2014
Cedrela fissilis Vell.	Bark	Decoction	One handful, grated, decocted in 2 liters of water until reduced to 1 liter Dosage regimen: 3 cups/day until resolution of symptoms	Bolivia	Muñoz, Sauvain, Bourdy, Callapa, Bergeron et al. 2000
Khaya anthotheca C.D.C.	Leaf	Decoction	Decoction drank for a week as required with leaves of *Musa paradisiaca* L., whole plant of *Sarcophrynium brachystachys* Schumann., leaves, young stem, and roots of *Bambusa vulgaris* Schreber., fruit of *Citrus aurantiifolia* (Christm.) Swingle, leaves and bark of three other unknown species	Ghana	Asase and Kadera 2014
Khaya senegalensis A. Juss.	Bark	Maceration	About 300 ml of filtrate after maceration of fresh bark in water taken BD–TDS until symptoms resolve	Tiv, Nigeria	Tor-anyiin, Sha'ato, and Oluma 2003
	Stem	Decoction	Alcohol extract of stem bark, 1 cup before food; not to be given to children	West Ghana	Asase, Akwetey, and Achel 2010

(*Continued*)

Table A.2 (Continued) Medicinal Plants Used for Malaria from Ethnobotanical Surveys That Provided Details on Preparation

Species	Part Used	Preparation	Remark	Country	Reference
			Menispermaceae		
	Root	Decoction	PO OD	Assam, Northeast India	Namsa, Mandal, and Tangjang 2011
Cissampelos mucronata	Aerial part	Decoction	PO QDS; may be used for bathing	Tiv, Nigeria	Tor-anyiin, Sha'ato, and Oluma 2003
Stephania japonica Miers.	Tuber	Not stated	Sun-dried tuber powder taken with boiled water BD for more than 4 days until cured	Northeast India	Shankar, Sharma, and Deb 2012
Stephania hermandifolia (Willd) Walp.	Root	Decoction	PO OD	Assam, Northeast India	Namsa, Mandal, and Tangjang 2011
Tinospora cordifolia Miers.	Stem	Not stated	50 ml BD	India	Poonam and Singh 2009
Tinospora crispa (L.) Miers	Stem	Decoction	15 cm of stem cut into smaller pieces and decocted in 500 ml water for 15 min	French Guiana	Bertani et al. 2005
			Monimiaceae		
Siparuna guianensis (Aubl.)	Leaf	Decoction	200 g fresh leaves decocted for 15 min and left to cool	French Guiana	Bertani et al. 2005
	Leaf	Juice	80 g pounded with 100 ml water and filtered	French Guiana	Bertani et al. 2005
			Moraceae		
Ficus bussei Warp ex Mildbr and Burret.	Root, leaf	Decoction	1 cup TDS for 4–5 days	Msambweni, Kenya	Nguta et al. 2010a
Maquira coriacea (Karst.) C. C. Berg	Root, bark	Decoction	Decoct one handful each of trunk and root barks in 2 liters of water until volume reduced to 1 liter Dosage regimen: half cup TDS until symptoms resolve	Bolivia	Muñoz, Sauvain, Bourdy, Callapa, Bergeron et al. 2000

(Continued)

Table A.2 (Continued) Medicinal Plants Used for Malaria from Ethnobotanical Surveys That Provided Details on Preparation

Species	Part Used	Preparation	Remark	Country	Reference
Moringa oleifera Lam.	Leaf	Decoction	1 cup TDS for 3–5 days Alternative: mash leaves in water, taken TDS	West Ghana	Asase, Akwetey, and Achel 2010
	Leaf	Decoction	Taken for 3 days, dosed by cups	Ethiopia	Suleman et al. 2009
			Musaceae		
Musa paradisiaca L.	Root	Decoction	Consume as required until recovery	West Ghana	Asase, Akwetey, and Achel 2010
			Myrsinaceae		
Myrsine africana	Leaf	Decoction	Taken for 3–5 days, dosed by cups	Ethiopia	Suleman et al. 2009
			Myrtaceae		
Eucalyptus sp. L'Her.	Leaf	Decoction	Decocted with leaves of *A. indica* for purging; decoction also used in bathing	Ghana	Asase, Hesse, and Simmonds 2012
Psidium guajava L.	Leaf	Decoction	Decoct with cut fruits of *Citrus aurantiifolia* and take 1 cup BD	West Ghana	Asase, Akwetey, and Achel 2010
Psidium personii McVaugh	Leaf	Decoction	150 g decocted in 1 liter of water for 15 min and left to cool	French Guiana	Bertani et al. 2005
			Orchidaceae		
Satyrium nepalense D. Don	Tuber	Not stated	Tonic	Northeast India	Shankar, Sharma, and Deb 2012
			Oleaceae		
Olea europa	Bark	Decoction	Taken for 3–5 days, dosed by cups	Ethiopia	Suleman et al. 2009

(*Continued*)

Table A.2 (Continued) Medicinal Plants Used for Malaria from Ethnobotanical Surveys That Provided Details on Preparation

Species	Part Used	Preparation	Remark	Country	Reference
			Papaveraceae		
Argemone mexicana Linn.	Not stated	Latex	2 teaspoons in 1:1 (lemon juice) BD	India	Poonam and Singh 2009
			Papilionaceae		
Bowdichia virgilioides Kunth.	Bark	Decoction	Decoct a piece (ca. 5 × 10 cm) in 2 liters of water until 1 liter	Bolivia	Deharo et al. 2001
			Dosage regimen: 3 small glasses a day for 3 days		
Securidaca longepedunculata Fres.	Root, leaf, stem bark	Decoction	1 cup TDS for 3–4 days	Msambweni, Kenya	Nguta et al. 2010a
			Piperaceae		
Piper longum L.	Fruit, root	Decoction	TDS	Assam, Northeast India	Namsa, Mandal, and Tangjang 2011
	Fruit	Juice	1 glass 1–2 times daily for 4 weeks	India	Prabhu et al. 2014
Piper marginatum Jacq.	Leaf	Decoction	10 fresh leaves decocted for 10 min in 500 ml water, and left to cool	French Guiana	Bertani et al. 2005
Piper mullesua Buch. Ham.	Leaf, fruit	Not stated	Dried plant consumed	Northeast India	Shankar, Sharma, and Deb 2012
			Plumbaginaceae		
Plumbago zeylanica Linn.	Root	Infusion	100–150 ml BD	India	Poonam and Singh 2009

(Continued)

Table A.2 (Continued) Medicinal Plants Used for Malaria from Ethnobotanical Surveys That Provided Details on Preparation

Species	Part Used	Preparation	Remark	Country	Reference
			Poaceae		
Bambusa vulgaris Schrad ex J. C. Wendl.	Leaf	Decoction	Consume as required until recovery	West Ghana	Asase, Akwetey, and Achel 2010
Cymbopogon citratus Stapf.	Leaf	Decoction	Boiled with stems of *Sacharrum officinale* L. and fruit peels of *Ananas comosus*; take decoction as desired	Ghana	Asase, Hesse, and Simmonds 2012
	Leaf	Infusion	1 cup BD until recovery	West Ghana	Asase, Akwetey, and Achel 2010
	Aerial part	Decoction	About 300 ml taken QDS for 7 days	Tiv, Nigeria	Tor-anyiin, Sha'ato, and Oluma 2003
Saccharum officinarum L.	Stem	Decoction	Decoct with fruit peels of *Ananas comosus* and leaves of *A. indica*; 1 cup TDS taken	West Ghana	Asase, Akwetey, and Achel 2010
Vetiveria zizanoides Nosk.	Root	Powder	1 bunch of powdered root taken once daily for 8 weeks	India	Prabhu et al. 2014
			Rananculaceae		
Coptis teeta Wall	Root, rhizome	Not stated	PO 150 g TDS	Northeast India	Shankar, Sharma, and Deb 2012
	Seed, root, rhizome	Decoction	Decoction taken with honey TDS	Assam, Northeast India	Namsa, Mandal, and Tangjang 2011
Nigella sativa	Seed	Paste	Taken for 3 days, dosed by teaspoons	Ethiopia	Suleman et al. 2009
Thalictrum foliolosum L.	Root, rhizome	Not stated	Tonic	Northeast India	Shankar, Sharma, and Deb 2012
			Rosaceae		
Randia fasciculata (Roxb.) D.C.	Leaf	Decoction	Leaf mixed with *Piper nigrum*	Northeast India	Shankar, Sharma, and Deb 2012

(*Continued*)

Table A.2 (Continued) Medicinal Plants Used for Malaria from Ethnobotanical Surveys That Provided Details on Preparation

Species	Part Used	Preparation	Remark	Country	Reference
			Rubiaceae		
Canthium glaucum Hiern.	Fruit	Decoction	1 cup TDS for 4–5 days	Msambweni, Kenya	Nguta et al. 2010a
Chimarrhis turbinata D.C.	Leaf	Decoction	50 g placed in 2 liters of water, decocted for 15 min, and left to cool	French Guiana	Bertani et al. 2005
	Bark	Decoction	200 g placed in 1 liter of water, decocted for 15 min, and left to cool	French Guiana	Bertani et al. 2005
Cinchona officinalis Linn. F.	Bark	Decoction	Bark ground into powder, then decocted	Northeast India	Shankar, Sharma, and Deb 2012
Coutarea hexandra (Jacq.) Schum.	Bark	Decoction	Decoct one handful of grated trunk bark in 2 liters of water until volume reduced to 1 liter Dosage regimen: half cup TDS until symptoms resolve	Bolivia	Muñoz, Sauvain, Bourdy, Callapa, Bergeron et al. 2000
Gardenia gummifera Linn.	Root, leaf	Infusion	50–70 ml BD	India	Poonam and Singh 2009
Mitragyna inermis	Leaf	Maceration	About 300 ml of filtrate is taken TDS for 7 days	Tiv, Nigeria	Tor-anyiin, Sha'ato, and Oluma 2003
Morinda lucida Benth.	Leaf, stem	Decoction	Decoction of leaves taken BD; if condition is serious, decoction of stem bark is taken	Ghana	Asase, Hesse, and Simmonds 2012
	Leaf	Decoction	Consume as required until recovery	West Ghana	Asase, Akwetey, and Achel 2010
	Leaf	Decoction	Decocted with fresh leaves of *Crossopteryx febrifuga*, *Newbouldia laevis*, and about 300 ml is taken QDS for 5 days	Tiv, Nigeria	Tor-anyiin, Sha'ato, and Oluma 2003
Nauclea latifolia Sm.	Root	Decoction	Consume as required until recovery	West Ghana	Asase, Akwetey, and Achel 2010

(Continued)

Table A.2 (Continued) Medicinal Plants Used for Malaria from Ethnobotanical Surveys That Provided Details on Preparation

Species	Part Used	Preparation	Remark	Country	Reference
Otiophora pauciflora Baker	Leaf	Maceration	Handful of crushed leaves mixed with 250 ml water; 1/2 glass OD for 2 days	Congo	Chifundera 2001
Rothmannia hispido K. Schum.	Leaf, root	Decoction	Decocted with seven *C. annuum*; drunk or used as enema	Ghana	Asase, Hesse, and Simmonds 2012
Sarcocephalus latifolius	Leaf	Juice	Fresh leaves are squeezed with a little water between palms, about 250 ml of exudates taken BD until resolution of symptoms	Tiv, Nigeria	Tor-anyiin, Sha'ato, and Oluma 2003
			Rutaceae		
Aegle mamelos (L.) Correa ex. Roxb	Bark	Juice	Juice taken with honey PO once	Assam, Northeast India	Namsa, Mandal, and Tangjang 2011
Citrus sp.	Root	Decoction	100 g placed in 500 ml of water, decocted for 15 min and left to cool	French Guiana	Bertani et al. 2005
Citrus aurantiifolia L.	Leaf, fruit	Decoction	Decoct either fruits or leaves with leaves of *A. indica* and take cupfuls TDS until recovery	West Ghana	Asase, Akwetey, and Achel 2010
Fagaropsis angolensis (Engl.) Del.	Leaf	Decoction	1 cup TDS for 3–4 days	Msambweni, Kenya	Nguta et al. 2010a
Monniera trifolia Rich.	Leaf	Decoction	300 g fresh leaves decocted for 10 min and left to cool	French Guiana	Bertani et al. 2005
Teclea simplicifolia (Eng) Verdoon	Root	Decoction	1/4 cup TDS for 2–3 days	Msambweni, Kenya	Nguta et al. 2010a
Zanthoxylum chalybeum (Eng) Engl.	Root bark	Decoction	1/2 glass TDS for 3–4 days	Msambweni, Kenya	Nguta et al. 2010a
Zanthoxylum hamiltonianum Wall.	Root	Decoction	Decoction in water taken OD	Assam, Northeast India	Namsa, Mandal, and Tangjang 2011
Zanthoxylum rhoifolium Lam.	Bark	Decoction	400 g inner bark decocted in 1.5 liters of water until volume reduced by half, and left to cool	French Guiana	Bertani et al. 2005

(Continued)

Table A.2 (Continued) Medicinal Plants Used for Malaria from Ethnobotanical Surveys That Provided Details on Preparation

Species	Part Used	Preparation	Remark	Country	Reference
			Salvadoraceae		
Salvadora persica L.	Leaf	Decoction	Leaves powdered, boiled, and consumed after addition of sugar	Ethiopia	Mesfin et al. 2012
			Sapindaceae		
Deinbollia pinnata Schum. & Thonn.	Leaf	Decoction	Consume as required until recovery	West Ghana	Asase, Akwetey, and Achel 2010
			Saxifragaceae		
Hydrangea macrophylla (Thunb.) Ser.	Leaf, root	Not stated	Said to be more potent than quinine	Northeast India	Shankar, Sharma, and Deb 2012
			Scrophulariaceae		
Brucea antidysenterica	Leaf	Decoction	Taken for 3–5 days, dosed by cups	Ethiopia	Suleman et al. 2009
Picrorhiza kurrooa Benth.	Root	Not stated	Pounded in water and taken	Northeast India	Shankar, Sharma, and Deb 2012
			Simaroubaceae		
Harrisonia abyssinica Oliv.	Root bark, leaf	Decoction	1 cup TDS for 2–3 days	Msambweni, Kenya	Nguta et al. 2010a
Picrasma javanica Bl.	Bark	Infusion	Inner coat of bark infused and taken PO in lieu of quinine at 10 ml BD	Northeast India	Shankar, Sharma, and Deb 2012
Picrolemma pseudocoffea Ducke (used both as prevention and cure in French Guiana)	Leaf	Infusion	20 g fresh leaves infused for 10 min in 1 liter of water and left to cool	French Guiana	Bertani et al. 2005

(*Continued*)

Table A.2 (Continued) Medicinal Plants Used for Malaria from Ethnobotanical Surveys That Provided Details on Preparation

Species	Part Used	Preparation	Remark	Country	Reference
Quassia amara L. (used both as prevention and cure in French Guiana)	Leaf	Decoction	10 fresh leaves (20 g) decocted for 10 min in 1 liter of water and left to cool	French Guiana	Bertani et al. 2005
	Stem	Decoction	100 g cut into smaller pieces and decocted for 15 min in 800 ml of water and left to cool	French Guiana	Bertani et al. 2005
	Leaf	Decoction	12 fresh leaves decocted with 3 fresh pieces (4 × 10 cm) of lemon root, 10 fresh roots (10 cm) of *E. precatoria* and 3 dry yellow leaves of *C. papaya* in 6 liters of water for 30 min and left to cool	French Guiana	Bertani et al. 2005
	Leaf	Decoction	8 fresh leaves decocted with 20 g fresh leaves of *P. pseudocoffea*, 40 g bark of *G. laevis* in 1.5 liters of water for 15 min and left to cool	French Guiana	Bertani et al. 2005
			Solanaceae		
Capsicum annum L.	Fruit	Eaten	Swallow 4 fruits 3 times daily for prevention	Zimbabwe, Africa	Ngarivhume et al. 2015
Datura metel L.	Seed, leaf, root	Not stated	Taken in malarial fever with catarrhal and cerebral involvement	Northeast India	Shankar, Sharma, and Deb 2012
Nicotiana tabacum L.	Leaf	Maceration	Leaves crushed in water, sometimes mixed with chicha; remedy drunk and then vomited	Amazonian Ecuador	Giovannini 2015
Solanum sp.	Root, root bark	Decoction	Fresh root crushed and likely boiled	Ethiopia	Mesfin et al. 2012
Solanum incanum	Root, leaf	Decoction	1 cup TDS for 3–4 days	Msambweni, Kenya	Nguta et al. 2010a
Solanum syzymbrifolium L.	Fruit	Infusion	Drunk or applied rectally OD	Congo	Chifundera 2001

(*Continued*)

Table A.2 (Continued) Medicinal Plants Used for Malaria from Ethnobotanical Surveys That Provided Details on Preparation

Species	Part Used	Preparation	Remark	Country	Reference
Solanum torvum Sw.	Fruit	Not stated	Burnt fruits taken	Northeast India	Shankar, Sharma, and Deb 2012
	Fruit, leaf	Decoction	Decoct fruits with leaves and consume before food	West Ghana	Asase, Akwetey, and Achel 2010
Withania somnifera (L.) Dunal	Whole plant	Decoction	Decoction taken TDS with honey	Assam, Northeast India	Namsa, Mandal, and Tangjang 2011
	Root	Decoction	Dried roots grounded, boiled, and consumed after addition of goat/camel milk	Ethiopia	Mesfin et al. 2012
			Tiliaceae		
Grewia hexamita Burret.	Root, leaf	Decoction	1 cup TDS for 3–4 days	Msambweni, Kenya	Nguta et al. 2010a
			Thymelaeaceae		
Aquilaria agallocha Roset.	Root, stem	Infusion, decoction	Thrice daily	India	Kichu et al. 2015
			Umbelliferae		
Centella asiatica L.	Whole plant	Decoction	PO TDS with honey	Assam, Northeast India	Namsa, Mandal, and Tangjang 2011
			Verbenaceae		
Clerodendrum infortunatum Gaertn.	Leaf	Decoction	OD	Assam, Northeast India	Namsa, Mandal, and Tangjang 2011

(Continued)

Table A.2 (Continued) Medicinal Plants Used for Malaria from Ethnobotanical Surveys That Provided Details on Preparation

Species	Part Used	Preparation	Remark	Country	Reference
Lantana camara L.	Bark	Decoction	Decoction taken OD; also used as insect repellent	Assam, Northeast India	Namsa, Mandal, and Tangjang 2011
	Whole plant	Decoction	30–50 ml BD	India	Poonam and Singh 2009
	Leaf	Decoction	1 cup TDS for 3–4 days	Msambweni, Kenya	Nguta et al. 2010a
Lippia multiflora Moldenke.	Leaf	Infusion	Boil dried leaves for 5 min, 1 cup TDS until recovery	West Ghana	Asase, Akwetey, and Achel 2010
Tectona grandis Lour.	Leaf	Decoction	Dried leaves decocted; consume as required until recovery	West Ghana	Asase, Akwetey, and Achel 2010
Verbena litoralis Kunth	Leaf	Not stated	Leaves crushed and mixed with warm water	Peruvian Amazon	Sanz-Biset et al. 2009
Vitex peduncularis Wall.	Bark, leaf, stem	Decoction, infusion	Bark crushed and boiled, steam inhaled for fever; infusion of leaves, root bark, or young stem bark for malaria, taken BD	Assam and Northeast India	Namsa, Mandal, and Tangjang 2011; Shankar, Sharma, and Deb 2012
			Vitaceae		
Cissus rotundifolia (Forssk.) Vahl	Root, root bark	Decoction	Fresh or dried root crushed or pounded, then boiled and consumed	Ethiopia	Mesfin et al. 2012
			Xanthorrhoeaceae		
Aloe sp.	Leaf	Juice	Fresh leaves squeezed and diluted with water and consumed; syrup prepared from dried leaves of the plant and that of *Asparagus africanus* and *Senna italica* is also consumed	Ethiopia	Mesfin et al. 2012
Aloe deserti	Leaf	Not stated	Leaves are dissolved in water for some time, 1/4 glass TDS	Msambweni, Kenya	Nguta et al. 2010a

(Continued)

Table A.2 (Continued) Medicinal Plants Used for Malaria from Ethnobotanical Surveys That Provided Details on Preparation

Species	Part Used	Preparation	Remark	Country	Reference
Aloe macrosiphon	Leaf	Not stated	Leaves are dissolved in water for some time, 1/4 glass TDS	Msambweni, Kenya	Nguta et al. 2010a
Aloe vera	Leaf	Not stated	Leaves are dissolved in water for some time, 1/4 glass TDS	Msambweni, Kenya	Nguta et al. 2010a
			Zingiberaceae		
Aframomum melegueta K. Schum	Leaf, stem	Not stated	Leaves and stem bark of this herb and those of *Erythrophleum ivorense, Anopyxis klaineana, Cocos nucifera, Turraeanthus africanus, Alstonia boonei*, and root of *Annona muricata* and *Thaumatococcus* sp. are pounded; mixture is applied on the body	Ghana	Asase, Hesse, and Simmonds 2012
Zingiber officinale Roscoe	Fruit	Not stated	Ground with fruits of *Citrus aurantiifolia* and *C. sinensis*, taken BD	Ghana	Asase, Hesse, and Simmonds 2012
	Rhizome	Decoction	About 300 ml of aqueous decoction of dry, powdered rhizome taken TDS for 6 days	Tiv, Nigeria	Tor-anyiin, Sha'ato, and Oluma 2003

Note: BD, twice daily; OD, once daily; PO, per os (orally); QDS, four times daily; TDS, thrice daily.

Table A.3 Medicinal Plant Species Listed by Family

Family	Plant Name
Acanthaceae	*Acanthus polystachyus*
	Adhatoda schimperiana
	Adhatoda zeylanica
	Andrographis paniculata
	Barleria acanthoides
	Brillantaisia patula
	Crabbea velutina
	Dicliptera paniculata
	Dyschoriste perrottetii
	Elytraria marginata
	Eremomastax speciosa
	Gendarussa vulgaris
	Justicia betonica
	Justicia flava
	Justicia insularis
	Justicia schimperiana
	Lepidagathis anobrya
	Monechma subsessile
	Phlogacanthus thyrsiflorus
	Thomandersia hensii
	Thunbergia alata
	Trichanthera gigantea
Acoraceae	*Acorus gramineus*
Agavaceae	*Dracaena reflexa*
	Sansevieria trifasciata
Aizoaceae	*Glinus oppositifolius*
	Limeum pterocarpum
	Mollugo nudicaulis
Amaranthaceae	*Achyranthe bidentata*
	Alternanthera brasiliana
	Alternanthera pungens
	Alternanthera sessilis
	Alternanthera tenella
	Amaranthus hybridus
	Cyathula cylindrica
	Cyathula prostrata
	Iresine calea
	Iresine diffusa
	Pupalia lappacea
	Pupalia micrantha
	Sericocomposis hildebrandtii

(*Continued*)

Table A.3 (Continued) Medicinal Plant Species Listed by Family

Family	Plant Name
Amaryllidaceae	*Allium cepa*
	Allium sativum
	Brunsvigia littoralis
	Crinum zeylanicum
	Hippeastrum puniceum
	Hymenocallis sp.
	Philodendron sp.
Anacardiaceae	*Amaranthus viridis*
	Anacardium occidentale
	Haematostaphis barteri
	Heeria insignis
	Juliania adstringens
	Lannea acida
	Lannea microcarpa
	Lannea schweinfurthii
	Mangifera caesia
	Mangifera indica
	Mangifera minor
	Ozoroa insignis
	Pseudoprotorhus longifolius
	Pseudospondias microcarpa
	Rhus natalensis
	Rhus semialata
	Rhus taratana
	Schinus molle
	Sclerocarya caffra
	Sorindeia juglandifolia
	Spondias mombin
	Spondias pinnata
	Spondias purpurea
	Tricoscypha ferruginea
Ancistrocladaceae	*Ancistrocladus extensus*
Anisophylleaceae	*Anisophyllea laurina*
Annonaceae	*Annickia affinis*
	Annickia chlorantha
	Annona muricata
	Annona reticulata
	Annona senegalensis
	Annona squamosa
	Artabotrys odoratissimus
	Cananga odorata
	Canna odorata

(Continued)

Table A.3 (Continued) Medicinal Plant Species Listed by Family

Family	Plant Name
	Cleistopholis patens
	Cremastosperma cauliflorum
	Enantia clorantha
	Greenwayodendron sp.
	Guatteria megalophylla
	Guatteria sp.
	Hexalobus crispiflorus
	Isolona hexaloba
	Monanthotaxis sp.
	Monodora myristica
	Pachypodanthium confine
	Polyalthia suaveolens
	Polyceratocarpus sp.
	Pseudoxandra uspidata
	Unonopsis floribunda
	Unonopsis spectabilis
	Uraria chamae
	Uvaria acuminata
	Uvaria afzelii
	Uvaria ambongoensis
	Uvaria chamae
	Uvaria leptocladon
	Uvaria sp.
	Uvariodendron spp.
	Xylopia spp.
	Xylopia aethiopica
	Xylopia parviflora
	Xylopia phloiodora
	Xylopia vielana
	Xymolox monosperma
Apiaceae	*Anethum graveolens*
	Bupleurum chinense
	Bupleurum falcatum
	Bupleurum longicaule
	Bupleurum scorzoneraefolium
	Daucus carota
	Eryngium amethystinum
	Foeniculum vulgare
	Heteromorpha trifoliata
	Peucedanum japonicum
	Sanicula marilandica
Apocynaceae	*Acokanthera schimperi*

(*Continued*)

Table A.3 (Continued) Medicinal Plant Species Listed by Family

Family	Plant Name
	Aganosma marginata
	Alstonia boonei
	Alstonia congensis
	Alstonia constricta
	Alstonia macrophylla
	Alstonia scholaris
	Alyxia spp.
	Apocynum cannabinum
	Aspidosperma discolor
	Aspidosperma excelsum
	Aspidosperma illustris
	Aspidosperma quebracho-blanco
	Aspidosperma rigidum
	Baissea multiflora
	Carissa carandas
	Carissa edulis
	Catharanthus roseus
	Cryptolepis sanguinolenta
	Diplorhynchus condylocarpon
	Funtumia africana
	Funtumia elastica
	Geissospermum argenteum
	Geissospermum laeve
	Geissospermum sericeum
	Himatanthus articulatus
	Himatanthus sucuuba
	Holarrhena floribunda
	Holarrhena pubescens
	Landolphia heudelotii
	Landolphia lanceolata
	Landolphia sp.
	Laudolphia buchananii
	Ludolphia buchananii
	Melodinus monogynus
	Motandra guineensis
	Ochrosia elliptica
	Picralima nitida
	Plumeria rubra
	Psiadia punctulata
	Rauvolfia duckei
	Rauvolfia macrophylla
	Rauvolfia obscura

(*Continued*)

Table A.3 (Continued) Medicinal Plant Species Listed by Family

Family	Plant Name
	Rauvolfia samarensis
	Rauvolfia serpentina
	Rauvolfia tetraphylla
	Rauvolfia vomitoria
	Rauwolfia sp.
	Saba senegalensis
	Strophanthus hispidus
	Tabernaemontana crassa
	Tabernaemontana elegans
	Tabernaemontana pachysiphon
	Tabernaemontana penduliflora
	Tabernaemontana sp.
	Thevetia peruviana
	Trachelospermum lucidum
	Voacanga africana
	Willughbeia sp.
	Wrightia dubia
Aquifoliaceae	*Ilex aquifolium*
	Ilex guayusa
Araceae	*Acorus calamus*
	Amorphophallus angolensis
	Anadendrum montanum
	Arisaema sp.
	Dieffenbachia sp.
	Philodendron cf. *uleanum*
	Pinellia tuberifera
	Pistia stratiotes
Araliaceae	*Cussonia arborea*
	Panax ginseng
	Polyscias filicifolia
Arecaceae	*Areca catechu*
	Astrocaryum chonta
	Bactris gasipaes
	Borassus aethiopum
	Borassus flabellifer
	Calamus salicifolius
Aristolochiaceae	*Aristolochia acuminata*
	Aristolochia albida
	Aristolochia bracteata
	Aristolochia brasiliensis
	Aristolochia brevipes
	Aristolochia elegans

(*Continued*)

Table A.3 (Continued) Medicinal Plant Species Listed by Family

Family	Plant Name
	Aristolochia heppii
	Aristolochia indica
	Aristolochia serpentaria
	Aristolochia spp.
	Aristolochia stahelii
	Aristolochia trilobata
Asclepiadaceae	*Calotropis gigantea*
	Calotropis procera
	Caralluma dalzielii
	Curroria volubilis
	Fischeria cf. *martiana*
	Hemidesmus indicus
	Leptadenia hastata
	Leptadenia madagascariensis
	Oxystelma secamone
	Parquetina nigrescens
	Pergularia daemia
	Periploca linearifolia
	Secamone africana
	Vincetoxicum atratum
Asparagaceae	*Asparagus africanus*
	Cordyline fruticosa
Aspleniaceae	*Asplenium adiantoides*
	Ceterach officinarum
Avicenniaceae	*Avicennia basilicum*
	Avicennia marina
Balanitaceae	*Balanites aegyptiaca*
	Balanites rotundifolia
Balsaminaceae	*Impatiens angustifolia*
Bambusaceae	*Cephalostachyum* sp.
Basellaceae	*Basella alba*
Begoniaceae	*Begonia inflata*
	Begonia parviflora
Berberidaceae	*Berberis aristata*
	Berberis goudotii
	Berberis lutea
	Berberis rigidifolia
	Berberis ruscifolia
	Berberis sp.
	Berberis vulgaris
	Mahonia aquifolia
	Mahonia bealei

(*Continued*)

Table A.3 (Continued) Medicinal Plant Species Listed by Family

Family	Plant Name
	Podophyllum versipelle
Betulaceae	*Alnus glutinosa*
	Alnus serrulata
Bignoniaceae	*Cresentia cujete*
	Doxantha unguis-cati
	Fernandoa sp.
	Jacaranda copaia
	Jacaranda spp.
	Kigelia africana
	Kigelianthe madagascariensis
	Mansoa alliacea
	Markhamia gellatiana
	Markhamia lutea
	Markhamia sessilis
	Markhamia tomentosa
	Newbouldia laevis
	Oroxylum indicum
	Pleonotoma melioides
	Spathodea campanulata
	Stereospermum suaveolens
	Sterospermum kunthiamum
	Tabebuia aurea
	Tabebuia impetiginosa
	Tabebuia ochracea
	Tabebuia rosea
Bixaceae	*Bixa orellana*
	Cochlospermum planchonii
Bombacaceae	*Adansonia digitata*
	Bombax buonopozense
	Bombax costatum
	Ceiba pentandra
	Cynoglossum glochidion
	Durio oxleyianus
	Pseudobombax munguba
Boraginaceae	*Chretia cymosa*
	Cordia curassavica
	Cordia riparia
	Cordia sinensis
	Ehretia buxifolia
	Ehretia cymosa
	Heliotropium arborecens
	Heliotropium europaeum

(Continued)

Table A.3 (Continued) Medicinal Plant Species Listed by Family

Family	Plant Name
	Heliotropium indicum
	Heliotropium peruvianum
Brassicaceae	*Brassica nigra*
	Chartoloma chartacea
	Lepidium sativum
Bromeliaceae	*Ananas comosus*
Burseraceae	*Boswellia dalzielii*
	Bursera simaruba
	Canarium schweinfurthii
	Commiphora africana
	Commiphora schimperi
	Dacryodes edulis
Buxaceae	*Buxus hyrcana*
	Buxus sempervirens
Campanulaceae	*Lobelia chinensis*
	Lobelia sp.
Canellaceae	*Cinnamosma fragrans*
	Warburgia ugandensis
Cannabaceae	*Humulus japonicus*
Cannaceae	*Canna bidentata*
	Canna indica
Capparaceae	*Bocia senegalensis*
	Boscia coriacea
	Buchholzia coriacea
	Buchholzia macrophylla
	Cadaba farinosa
	Capparis decidua
	Capparis sepiaria
	Capparis sp.
	Cleome arborea
	Cleome ciliata
	Cleome rutidosperma
	Crataeva adansonii
	Crateva adansonii
	Crateva nurvala
	Crateva religiosa
	Gynandropsis gynandra
	Maerua oblongifolia
Caprifoliaceae	*Sambucus australis*
	Viburnum nudum
	Viburnum obovatum
Cardiopteridaceae	*Citronella* cf. *melliodora or incanum*

(Continued)

Table A.3 (Continued) Medicinal Plant Species Listed by Family

Family	Plant Name
Caricaceae	*Carica papaya*
	Polycarpa glabrifolia
Caryophyllaceae	*Dianthus anatolicus*
	Drymaria cordata
	Silene flos-cuculi
Celastraceae	*Cheiloclinium cognatum*
	Euonymus thunbergianus
	Euonynmus europaeus
	Hippocratea africana
	Hua gabonii
	Loeseneriella africana
	Maytenus acuminata
	Maytenus arbutifolia
	Maytenus heterophylla
	Maytenus krukovii
	Maytenus macrocarpa
	Maytenus putterlickioides
	Maytenus senagalensis
	Maytenus spp.
	Maytenus undata
	Moya davana-dependens
Chenopodiaceae	*Chenopodium ambrosioides*
	Chenopodium opulifolium
	Halothamnus somalensis
	Salsola somalensis
Chrysobalanaceae	*Licania hebantha*
	Licania parviflora
	Parinari benna
	Parinari excelsa
Clusiaceae	*Allanblackia monticola*
	Garcinia kola
	Garcinia luzonensis
	Mammea africana
	Psorospermum senegalense
Cochlospermaceae	*Cochlospermum tinctorium*
Combretaceae	*Anogeissus leiocarpa*
	Calycopteris floribunda
	Combretum adenogonium
	Combretum decandrum
	Combretum ghasalense
	Combretum glutinosum
	Combretum illairii

(*Continued*)

Table A.3 (Continued) Medicinal Plant Species Listed by Family

Family	Plant Name
	Combretum latiatum
	Combretum lecardii
	Combretum micranthum
	Combretum molle
	Combretum nigricans
	Combretum padoides
	Combretum paniculatum
	Combretum platysterum
	Combretum raimbaulti
	Combretum sp.
	Combretum spinesis
	Guiera senegalensis
	Pteleopsis suberosa
	Terminalia albida
	Terminalia avicennoides
	Terminalia boivinii
	Terminalia brevipes
	Terminalia catappa
	Terminalia chebula
	Terminalia glaucescens
	Terminalia ivorensis
	Terminalia latifolia
	Terminalia macroptera
	Terminalia superba
Commelinaceae	*Commelina benghalensis*
	Commelina communis
	Commelina erecta
	Dichorisandra hexandra
	Palisota hirsuta
	Pollia condensata
Compositae	*Acanthospermum australe*
	Acanthospermum hispidum
	Acanthospermum hispidus
	Ageratum conyzoides
	Ageratum echioides
	Amberboa ramosa
	Ambrosia artemisiifolia
	Argyrovernonia martii
	Artemisia absinthium
	Artemisia afra
	Artemisia annua
	Artemisia apiacea

(*Continued*)

Table A.3 (Continued) Medicinal Plant Species Listed by Family

Family	Plant Name
	Artemisia argyi
	Artemisia capillaris
	Artemisia nilagirica
	Artemisia scoparia
	Artemisia spp.
	Artemisia tridentata
	Artemisia vulgaris
	Aspilia africana
	Aster amellus
	Aster trinervius
	Atractylis sinensis
	Ayapana lanceolata
	Ayapana triplinervis
	Baccharis genistelloides
	Baccharis lanceolata
	Baccharis trimera
	Bidens bipinata
	Bidens cynapiifolia
	Bidens grantii
	Bidens pilosa
	Blumea balsamifera
	Bothriocline longipes
	Brachylaena huillensis
	Brachylaena ramiflora
	Brickellia cavanillesi
	Calea berteriana
	Calea zacatechichi
	Centaurea calcitrapa
	Centaurea solstitialis
	Centipeda minima
	Centipeda minuta
	Centratherum punctatum
	Chromolaena odorata
	Chuquiraga jussieui
	Cichorium intybus
	Conyza aegyptiaca
	Conyza bonariensis
	Conyza lyrata
	Conyza sp.
	Conyza sumatrensis
	Crassocephalum biafrae
	Cynara cardunculus

(Continued)

Table A.3 (Continued) Medicinal Plant Species Listed by Family

Family	Plant Name
	Dicoma tomentosa
	Echinops hoehnelii
	Eclipta prostata
	Emilia coccinea
	Enanthia chlorantha
	Erlangea cordifolia
	Erlangea tomentosa
	Erythrocephalum zambesianum
	Eupatorium perfoliatum
	Gnaphalium leuteo-album
	Guizotia scabra
	Gutenbergia cordifolia
	Gutierrezia sarothrae
	Gynura
	Helianthus annuus
	Helichrysum faradifani
	Helichrysum sp.
	Inula perrieri
	Inula viscosa
	Isocarpha divaricata
	Isocarpha microcephala
	Laggera alata
	Launea cornuta
	Macroclinidium verticillatum
	Matricaria chamomilla
	Matricaria recutita
	Melanthera scandens
	Microglossa angolensis
	Microglossa pyrifolia
	Microglossa sp.
	Mikania cordifolia
	Mikania guaco
	Mikania hookeriana
	Mikania micrantha
	Mikania officinalis
	Mikania scandens
	Neurolaena lobata
	Parthenium hysterophorus
	Pectis capillaris
	Picris repens
	Pluchea ovalis
	Polydora serratuloides

(Continued)

Table A.3 (Continued) Medicinal Plant Species Listed by Family

Family	Plant Name
	Polymnia uvedalia
	Porophyllum gracile
	Porophyllum seemannii
	Schkuhria pinnata
	Senecio hoffmannii
	Senecio ompricaefolius
	Soncoio salignus
	Senecio scandens
	Senecio syringitolius
	Senencio nandensis
	Senencio petitianus
	Siegesbeckia octoaristata
	Siegesbeckia orientalis
	Solanecio manii
	Solidago virgoaurea
	Sonchus oleraceus
	Sphaeranthus suaveolens
	Spilanthes oleraceae
	Stenocline inuloides
	Stevia eupatoria
	Struchium sparganophorum
	Synedrela nodiflora
	Tagetes erecta
	Tagetes filifolia
	Tagetes florida
	Tagetes lucida
	Tagetes patula
	Tanacetum parthenium
	Taraxacum officinale
	Taraxacum panalptnum
	Tessaria integrifolia
	Tithonia diversifolia
	Tithonia rotundifolia
	Tricholepis furcata
	Tridax procumbens
	Triplotaxis stellulifera
	Tussilago farfara
	Vernonia adoensis
	Vernonia ampandrandavensis
	Vernonia amygdalina
	Vernonia auriculifera
	Vernonia brachycalyx

(Continued)

Table A.3 (Continued) Medicinal Plant Species Listed by Family

Family	Plant Name
	Vernonia brachiata
	Vernonia brazzavillensis
	Vernonia campanea
	Vernonia chapelieri
	Vernonia cinerea
	Vernonia colorata
	Vernonia condensata
	Vernonia conferta
	Vernonia jugalis
	Vernonia lasiopus
	Vernonia pectoralis
	Vernonia sp.
	Vernonia trichodesma
	Wedelia biflora
	Wedelia mossambicensis
	Xanthium spinosum
	Xanthium strumarium
	Zinnia peruviana
Connaraceae	*Cnestis palala*
	Rourea coccinea
Convolvulaceae	*Astripomoea malvacea*
	Hewittia malabarica
	Ipomoea batatas
	Ipomoea cairica
	Ipomoea hederifolia
	Ipomoea pes-caprae
	Ipomoea spathulala
Cornaceae	*Cornus florida*
	Cornus officinalis
Costaceae	*Costus spectabilis*
Crassulaceae	*Kalanchoe densiflora*
Cruciferae	*Capsella bursa-pastoris*
	Coronopus didymus
	Sinapsis arvensis
Cucurbitaceae	*Citrullus lanatus*
	Cogniauxia podolaena
	Cucumis ficifolius
	Ecballium elaterium
	Gerranthus lobatus
	Gymnopetalum cochinchinensis
	Luffa acutangula
	Momordica balsamina

(*Continued*)

Table A.3 (Continued) Medicinal Plant Species Listed by Family

Family	Plant Name
	Momordica charantia
	Momordica cochinchinensis
	Momordica condensata
	Momordica foetida
	Oreosyce africana
	Trichosanthes cucumerina
	Zehneria scabra
Cupressaceae	*Thujia occidentalis*
Cyclanthaceae	*Carludovica palmata*
Cyperaceae	*Cyperus articulatus*
	Cyperus esculentus
	Cyperus papyrus
	Scleria barteri
	Scleria hirtella
Davalliaceae	*Nephrolepis biserrata*
Dichapetalaceae	*Dichapetalum madagascariense*
Dilleniaceae	*Dillenia indica*
	Doliocarpus dentatus
	Tetracera alnifolia
	Tetracera spp.
	Tetracera potatoria
	Tetracera volubilis
Dioscoreaceae	*Dioscorea dumetorum*
Dipterocarpaceae	*Monotes kerstingii*
Dracaenaceae	*Sansevieria liberica*
Dryopteridaceae	*Arcypteris difformis*
Ebenaceae	*Diospyros ebenaster*
	Diospyros malabarica
	Diospyros mespiliformis
	Diospyros revoluta
	Diospyros zombensis
	Euclea divinorum
	Euclea natalensis
Elaeocarpaceae	*Elaeocarpus kontumensis*
Elatinaceae	*Bergia suffruticosa*
Ephedraceae	*Ephedra vulgaris*
Ericaceae	*Rhododendron molle*
Erythroxylaceae	*Erythroxylum spp.*
	Erythroxylum tortuosum
Euphorbiaceae	*Acalypha indica*
	Alchornea cordifolia
	Alchornea triplinervia

(*Continued*)

Table A.3 (Continued) Medicinal Plant Species Listed by Family

Family	Plant Name
	Alchromanes difformis
	Amanoa sp.
	Antidesma laciniatum
	Antidesma montanum
	Breynia vitis-idaea
	Bridelia ferruginea
	Bridelia micrantha
	Clutia abyssinica
	Croton spp.
	Croton cajucara
	Croton eleuteria
	Croton goudoti
	Croton gratissimus
	Croton guatemalensis
	Croton humilis
	Croton lechleri
	Croton leptostachyus
	Croton macrostachis
	Croton macrostachyus
	Croton megalocarpus
	Croton niveus
	Croton reflexifolius
	Croton tiglium
	Croton xalapensis
	Croton zambesicus
	Denteromallotus acuminatus
	Drypetes natalensis
	Erythrococca anomala
	Euphorbia abyssinica
	Euphorbia calyculata
	Euphorbia helioscopia
	Euphorbia heterophylla
	Euphorbia hirta
	Euphorbia poinsonii
	Excoecaria grahamii
	Flueggea microcarpa
	Flueggea virosa
	Glochidion puberum
	Glochidion sp.
	Homonoia riparia
	Hura crepitans
	Hymenocardia acida

(*Continued*)

Table A.3 (Continued) Medicinal Plant Species Listed by Family

Family	Plant Name
	Hymenocardia ulmoides
	Jatropha curcas
	Jatropha gossypifolia
	Jatropha podagrica
	Macaranga populifolia
	Mallotus oppositofolius
	Mallotus paniculatus
	Manihot esculenta
	Manihot utilisma
	Manniophyton fulvum
	Mareya micrantha
	Margaritaria discoidea
	Millettia griffoniana
	Neoboutonia macrocalyx
	Neoboutonia velutina
	Phyllanthus amarus
	Phyllanthus corcovadensis
	Phyllanthus emblica
	Phyllanthus muellerianus
	Phyllanthus niruri
	Phyllantus reticulatus
	Phyllanthus sp.
	Phyllanthus stipulatus
	Ricinus communis
	Tetrorchidium didymostemon
	Triadica cochinchinensis
	Uapaca paludosa
	Uapaca togoensis
Fabaceae	*Abarema laeta*
	Abrus precatorius
	Acacia catechu
	Acacia concinna
	Acacia dudgeoni
	Acacia farnesiana
	Acacia foetida
	Acacia hockii
	Acacia macrostachya
	Acacia mellifera
	Acacia nilotica
	Acacia pennata
	Acacia robusta
	Acacia senegal

(*Continued*)

Table A.3 (Continued) Medicinal Plant Species Listed by Family

Family	Plant Name
	Acacia seyal
	Acacia sieberiana
	Acacia sp.
	Acacia tortilis
	Acosmium panamense
	Acosmium sp.
	Adenanthera microsperma
	Afzelia afranica
	Albizia amara
	Albizia anthelmintica
	Albizia chevalieri
	Albizia coriaria
	Albizia ferruginea
	Albizia grandibracteata
	Albizia gummifera
	Albizia zygia
	Albizzia lebbek
	Andira inermis
	Andira surinamensis
	Anthonotha macrophylla
	Arachis hypogaea
	Astragalus hoantchy
	Bauhinia guianensis
	Bauhinia rufescens
	Bauhinia sp.
	Bauhinia ungulata
	Bowdichia virgilioides
	Caesalpinia bonduc
	Caesalpinia coriaria
	Caesalpinia ferrea
	Caesalpinia pulcherrima
	Caesalpinia sepiara
	Caesalpinia volkensii
	Cajanus cajan
	Calliandra anomala
	Calliandra clavellina
	Calliandra houstoniana
	Campsiandra angustifolia
	Campsiandra comosa
	Campsiandra comosa var. *laurifolia*
	Capsiandra angustifolia
	Cassia aff. *abbreviata*

(*Continued*)

Table A.3 (Continued) Medicinal Plant Species Listed by Family

Family	Plant Name
	Cassia abbreviata
	Cassia arereh
	Cassia fistula
	Cassia mimosoides
	Cassia nigricans
	Cassia sieberiana
	Chamaecrista rotundifolia
	Christia vespertilionis
	Copaifera pauperi
	Crotalaria naragutensis
	Crotolaria occulta
	Crotolaria spinosa
	Cylicodiscus gabunensis
	Cynometra sphaerocarpa
	Dalbergia pinnata
	Daniellia ogea
	Daniellia oliveri
	Delonix elata
	Delonix regia
	Desmodium gangeticum
	Desmodium hirtum
	Desmodium incanum
	Desmodium ramosissimum
	Detarium microcarpum
	Detarium senegalense
	Dialium guineense
	Dichrostachys cinerea
	Dioclea sp.
	Disthmonanthus benthamianus
	Dolichos schweinfurthii
	Elephantorrhiza goetzei
	Entada abyssinica
	Entada africana
	Entada phaseoloides
	Eriosema stanerianum
	Erythrina abyssinica
	Erythrina fusca
	Erythrina sacleuxii
	Erythrina senegalensis
	Erythrina sigmoidea
	Erythrophleum suaveolens
	Erythrynia indica

(*Continued*)

Table A.3 (Continued) Medicinal Plant Species Listed by Family

Family	Plant Name
	Gliricidia sepium
	Glycine max
	Glycyrrhiza glabra
	Guibourtia tessmannii
	Hymenaea courbaril
	Indigofera arrecta
	Indigofera articulata
	Indigofera coerulea
	Indigofera congesta
	Indigofera emarginella
	Indigofera erecta
	Indigofera lupatana
	Indigofera nigritana
	Indigofera pulchra
	Indigofera sp.
	Indigofera spicata
	Indigofera suffruticosa
	Inga acreana
	Inga sessilis
	Inga setulifera
	Inga spp.
	Isoberlinia doka
	Lonchocarpus cyanescens
	Machaerium floribunda
	Macrotyloma axillare
	Mezoneuron benthamianum
	Millettia diptera
	Millettia laurentii
	Millettia versicolor
	Millettia zechiana
	Mimosa flexuosa
	Mimosa pudica
	Myroxylon balsamum
	Otholobium mexicanum
	Parkia biglobosa
	Parkinsonia aculeata
	Pericopsis elata
	Pericopsis laxiflora
	Piliostigma reticulatum
	Piliostigma thonningii
	Pithecellobium (Albizia) panurense
	Pithecellobium laetum

(Continued)

Table A.3 (Continued) Medicinal Plant Species Listed by Family

Family	Plant Name
	Pithecellobium sp.
	Pongamia pinnata
	Prosopis africana
	Pseudarthria hookeri
	Psoralea pentaphylla
	Pterocarpus erinaceus
	Pterocarpus macrocarpus
	Pterocarpus rohrii
	Pterocarpus santalinoides
	Pterocarpus soyauxii
	Rhynchosia sp.
	Rhynchosia viscosa
	Senna alata
	Senna didymobotrya
	Senna hirsuta
	Senna italica
	Senna longiracemosa
	Senna obtusifolia
	Senna occidentalis
	Senna reticulata
	Senna septemtrionalis
	Senna siamea
	Senna singueana
	Senna spp.
	Senna tora
	Sesbania pachycarpa
	Sesbania sesban
	Stylosanthes erecta
	Sweetia panamensis
	Tamarindus indica
	Tephrosia bracteolata
	Tephrosia purpurea
	Tetrapleura tetraptera
	Uraria lagopodioides
	Uraria picta
	Uraria prunellaefolia
	Vicia hirsuta
	Vigna subterranea
	Vigna unguiculata
	Xeroderris stuhlmannii
	Zygia longifolia
Fagaceae	*Fagus sylvatica*

(*Continued*)

Table A.3 (Continued) Medicinal Plant Species Listed by Family

Family	Plant Name
Flacourtiaceae	*Casearia* aff. *spruceana*
	Casearia sp.
	Flacourtia indica
	Homalium involucratum
	Homalium letestui
	Homalium sp.
	Lindackeria paludosa
	Trimeria bakeri
	Trimeria grandifolia
Garryaceae	*Garrya fremontii*
Gentianaceae	*Coutoubea racemosa*
	Coutoubea spicata
	Enicostema hyssopitolium
	Enicostema littorale
	Enicostema verticillatum
	Fagraea fragans
	Gentiana lutea
	Gentiana quinquefolia
	Irlbachia alata
	Irlbachia speciosa
	Sabatia elliottii
	Schultesia guianensis
	Schultesia lisianthoides
	Swertia angustifolia
	Swertia chirata
	Swertia chirayata
	Swertia chirayita
	Swertia nervosa
	Swertia racemosa
	Tachia guianensis
Gnetaceae	*Gnetum* spp.
	Gnetum ula
Gramineae	*Andropogon schireusis*
	Andropogon schoenanthus
	Bambusa vulgaris
	Cenchrus echinatus
	Cymbopogon citratus
	Panicum maximum
	Phragmites mauritianus
	Rottboelia exaltata
	Saccharum officinarum
Grossulariaceae	*Ribes nigrum*

(*Continued*)

Table A.3 (Continued) Medicinal Plant Species Listed by Family

Family	Plant Name
Guttiferae	*Allanblackia floribunda*
	Chrysochlamys sp.
	Harungana madagascariensis
	Mammeaamericana
	Vismia guianensis
	Vismia sessilifolia
Hamamelidaceae	*Liquidambar orientalis*
Hernandiaceae	*Hernandia voyroni*
Hydnoraceae	*Hydnora johannis*
Hydrangeaceae	*Hydrangea paniculata*
	Hydrangea strigosa
	Hydrangea umbellata
Hypericaceae	*Psorospermum febrifugum*
	Vismia guineensis
Hypoxidaceae	*Curculigo pilosa*
Icacinaceae	*Cassinopsis madagascariensis*
	Leretia cordata
	Poraqueiba sericea
	Rhaphiostylis beninensis
Iridaceae	*Belamcanda chinensis*
	Crocus sativus
	Watsonia sp.
Irvingiaceae	*Irvingia gabonensis*
	Irvingia malayana
Labiatae	*Ajuga remota*
	Anisomeles indica
	Clerodendrum myricoides
	Clerodendrum rotundifolium
	Clinopodium laevigatum
	Clinopodium taxifolium
	Cornutia sp.
	Fuerstia africana
	Gomphostemma parviflora
	Hiptis sidifolia
	Hoslundia opposita
	Hyptis crenata
	Hyptis pectinata
	Hyptis spicigera
	Hyptis suaveolens
	Iboza multiflora
	Leonotis nepetifolia
	Leonurus japonicus

(*Continued*)

Table A.3 (Continued) Medicinal Plant Species Listed by Family

Family	Plant Name
	Leucas martinicensis
	Marrubium incanum
	Marrubium vulgare
	Mentha sp.
	Mentha spicata
	Mentha sylvestris
	Mentha viridis
	Minthostachys mollis
	Ocimum americanum
	Ocimum campechianum
	Ocimum gratissimum
	Ocimum sanctum
	Ocimum spicatum
	Otostegia integrifolia
	Otostegia michauxii
	Otostegia persica
	Perilla frutescens
	Plectranthus amboinicus
	Plectranthus barbatus
	Plectranthus cf. *forskohlii*
	Plectranthus sp.
	Pogostemon glaber
	Pycnostachys eminii
	Roylea calycina
	Salvia lavanduloides
	Salvia shannonii
	Scutellaria galericulata
	Solenostemon latifolius
	Solenostemon monostachyus
	Tetradenia sp.
	Teucrium chamaedrys
	Teucrium cubense
	Vitex cannabifolia
Lacistemataceae	*Lacistema aggregatum*
Lamiaceae	*Ocimum angustifolium*
	Ocimum canum
	Ocimum lamiifolium
	Vitex madiensis
Lauraceae	*Cassytha filiformis*
	Chlorocardium rodiaei
	Cinnamomum bejolghota
	Cinnamomum burmanni

(*Continued*)

Table A.3 (Continued) Medicinal Plant Species Listed by Family

Family	Plant Name
	Cinnamomum camphora
	Cinnamomum iners
	Cinnamomum zeylanicum
	Laurus nobilis
	Lindera strychnifolia
	Litsea bubeba
	Litsea floribunda
	Litsea semecarpifolia
	Litsea sp.
	Mesona wallichiana
	Nectandra rodiaei
	Neocinnamomum hainanianum
	Ocotea usambarensis
	Persea americana
Lecythidaceae	*Barringtonia acutangula*
	Bertholletia excelsa
	Couroupita guianensis
	Grias neuberthii
	Grias peruviana
	Gustavia longifolia
	Napoleona vogelli
	Petersianthus macrocarpus
Lemnaceae	*Lemna minor*
Liliaceae	*Crinum erubescens*
	Gloriosa superba
	Narcissus sp.
	Polygonatum officinale
Linaceae	*Roucheria calophylla*
	Roucheria punctata
Loganiaceae	*Anthocleista amplexicaulus*
	Anthocleista djalonensis
	Anthocleista nobilis
	Anthocleista rhizophoroides
	Anthocleista schweinfurthii
	Anthocleista vogelii
	Buddleja asiatica
	Fagraea cochinchinensis
	Gelsemium sempervirens
	Potalia amara
	Potalia resinifera
	Spigelia marilandica
	Strychnos colubrina

(Continued)

Table A.3 (Continued) Medicinal Plant Species Listed by Family

Family	Plant Name
	Strychnos fendleri
	Strychnos henningsii
	Strychnos icaja
	Strychnos innocua
	Strychnos mostuoides
	Strychnos potatorum
	Strychnos pseudoquina
	Strychnos spinosa
	Strychnos spp.
	Strychnos wallichiana
Loranthaceae	*Loranthus ferrugineus*
	Loranthus kaempferi
	Psittacanthus cordatus
	Tacillus umbellifer
	Tapinanthus dodoneifolius
	Tapinanthus paradoxa
	Tapinanthus sessilifolius
Lycopodiaceae	*Lycopodium crasum*
	Lyopodium saururus
Lythraceae	*Cuphea carthagenensis*
	Cuphea glutinosa
	Lagerstroemia speciosa
	Lawsonia inermis
	Punica granatum
Maesaceae	*Maesa lanceolata*
Magnoliaceae	*Liriodendron tulipifera*
	Magnolia grandiflora
	Magnolia officinalis
	Magnolia virginiana
	Michelia alba
	Michelia champaca
	Talauma mexicana
Malpighiaceae	*Banisteriopsis caapi*
	Burdachia prismatocarpa
	Byrsonima crassa
	Byrsonima spp.
	Mascagnia benthamiana
	Stigmaphyllon convolvulifolium
	Stigmaphyllon sinuatum
	Tetrapterys discolor
	Tristellateia madagascariensis
Malvaceae	*Althaea rosea*

(*Continued*)

Table A.3 (Continued) Medicinal Plant Species Listed by Family

Family	Plant Name
	Briquetia spicata
	Cola acuminata
	Cola millenii
	Cola nitida
	Dombeya shupangae
	Gossypium arboreum
	Gossypium barbadense
	Gossypium herbaceum
	Gossypium hirsutum
	Gossypium spp.
	Guazuma ulmifolia
	Hibiscus asper
	Hibiscus cannabinus
	Hibiscus esculentus
	Hibiscus rosa-sinensis
	Hibiscus sabdariffa
	Hibiscus surattensis
	Hibiscus tiliaceus
	Malachra alceifolia
	Sabdariffa rubra
	Sida acuta
	Sida rhombifolia
	Sida urens
	Sterculia apetala
	Sterculia setigera
	Theobroma bicolor
	Theobroma cacao
	Thespesia populnea
	Urena lobata
	Waltheria indica
	Wissadula amplissima
Marantaceae	*Maranta arundinacea*
	Thalia geniculata
Melastomataceae	*Anplectrum glaucum*
	Arthrostemma macrodesmum
	Arthrostemma volubile
	Clidemia hirta
	Heterotis rotundifolia
	Miconia spp.
	Phyllagathis griffithii
	Phyllagathis rotundifolia
Meliaceae	*Azadirachta indica*

(*Continued*)

Table A.3 (Continued) Medicinal Plant Species Listed by Family

Family	Plant Name
	Cabralea canjerana
	Carapa guianensis
	Carapa procera
	Cedrela fissilis
	Cedrela odorata
	Cedrela toona
	Ekebergia capensis
	Enthadrophragma angolense
	Khaya anthotheca
	Khaya grandifoliola
	Khaya senegalensis
	Lansium domesticum
	Melia azedarach
	Melia indica
	Munronia henryi
	Naregamia alata
	Pseudocedrela kotschyi
	Soymida febrifuga
	Swietenia macrophylla
	Swietenia mahagoni
	Trichilia cipo
	Trichilia emetica
	Trichilia gilletii
	Trichilia havanensis
	Trichilia monadelpha
	Trichilia prieureana
	Turraea mombassana
	Turreanthus africanus
Melianthaceae	*Bersama abyssinica*
	Bersama engleriana
Menispermaceae	*Abuta grandifolia*
	Abuta rufescens
	Abuta sp.
	Arcangelisia flava
	Burasaia australis
	Burasaia congesta
	Burasaia gracilis
	Burasaia madagascariensis
	Burasaia nigrescens
	Chasmanthera dependens
	Chasmanthera uviformis
	Chondrodendron platyphyllum

(Continued)

Table A.3 (Continued) Medicinal Plant Species Listed by Family

Family	Plant Name
	Cissampelos glaberrima
	Cissampelos mucronata
	Cissampelos ovalifolia
	Cissampelos pareira
	Cocculus leaeba
	Curarea sp.
	Curarea tecunarum
	Cyclea peltata
	Fibraurea tinctoria
	Penianthus longifolius
	Rhigiocarya racemifera
	Sphenocentrum jollyanum
	Spirospermum penduliflorum
	Stephania hermandifolia
	Stephania japonica
	Strychnopsis thouarsii
	Tinospora cordifolia
	Tinospora crispa
	Tinospora sinensis
	Triclisia dictyophylla
	Triclisia gelletii
	Triclisia macrocarpa
Molluginaceae	*Mollugo pentaphylla*
Monimiaceae	*Glossocalyx brevipes*
	Siparuna guianensis
	Siparuna pauciflora
	Siparuna poeppigii
	Tambourissa leptophylla
Moraceae	*Brosimum lactescens*
	Brosimum rubescens
	Cecropia peltata
	Cecropia spp.
	Cudrania triloba
	Ficus adenosperma
	Ficus bussei
	Ficus carica
	Ficus dicranostyla
	Ficus exasperata
	Ficus glabrata
	Ficus gnaphalocarpa
	Ficus hispida
	Ficus insipida

(Continued)

Table A.3 (Continued) Medicinal Plant Species Listed by Family

Family	Plant Name
	Ficus ischnopoda
	Ficus megapoda
	Ficus ovata
	Ficus platyphylla
	Ficus polita
	Ficus racemosa
	Ficus ribes
	Ficus sur
	Ficus sycomorus
	Ficus thonningii
	Ficus umbellata
	Ficus vallis-choudae
	Maclura sp.
	Maclura tinctoria
	Maquira coriacea
	Milicia excelsa
	Moringa oleifera
	Morus alba
	Pseudolmedia spuria
Musaceae	*Musa paradisiaca*
	Musa spp.
Myricaceae	*Myrica kandtiana*
	Myrica salicifolia
Myristicaceae	*Myristica fragrans*
	Pycnanthus angolensis
	Virola calophylla
	Virola carinata
	Virola sp.
Myrsinaceae	*Ardisia* sp.
	Ardisia virens
	Labisia pumila
	Maesa doraena
	Myrsine africana
	Myrsine latifolia
Myrtaceae	*Campomanesia aromatica*
	Campomanesia grandiflora
	Eucalyptus camaldulensis
	Eucalyptus citriodora
	Eucalyptus globulus
	Eucalyptus grandis
	Eucalyptus robusta
	Eucalyptus sp.

(Continued)

Table A.3 (Continued) Medicinal Plant Species Listed by Family

Family	Plant Name
	Eugenia pitanga
	Eugenia sulcata
	Melaleuca leucadendron
	Myrciaria dubia
	Myrtus communis
	Psidium guajava
	Psidium personii
	Psidium sp.
	Pycananthus angolensis
	Syzygium aromaticum
	Syzygium cordatum
	Syzygium guineense
Nyctaginaceae	*Boerhaavia diffusa*
	Boerhavia elegans
	Boerhaavia erecta
	Boerhavia coccinea
	Boerhavia hirsuta
Nymphaeaceae	*Nymphaea lotus*
Ochnaceae	*Lophira alata*
	Lophira lanceolata
	Polyouratea hexasperma
	Quiina guianensis
Olacaceae	*Minquartia guianensis*
	Olax gambecola
	Ximenia americana
	Ximenia caffra
Oleaceae	*Chionanthus virginicus*
	Fraxinus excelsior
	Jasminum fluminense
	Jasminum syringifolia
	Linociera ramiflora
	Nyctanthes arbor-tristis
	Olea capensis
	Olea europaea
	Olea europaea ssp. *africana*
	Olea glandulifera
	Syringa vulgaris
Oliniaceae	*Olinia macrophylla*
Onagraceae	*Ludwigia erecta*
	Ludwigia peruviana
Ophioglossaceae	*Helminthostachys zeylanica*
Opiliaceae	*Opilia celtidifolia*

(*Continued*)

Table A.3 (Continued) Medicinal Plant Species Listed by Family

Family	Plant Name
Orchidaceae	*Laelia* sp.
	Luicia teres
	Satyrium nepalense
	Tropidia cucurligioides
Oxalidaceae	*Biophytum petersianum*
	Biophytum umbraculum
	Oxalis corniculata
Palmae	*Arenga pinnata*
	Cocos nucifera
	Elaeis guineensis
	Euterpe edulis
	Euterpe oleracea
	Euterpe precatoria
	Euterpe sp.
	Mauritia flexuosa
	Oenocarpus bataua
	Oenocarpus mapora
Pandaceae	*Microdesmis keayana*
Papaveraceae	*Argemone mexicana*
	Corydalis govaniana
	Papaver somniferum
Parmeliaceae	*Usnea* spp.
Passifloraceae	*Adenia cissampeloides*
	Passiflora edulis
	Passiflora foetida
	Passiflora lauriflora
	Passiflora nepalensis
Pedaliaceae	*Ceratotheca sesamoides*
	Martynia annua
	Sesamum indicum
Pentadiplandraceae	*Pentadiplandra brazzeana*
Phytolaccaceae	*Petiveria alliacea*
	Phytolacca dodecandra
Piperaceae	*Peperomia nigropunctata*
	Peperomia spp.
	Peperomia trifolia
	Peperomia vulcanica
	Piper aduncum
	Piper betel
	Piper capense
	Piper guineense
	Piper hispidum

(*Continued*)

Table A.3 (Continued) Medicinal Plant Species Listed by Family

Family	Plant Name
	Piper longum
	Piper marginatum
	Piper mullesua
	Piper nigrum
	Piper peltatum
	Piper pyrifolium
	Piper sarmentosum
	Potomorphe sp.
Pittosporaceae	*Pittosporum ferrugineum*
	Pittosporum viridiflorum
Plantaginaceae	*Plantago amplexicaulis*
	Plantago intermedia
	Plantago major
Plumbaginaceae	*Plumbago zeylanica*
Poaceae	*Axonopus compressus*
	Cymbopogon giganteus
	Cymbopogon nardus
	Cymbopogon proximus
	Digitaria abyssinica
	Eleusine indica
	Hymenachne donacifolia
	Imperata cylindrica
	Oryza sativa
	Panicum subalbidum
	Pennisetum pedicellatum
	Pennisetum purpureum
	Polygala persicariaefolia
	Setaria megaphylla
	Setaria sp.
	Sorghum guineense
	Sporobolus helveolus
	Vetiveria zizanoides
	Zea mays
Podocarpaceae	*Podocarpus latifolius*
	Podocarpus sprucei
Polemoniaceae	*Cantua quercifolia*
	Loeselia ciliata
Polygalaceae	*Polygala micrantha*
	Polygala paniculata
	Securidaca longipedunculata
Polygonaceae	*Polygonum aviculare*
	Polygonum bistorta

(*Continued*)

Table A.3 (Continued) Medicinal Plant Species Listed by Family

Family	Plant Name
	Polygonum punctatum
	Rheum officinale
	Rumex abyssinicus
	Rumex crispus
	Rumex obtusifolius
	Rumex usambarensis
	Triplaris americana
	Triplaris cf. *poeppigiana*
	Triplaris weigeltiana
Polypodiaceae	*Drynaria quercifolia*
	Polypodium decumanum
Pontederiaceae	*Monochoria hastata*
Portulacaceae	*Portulaca oleracea*
	Portulaca sp.
	Talinum portulacifolium
Potamogetonaceae	*Potamogeton javanicus*
Pteridaceae	*Acrostichum aureum*
Ranunculaceae	*Aconitum deinorhizum*
	Aconitum heterophyllum
	Anemone vitifolia
	Cimicifuga foetida
	Cimicifuga racemosa
	Clematis brachiata
	Clematis mauritiana
	Clematis minor
	Coptis teeta
	Nigella sativa
	Ranunculus japonicus
	Ranunculus muricatus
	Ranunculus trichophyllus
	Ranunculus zuccarinii
	Thalictrum foliolosum
Rhamnaceae	*Ampelozizyphus amazonicus*
	Colubrina glomerata
	Colubrina guatemalensis
	Discaria febrifuga
	Rhamnus prinoides
	Rhamnus staddo
	Ziziphus jujuba
	Ziziphus mucronata
	Ziziphus mauritania
	Ziziphus spina-christi

(*Continued*)

Table A.3 (Continued) Medicinal Plant Species Listed by Family

Family	Plant Name
Rhizophoraceae	*Carallia brachiata*
	Ceriops tagal
	Rhizophora mangle
Rosaceae	*Amygdalus davidiana*
	Fragaria indica
	Geum urbanum
	Malus sylvestris
	Parinari polyandra
	Potentilla discolor
	Potentilla reptans
	Prunus africana
	Prunus cerasoides
	Prunus domestica
	Prunus mume
	Prunus persica
	Prunus sphaerocarpa
	Prunus virginiana
	Randia fasciculata
	Rosa canina
	Rubus ellipticus
Rubiaceae	*Aganthesanthemum bojeri*
	Agathisanthenum globosum
	Borreria ocimoides
	Boscia angustifolia
	Breonadia salicina
	Calycophylum acreanum
	Canthium glaucum
	Cephalanthus occidentalis
	Cephalanthus spathelliferus
	Chasallia chartacea
	Chimarrhis turbinata
	Chiococcca alba
	Cinchona barbacoensis
	Cinchona henleana
	Cinchona ledgeriana
	Cinchona officinalis
	Cinchona pitayensis
	Cinchona pubescens
	Cinchona sp.
	Coffea arabica
	Coffea canephora
	Condaminea sp.

(*Continued*)

Table A.3 (Continued) Medicinal Plant Species Listed by Family

Family	Plant Name
	Coutarea hexandra
	Craterispermum laurinum
	Crossopteryx febrifuga
	Danais breviflora
	Danais cernua
	Danais fragrans
	Danais gerrardii
	Danais verticillata
	Exostema australe
	Exostema cuspidatum
	Exostema caribaeum
	Exostema mexicanum
	Exostemma peruviana
	Fadogia agrestis
	Feretia apodenthera
	Gardenia aqualla
	Gardenia erubescens
	Gardenia gummifera
	Gardenia lutea
	Gardenia sokotensis
	Gardenia sp.
	Gardenia ternifolia
	Gardenia vogelii
	Geophila obvallata
	Guarea grandifolia
	Guettarda speciosa
	Hallea rubrostipulata
	Hamelia patens
	Hedyotis biflora
	Hedyotis diffusa
	Hedyotis scandens
	Hintonia latiflora
	Hymenodictyon excelsum
	Hymenodyction lohavato
	Isertia haenkeana
	Joosia dichotoma
	Ladenbergia cujabensis
	Ladenbergia macrocarpa
	Ladenbergia magnifolia
	Ladenbergia malacophylla
	Ladenbergia moritziana
	Ladenbergia oblongifolia

(*Continued*)

Table A.3 (Continued) Medicinal Plant Species Listed by Family

Family	Plant Name
	Macrocnemum roseum
	Mitacarpus scaber
	Mitracarpus megapotamicus
	Mitragyna inermis
	Morinda citrifolia
	Morinda confusa
	Morinda geminata
	Morinda lucida
	Morinda morindoides
	Mussaenda glabra
	Myragina stipulosa
	Mytragina ciliata
	Mytragina stipulosa
	Nauclea diderrichii
	Nauclea latifolia
	Nauclea officinalis
	Nauclea pobeguinii
	Neolamarckia cadamba
	Neonauclea formicaria
	Oldenlandia corymbosa
	Oldenlandia herbacea
	Otiophora pauciflora
	Palicourea rigida
	Pavetta corymbosa
	Pavetta crassipes
	Pavetta schumanniana
	Pavetta ternifolia
	Pentanisia ouranogyne
	Pentas lanceolata
	Pentas longiflora
	Pogonopus speciosus
	Porterandia cladantha
	Psychotria sp.
	Psychotria ipecacuanha
	Psychotria kirkii
	Randia echinocarpa
	Randia spinosa
	Randia talangnigna
	Remijia amphithrix
	Remijia ferruginea
	Remijia firmula
	Remijia macrophylla

(*Continued*)

Table A.3 (Continued) Medicinal Plant Species Listed by Family

Family	Plant Name
	Remijia macrocnemia
	Remijia pedunculata
	Remijia peruviana
	Remijia purdieana
	Remijia sp.
	Remijia trianae
	Remijia ulei
	Remijia vellozii
	Rothmannia hispido
	Rustia formosa
	Rytigynia umbellulata
	Sabicea villosa
	Saldinia sp.
	Sarcocephalus officinalis
	Schismatoclada concinna
	Schismatoclada farahimpensis
	Schismatoclada viburnoides
	Schumanniophyton magnificum
	Timonius timon
	Tricalysia cryptocalyx
	Uncaria guianensis
	Urophyllum lyallii
	Vangueria madagascariensis
Rutaceae	*Aegle mamelos*
	Afraegle paniculata
	Angostura spp.
	Araliopsis tabuensis
	Chloroxylon falcatum
	Citrus aurantiifolia
	Citrus aurantium
	Citrus limon
	Citrus maxima
	Citrus medica
	Citrus nobilis
	Citrus paradisi
	Citrus sinensis
	Citrus sp.
	Clausena anisata
	Clausena excavata
	Dictamnus albus
	Esenbeckia febrifuga
	Evodia fatraina

(*Continued*)

Table A.3 (Continued) Medicinal Plant Species Listed by Family

Family	Plant Name
	Fagara macrophylla
	Fagaropsis angolensis
	Feroniella lucida
	Galipea jasminiflora
	Hortia sp.
	Hortia brasiliana
	Melicope pteleifolia
	Monniera trifolia
	Orixa japonica
	Pilocarpus jaborandi
	Ptelea trifoliata
	Ruta graveolens
	Teclea nobilis
	Teclea simplicifolia
	Toddalia aculeata
	Toddalia asiatica
	Vepris ampody
	Vepris lanceolata
	Zanthoxylum avicennae
	Zanthoxylum chalybeum
	Zanthoxylum gilletti
	Zanthoxylum hamiltonianum
	Zanthoxylum hermaphroditum
	Zanthoxylum lemarei
	Zanthoxylum leprieurii
	Zanthoxylum pentandrum
	Zanthoxylum perrotteti
	Zanthoxylum piperitum
	Zanthoxylum rhoifolium
	Zanthoxylum tsihanimpotsa
	Zanthoxylum usambarense
	Zanthoxylum xanthoxyloides
	Zanthoxylum zanthoxyloides
Salicaceae	*Populus alba*
	Salix alba
	Salix babylonica
	Salix fragilis
	Salix humboldtiana
	Salix purpurea
	Salix taxifolia
Salvadoraceae	*Salvadora persica*
Samydaceae	*Trimelia bakeri*

(*Continued*)

Table A.3 (Continued) Medicinal Plant Species Listed by Family

Family	Plant Name
Sapindaceae	*Allophylus pervillei*
	Blighia sapida
	Blighia unijugata
	Dedonaea viscosa
	Deinbollia pinnata
	Dodonaea madagascariensis
	Dodonaea viscosa
	Dodonea angustifolia
	Erioglossum edule
	Lecaniodiscus cupanoids
	Pappea capensis
	Paullinia pinnata
	Paullinia splendida
	Paullinia yoco
	Schleichera oleosa
	Serjania racemosa
	Tristiropsis sp.
	Zanha golungensis
Sapotaceae	*Autranella congolensis*
	Chrysophyllum albidum
	Chrysophyllum perpulchrum
	Ecclinusa ramiflora
	Lucuma salicifolia
	Pouteria caimito
	Vitellaria paradoxa
Saururaceae	*Houttuynia cordata*
	Saururus loureiri
Saxifragaceae	*Dichroa febrifuga*
	Hydrangea macrophylla
Schizaeaceae	*Mohria caffrorum*
Scrophulariaceae	*Bacopa monniera*
	Brucea antidysenterica
	Picrorhiza kurrooa
	Picrorhiza scrophulariaefolia
	Scoparia dulcis
	Scrophularia dulcis
	Scrophularia oldhami
	Striga hermonthica
	Vandellia sessiliflora
	Verbascum thapsus
	Veronica cymbalaria
Selaginellaceae	*Selaginella pallescens*

(Continued)

Table A.3 (Continued) Medicinal Plant Species Listed by Family

Family	Plant Name
Simaroubaceae	*Ailanthus altissima*
	Ailanthus excelsa
	Brucea javanica
	Eurycoma longifolia
	Harrisonia abyssinica
	Harrisonia perforata
	Odyendyea gabonensis
	Picramnia antidesma
	Picramnia pentandra
	Picrasma excelsa
	Picrasma javanica
	Picrolemma pseudocoffea
	Quassia africana
	Quassia amara
	Quassia undulata
	Simaba cedron
	Simaba ferruginea
	Simaba morettii
	Simarouba amara
	Simarouba glauca
Siparunaceae	*Siparuna* sp.
Smilacaceae	*Smilax china*
	Smilax spinosa
	Smilax spp.
Solanaceae	*Brunfelsia grandiflora*
	Brunfelsia sp.
	Capsicum annuum
	Capsicum frutescens
	Cestrum euanthes
	Cestrum laevigatum
	Cestrum megalophyllum
	Cyphomandra pendula
	Cyphomandra sp.
	Datura inoxia
	Datura metel
	Datura stramonium
	Nicotiana tabacum
	Physalis alkekengi
	Physalis angulata
	Scopolia japonica
	Solanum dulcamara
	Solanum erianthum

(*Continued*)

Table A.3 (Continued) Medicinal Plant Species Listed by Family

Family	Plant Name
	Solanum grandiflorum
	Solanum incanum
	Solanum indicum
	Solanum leucocarpon
	Solanum micranthum
	Solanum nigrum
	Solanum nitidum
	Solanum paniculatum
	Solanum pseudoquina
	Solanum sp.
	Solanum syzymbrifolium
	Solanum torvum
	Solanum vairum
	Withania somnifera
Sterculiaceae	*Cola cordifolia*
Taccaceae	*Tacca chantrieri*
Tamaricaceae	*Tamarix chinensis*
Theaceae	*Anneslea fragrans*
	Bonnetia paniculata
	Camellia sinensis
Thelypteridaceae	*Thelypteris* sp.
Thymelaeaceae	*Aquilaria agallocha*
	Daphne genkwa
	Peddia involucrata
Tiliaceae	*Grewia cyclea*
	Grewia hexamita
	Grewia mollis
	Grewia paniculata
	Grewia plagiophylla
	Grewia trichocarpa
	Grewia villosa
Turneraceae	*Turnera diffusa*
	Turnera ulmifolia
Ulmaceae	*Celtis* cf. *tessmannii*
	Celtis durandii
	Trema commersonii
	Trema orientalis
Umbelliferae	*Angelica archangelica*
	Angelica sylvestris
	Apium graveolens
	Centella asiatica
	Coriandrum sativum

(*Continued*)

Table A.3 (Continued) Medicinal Plant Species Listed by Family

Family	Plant Name
	Eryngium foetidum
	Steganotaenia araliacea
Urticaceae	*Musanga cecropiodies*
	Myriocarpa longipes
	Urera baccifera
	Urera laciniata
	Urtica dioica
	Urtica massaica
Verbenaceae	*Bouchea prismatica*
	Cleodendrum colebrookianum
	Cleodendrum serratum
	Clerodendron indicum
	Clerodendron scandens
	Clerodendrum eriophyllum
	Clerodendrum inerme
	Clerodendrum infortunatum
	Clerodendrum serratum
	Clerodendrum splendens
	Clerodendrum wallii
	Gmelia arborea
	Lantana camara var. *aculeata*
	Lantana camara
	Lantana involucrata
	Lantana rhodesiensis
	Lantana trifolia
	Lippia alba
	Lippia chevalieri
	Lippia multiflora
	Lippia schomburgkiana
	Premna angolensis
	Premna chrysoclada
	Premna latifolia
	Premna serratifolia
	Stachytarpheta angustifolia
	Stachytarpheta cayennensis
	Stachytarpheta jamaicensis
	Stachytarpheta straminea
	Tectona grandis
	Verbena litoralis
	Verbena officinalis
	Vitex doniana
	Vitex negundo

(*Continued*)

Table A.3 (Continued) Medicinal Plant Species Listed by Family

Family	Plant Name
	Vitex peduncularis
	Vitex trifolia
Violaceae	*Viola odorata*
Viscaceae	*Viscum album*
Vitaceae	*Ampelocissus bombycina*
	Cissus aralioides
	Cissus flavicans
	Cissus populnea
	Cissus quadrangularis
	Cissus rotundifolia
	Cyphostemma digitatum
	Leea sp.
	Rhoicissus tridentata
	Vitis japonica
Winteraceae	*Tasmannia piperita*
Xanthorrhoeaceae	*Aloe kedongensis*
	Aloe volkensii
	Aloe deserti
	Aloe macrosiphon
	Aloe pirottae
	Aloe secundiflora
	Aloe sp.
	Aloe vera
Zingiberaceae	*Aframomum citratum*
	Aframomum latifoilum
	Aframomum melegueta
	Aframomum sceptrum
	Aframomum zambesiacum
	Alpinia officinarum
	Alpinia zerumbet
	Costus afer
	Costus arabicus
	Costus dubius
	Curcuma caesia
	Curcuma longa
	Elettaria cardamomum
	Hedychium cylindricum
	Kaempferia galanga
	Languas galanga
	Renealmia guianensis
	Siphonochilus aethiopicus
	Zingiber cassumunar

(*Continued*)

Table A.3 (Continued) Medicinal Plant Species Listed by Family

Family	Plant Name
	Zingiber mioga
	Zingiber officinale
Zygophyllaceae	*Peganum harmala*
	Tribulus cistoides

REFERENCES

Adamu, H.M., O.J. Abayeh, M.O. Agho, A.L. Abdullahi, A. Uba, H.U. Dukku, and B.M. Wufem. 2005. "An ethnobotanical survey of Bauchi State herbal plants and their antimicrobial activity." *Journal of Ethnopharmacology* 99:1–4.

Adebayo, J.O., and A.U. Krettli. 2011. "Potential antimalarials from Nigerian plants: A review." *Journal of Ethnopharmacology* 133:289–302.

Agra, M.F., G.S. Baracho, K. Nurit, I. Basílio, and V.P.M. Coelho. 2007. "Medicinal and poisonous diversity of the flora of 'Cariri Paraibano,' Brazil." *Journal of Ethnopharmacology* 111:383–395.

Ajibesin, K.K., B.A. Ekpo, D.N. Bala, E.E. Essien, and S.A. Adesanya. 2007. "Ethnobotanical survey of Akwa Ibom State of Nigeria." *Journal of Ethnopharmacology* 115:387–408.

Akendengue, B. 1992. "Medicinal plants used by the Fang traditional healers in Equatorial Guinea." *Journal of Ethnopharmacology* 37:165–173.

Al-adhroey, A.H., Z.M. Nor, H.M. Al-mekhlafi, and R. Mahmud. 2010. "Ethnobotanical study on some Malaysian anti-malarial plants: A community based survey." *Journal of Ethnopharmacology* 132:362–364.

Albuquerque, U.P.D., M.D. Medeiros, A.L.S.D. Almeida, M. Monteiro, E. Machado, D.F. Lins, and J. Patr. 2007. "Medicinal plants of the Caatinga (semi-arid) vegetation of NE Brazil: A quantitative approach." *Journal of Ethnopharmacology* 114:325–354.

Alfaro, M.A.M. 1984. "Medicinal plants used in a Totonac community of the Sierra Norte de Puebla: Tuzamapan de Galeana, Puebla, Mexico." *Journal of Ethnopharmacology* 11:203–221.

Allabi, A.C., K. Busia, V. Ekanmian, and F. Bakiono. 2011. "The use of medicinal plants in self-care in the Agonlin region of Benin." *Journal of Ethnopharmacology* 133: 234–243.

Altundag, E., and M. Ozturk. 2011. "Ethnomedicinal studies on the plant resources of east Anatolia, Turkey." *Procedia—Social and Behavioral Sciences* 19:756–777.

Asase, A., and G. Oppong-Mensah. 2009. "Traditional antimalarial phytotherapy remedies in herbal markets in southern Ghana." *Journal of Ethnopharmacology* 126:492–499.

Asase, A., and M.L. Kadera. 2014. "Herbal medicines for child healthcare from Ghana." *Journal of Herbal Medicine* 4:24–36.

Asase, A., A.A. Oteng-yeboah, G.T. Odamtten, and M.S.J. Simmonds. 2005. "Ethnobotanical study of some Ghanaian anti-malarial plants." *Journal of Ethnopharmacology* 99:273–279.

Asase, A., G.A. Akwetey, and D.G. Achel. 2010. "Ethnopharmacological use of herbal remedies for the treatment of malaria in the Dangme West District of Ghana." *Journal of Ethnopharmacology* 129:367–376.

Asase, A., D.N. Hesse, and M.S.J. Simmonds. 2012. "Uses of multiple plants prescriptions for treatment of malaria by some communities in southern Ghana." *Journal of Ethnopharmacology* 144:448–452.

Au, D.T., J. Wu, Z. Jiang, H. Chen, G. Lu, and Z. Zhao. 2008. "Ethnobotanical study of medicinal plants used by Hakka in Guangdong, China." *Journal of Ethnopharmacology* 117:41–50.

Bahadur, M., Z. Münzbergová, and B. Timsina. 2010. "Ethnobotanical study of medicinal plants from the Humla district of western Nepal." *Journal of Ethnopharmacology* 130:485–504.

Belayneh, A., Z. Asfaw, S. Demissew, and N.F. Bussa. 2012. "Medicinal plants potential and use by pastoral and agro-pastoral communities in Erer Valley of Babile Wereda, Eastern Ethiopia." *Journal of Ethnobiology and Ethnomedicine* 8:42.

Bertani, S., G. Bourdy, I. Landau, J.C. Robinson, P. Esterre, and E. Deharo. 2005. "Evaluation of French Guiana traditional antimalarial remedies." *Journal of Ethnopharmacology* 98:45–54.

Betti, J.L., and S.R.M. Yemefa. 2011. "An ethnobotanical study of medicinal plants used in the Kalamaloué National Park, Cameroon." *Journal of Medicinal Plants Research* 5 (8):1447–1458.

Beverly, C.D., and G. Sudarsanam. 2011. "Ethnomedicinal plant knowledge and practice of people of Javadhu hills in Tamilnadu." *Asian Pacific Journal of Tropical Biomedicine* 1 (1):79–81.

Bhandary, M.J., K.R. Chandrashekar, and K.M. Kaveriappa. 1995. "Medical ethnobotany of the Siddis of Uttara Kannada district, Karnataka, India." *Journal of Ethnopharmacology* 47:149–158.

Bosco, F.G., and R. Arumugam. 2012. "Ethnobotany of Irular tribes in Redhills, Tamilnadu, India." *Asian Pacific Journal of Tropical Disease* 2:S874–S877.

Bourdy, G., S.J. Dewalt, and L.R. Cha. 2000. "Medicinal plants uses of the Tacana, an Amazonian Bolivian ethnic group." *Journal of Ethnopharmacology* 70:87–109.

Bourdy, G., L.R. Chāvez de Michel, and A. Roca-Coulthard. 2004. "Pharmacopoeia in a shamanistic society: The Izoceño-Guaraní (Bolivian Chaco)." *Journal of Ethnopharmacology* 91:189–208.

Brandão, M.G., T.S. Grandi, E.M. Rocha, D.R. Sawyer, and A.U. Krettli. 1992. "Survey of medicinal plants used as antimalarials in the Amazon." *Journal of Ethnopharmacology* 36 (2):175–182.

Bruschi, P., M. Morganti, M. Mancini, and M.A. Signorini. 2011. "Traditional healers and laypeople: A qualitative and quantitative approach to local knowledge on medicinal plants in Muda (Mozambique)." *Journal of Ethnopharmacology* 138 (2):543–563.

Bulut, G., and E. Tuzlaci. 2013. "An ethnobotanical study of medicinal plants in Turgutlu (Manisa, Turkey)." *Journal of Ethnopharmacology* 149:633–647.

Cai, G. 2004. *A Record of Hunan Pharmacy*. China: Hunan Science and Technology Press.

Calderón, A.I., J. Simithy-Williams, and M.P. Gupta. 2012. "Antimalarial natural products drug discovery in Panama." *Pharmaceutical Biology* 50 (1):61–71.

Cano, J.H., and G. Volpato. 2004. "Herbal mixtures in the traditional medicine of eastern Cuba." *Journal of Ethnopharmacology* 90 (2–3):293–316.

Chander, M.P., C. Kartick, J. Gangadhar, and P. Vijayachari. 2014. "Ethno medicine and healthcare practices among Nicobarese of Car Nicobar—An indigenous tribe of Andaman and Nicobar Islands." *Journal of Ethnopharmacology* 158:18–24.

Chifundera, K. 2001. "Contribution to the inventory of medicinal plants from the Bushi area, South Kivu Province, Democratic Republic of Congo." *Fitoterapia* 72 (4):351–368.

Cosenza, G.P., N.S. Somavilla, C.W. Fagg, and M.G.L. Brandão. 2013. "Bitter plants used as substitute of *Cinchona* spp. (quina) in Brazilian traditional medicine." *Journal of Ethnopharmacology* 149:790–796.

de Madureira, M.d.C., A.P. Martins, M. Gomes, J. Paiva, A.P. da Cunha, and V. do Rosario. 2002. "Antimalarial activity of medicinal plants used in traditional medicine in S. Tome and Prncipe islands." *Journal of Ethnopharmacology* 81:23–29.

De Natale, A., and A. Pollio. 2007. "Plants species in the folk medicine of Montecorvino Rovella (Inland Campania, Italy)." *Journal of Ethnopharmacology* 109:295–303.

Deharo, E., G. Bourdy, C. Quenevo, V. Muñoz, G. Ruiz, and M. Sauvain. 2001. "A search for natural bioactive compounds in Bolivia through a multidisciplinary approach. Part V. Evaluation of the antimalarial activity of plants used by the Tacana Indians." *Journal of Ethnopharmacology* 77 (1):91–98.

di Tizio, A., Ł.J. Łuczaj, C.L. Quave, S. Redzic, and A. Pieroni. 2012. "Traditional food and herbal uses of wild plants in the ancient South-Slavic diaspora of Mundimitar/Montemitro (Southern Italy)." *Journal of Ethnobiology and Ethnomedicine* 8:21.

Diallo, D., B. Hveem, M.A. Mahmoud, G. Berge, B.S. Paulsen, and A. Maiga. 1999. "An ethnobotanical survey of herbal drugs of Gourma District, Mali."

Diarra, N., C. van't Klooster, A. Togola, D. Diallo, M. Willcox, and J.D. Jong. 2015. "Ethnobotanical study of plants used against malaria in Sélingué subdistrict, Mali." *Journal of Ethnopharmacology* 166:352–360.

Dike, I.P., O.O. Obembe, and F.E. Adebiyi. 2012. "Ethnobotanical survey for potential anti-malarial plants in south-western Nigeria." *Journal of Ethnopharmacology* 144 (3):618–626.

Dimayuga, R.E., and J. Agundez. 1986. "Traditional medicine of Baja California Sur (Mexico). I." *Journal of Ethnopharmacology* 17 (2):183–193.

Dimayuga, R.E., R.F. Murillo, and M.L. Pantoja. 1987. "Traditional medicine of Baja California Sur (Mexico). II." *Journal of Ethnopharmacology* 20:209–222.

Duke, J.A. 1996. "Dr. Duke's Phytochemical and Ethnobotanical Database." Accessed September 10, 2014. http://www.ars-grin.gov/duke/.

Elisabetsky, E., and P. Shanley. 1994. "Ethnopharmacology in the Brazilian Amazon." *Pharmacology and Therapeutics* 64 (94):201–214.

El-Kamali, H.H., and K.F. El-Khalifa. 1999. "Folk medicinal plants of riverside forests of the Southern Blue Nile district, Sudan." *Fitoterapia* 70 (5):493–497.

Ellena, R., C.L. Quave, and A. Pieroni. 2012. "Comparative Medical Ethnobotany of the Senegalese Community Living in Turin (Northwestern Italy) and in Adeane (Southern Senegal)." *Evidence-Based Complementary and Alternative Medicine* 2012 (4):1–30.

Elliott, S., and J. Brimacombe. 1987. "The medicinal plants of Gunung Leuser National Park, Indonesia." *Journal of Ethnopharmacology* 19 (3):285–317.

Esmaeili, S., F. Naghibi, M. Mosaddegh, S. Sahranavard, S. Ghafari, and N.R. Abdullah. 2009. "Screening of antiplasmodial properties among some traditionally used Iranian plants." *Journal of Ethnopharmacology* 121:400–404.

Fernandez, E.C., Y.E. Sandi, and L. Kokoska. 2003. "Ethnobotanical inventory of medicinal plants used in the Bustillo Province of the Potosi." *Fitoterapia* 74 (03):407–416.

Gakuya, D.W., S.M. Itonga, J.M. Mbaria, J.K. Muthee, and J.K. Musau. 2013. "Ethnobotanical survey of biopesticides and other medicinal plants traditionally used in Meru central district of Kenya." *Journal of Ethnopharmacology* 145 (2):547–553.

Gathirwa, J.W., G.M. Rukunga, P.G. Mwitari, N.M. Mwikwabe, C.W. Kimani, C.N. Muthaura, and S.A. Omar. 2011. "Traditional herbal antimalarial therapy in Kilifi district, Kenya." *Journal of Ethnopharmacology* 134 (2):434–442.

Geissler, P.W., S.A. Harris, R.J. Prince, A. Olsen, R.A. Odhiambo, H. Oketch-Rabah, and P. Mølgaard. 2002. "Medicinal plants used by Luo mothers and children in Bondo district, Kenya." *Journal of Ethnopharmacology* 83 (1–2):39–54.

Gessler, M.C., D.E. Msuya, M.H. Nkunya, L.B. Mwasumbi, A. Schär, M. Heinrich, and M. Tanner. 1995. "Traditional healers in Tanzania: The treatment of malaria with plant remedies." *Journal of Ethnopharmacology* 48 (3):131–44.

Ghimire, K., and R.R. Bastakoti. 2009. "Ethnomedicinal knowledge and healthcare practices among the Tharus of Nawalparasi district in central Nepal." *Forest Ecology and Management* 257 (10):2066–2072.

Ghorbani, A., G. Langenberger, L. Feng, and J. Sauerborn. 2011. "Ethnobotanical study of medicinal plants utilised by Hani ethnicity in Naban River Watershed National Nature Reserve, Yunnan, China." *Journal of Ethnopharmacology* 134 (3):651–667.

Giday, M., T. Teklehaymanot, A. Animut, and Y. Mekonnen. 2007. "Medicinal plants of the Shinasha, Agew-awi and Amhara peoples in northwest Ethiopia." *Journal of Ethnopharmacology* 110 (3):516–525.

Giday, M., Z. Asfaw, and Z. Woldu. 2009. "Medicinal plants of the Meinit ethnic group of Ethiopia: An ethnobotanical study." *Journal of Ethnopharmacology* 124:513–521.

Giday, M., Z. Asfaw, and Z. Woldu. 2010. "Ethnomedicinal study of plants used by Sheko ethnic group of Ethiopia." *Journal of Ethnopharmacology* 132 (1):75–85.

Giovannini, P. 2015. "Medicinal plants of the Achuar (Jivaro) of Amazonian Ecuador: Ethnobotanical survey and comparison with other Amazonian pharmacopoeias." *Journal of Ethnopharmacology* 164:78–88.

Githinji, C.W., and J.O. Kokwaro. 1993. "Ethnomedicinal study of major species in the family Labiatae from Kenya." *Journal of Ethnopharmacology* 39 (3):197–203.

Grønhaug, T.E., S. Glæserud, M. Skogsrud, N. Ballo, S. Bah, D. Diallo, and B.S. Paulsen. 2008. "Ethnopharmacological survey of six medicinal plants from Mali, West-Africa." *Journal of Ethnobiology and Ethnomedicine* 4:26.

Guarrera, P.M., G. Salerno, and G. Caneva. 2005. "Folk phytotherapeutical plants from Maratea area (Basilicata, Italy)." *Journal of Ethnopharmacology* 99:367–378.

Guédé, N.Z., K. N'guessan, T.E. Dibié, and P. Grellier. 2010. "Ethnopharmacological study of plants used to treat malaria, in traditional medicine, by Bete Populations of Issia (Côte d' Ivoire)." *Journal of Pharmaceutical Sciences and Research* 2 (4):216–227.

Gupta, M.P., M.D. Correa, P.N. Solís, A. Jones, C. Galdames, and F. Guionneau-Sinclair. 1993. "Medicinal plant inventory of Kuna Indians: Part 1." *Journal of Ethnopharmacology* 40 (2):77–109.

Gurib-Fakim, A., M. Sewraj, J. Gueho, and E. Dulloo. 1993. "Medicalethnobotany of some weeds of Mauritius and Rodrigues." *Journal of Ethnopharmacology* 39 (3):175–185.

Hajdu, Z., and J. Hohmann. 2012. "An ethnopharmacological survey of the traditional medicine utilized in the community of Porvenir, Bajo Paraguá Indian Reservation, Bolivia." *Journal of Ethnopharmacology* 139 (3):838–857.

Hamill, F.A., S. Apio, N.K. Mubiru, M. Mosango, O.W. Maganyi, and D.D. Soejarto. 2000. "Traditional herbal drugs of southern Uganda, I." *Journal of Ethnopharmacology* 70:281–300.

Hanlidou, E., R. Karousou, V. Kleftoyanni, and S. Kokkini. 2004. "The herbal market of Thessaloniki (N Greece) and its relation to the ethnobotanical tradition." *Journal of Ethnopharmacology* 91:281–299.

Hassan-Abdallah, A., A. Merito, S. Hassan, D. Aboubaker, M. Djama, Z. Asfaw, and E. Kelbessa. 2013. "Medicinal plants and their uses by the people in the Region of Randa, Djibouti." *Journal of Ethnopharmacology* 148 (2):701–713.

Hout, S., A. Chea, S.-S. Bun, R. Elias, M. Gasquet, P. Timon-David, and N. Azas. 2006. "Screening of selected indigenous plants of Cambodia for antiplasmodial activity." *Journal of Ethnopharmacology* 107 (1):12–18.

Ibrahim, J.A., I. Muazzam, I.A. Jegede, and O.F. Kunle. 2010. "Medicinal plants and animals sold by the "Yan-Shimfidas" of Sabo Wuse in Niger State, Nigeria." *African Journal of Pharmacy and Pharmacology* 4 (6):386–394.

Idolo, M., R. Motti, and S. Mazzoleni. 2010. "Ethnobotanical and phytomedicinal knowledge in a long-history protected area, the Abruzzo, Lazio and Molise National Park (Italian Apennines)." *Journal of Ethnopharmacology* 127:379–395.

Idowu, O.A., O.T. Soniran, O. Ajana, and D.O. Aworinde. 2010. "Ethnobotanical survey of antimalarial plants used in Ogun State, Southwest Nigeria." *African Journal of Pharmacy and Pharmacology* 4:55–60.

Inta, A., P. Trisonthi, and C. Trisonthi. 2013. "Analysis of traditional knowledge in medicinal plants used by Yuan in Thailand." *Journal of Ethnopharmacology* 149 (1):344–351.

Islam, M.K., S. Saha, I. Mahmud, K. Mohamad, K. Awang, S. Jamal Uddin, and J.A. Shilpi. 2014. "An ethnobotanical study of medicinal plants used by tribal and native people of Madhupur forest area, Bangladesh." *Journal of Ethnopharmacology* 151:921–930.

Jaganath, I.B., and L.T. Ng. 2000. *Herbs: The Green Pharmacy of Malaysia*. Kuala Lumpur: Vinpress.

Jain, S.P., and H.S. Puri. 1984. "Ethnomedicinal plants of Jaunsar-Bawar hills, Uttar Pradesh, India." *Journal of Ethnopharmacology* 12 (2):213–222.

Jamir, T.T., H.K. Sharma, and A.K. Dolui. 1999. "Folklore medicinal plants of Nagaland, India." *Fitoterapia* 70 (4):395–401.

Jansen, O., L. Angenot, M. Tits, J.P. Nicolas, P.D. Mol, J. Nikiéma, and M. Frédérich. 2010. "Evaluation of 13 selected medicinal plants from Burkina Faso for their antiplasmodial properties." *Journal of Ethnopharmacology* 130 (1):143–150.

Jeruto, P., C. Lukhoba, G. Ouma, D. Otieno, and C. Mutai. 2008. "An ethnobotanical study of medicinal plants used by the Nandi people in Kenya." *Journal of Ethnopharmacology* 116:370–376.

Joly, L.G., S. Guerra, R. Septimo, P.N. Solis, M. Correa, M. Gupta, and F. Sandberg. 1987. "Ethnobotanical inventory of medicinal plants used by the Guaymi Indians in western Panama. Part I." *Journal of Ethnopharmacology* 20:145–171.

Joly, L.G., S. Guera, R. Septimo, P.N. Solis, M.D.A. Correa, M.P. Gupta, and P. Perera. 1990. "Ethnobotanical inventory of medicinal plants used by the Guaymi Indians in Western Panama. Part II." *Journal of Ethnopharmacology* 28 (2):191–206.

Jorim, R.Y., S. Korape, W. Legu, M. Koch, L.R. Barrows, T.K. Matainaho, and P.P. Rai. 2012. "An ethnobotanical survey of medicinal plants used in the eastern highlands of Papua New Guinea." *Journal of Ethnobiology and Ethnomedicine* 8:47–47.

Junsongduang, A., H. Balslev, A. Inta, A. Jampeetong, and P. Wangpakapattanawong. 2014. "Karen and Lawa medicinal plant use: Uniformity or ethnic divergence?" *Journal of Ethnopharmacology* 151 (1):517–527.

Kadir, M.F., M.S. Bin Sayeed, N.I. Setu, A. Mostafa, and M.M.K. Mia. 2014. "Ethnopharmacological survey of medicinal plants used by traditional health practitioners in Thanchi, Bandarban Hill Tracts, Bangladesh." *Journal of Ethnopharmacology* 155:495–508.

Kamaraj, C., N.K. Kaushik, A.A. Rahuman, D. Mohanakrishnan, A. Bagavan, G. Elango, and D. Sahal. 2012. "Antimalarial activities of medicinal plants traditionally used in the villages of Dharmapuri regions of South India." *Journal of Ethnopharmacology* 141 (3):796–802.

Kantamreddi, V.S., and C.W. Wright. 2012. "Screening Indian plant species for antiplasmodial properties—Ethnopharmacological compared with random selection." *Phytotherapy Research* 26:1793–1799.

Kaou, A.M., V. Mahio-Leddet, S. Hutter, S. Ainouddine, S. Hassani, I. Yahaya, and E. Ollivier. 2008. "Antimalarial activity of crude extracts from nine African medicinal plants." *Journal of Ethnopharmacology* 116:74–83.

Karunamoorthi, K., and E. Tsehaye. 2012. "Ethnomedicinal knowledge, belief and self-reported practice of local inhabitants on traditional antimalarial plants and phytotherapy." *Journal of Ethnopharmacology* 141:143–150.

Katewa, S.S., B.L. Chaudhary, and A. Jain. 2004. "Folk herbal medicines from tribal area of Rajasthan, India." *Journal of Ethnopharmacology* 92 (1):41–46.

Katuura, E., P. Waako, J. Ogwal-Okeng, and R. Bukenya-Ziraba. 2007. "Traditional treatment of malaria in Mbarara District, western Uganda." *African Journal of Ecology* 45 (Suppl. 1):48–51.

Kayani, S., M. Ahmad, S. Sultana, Z.K. Shinwari, M. Zafar, G. Yaseen, and T. Bibi. 2015. "Ethnobotany of medicinal plants among the communities of Alpine and Sub-alpine regions of Pakistan." *Journal of Ethnopharmacology* 164:186–202.

Khajoei Nasab, F., and A.R. Khosravi. 2014. "Ethnobotanical study of medicinal plants of Sirjan in Kerman Province, Iran." *Journal of Ethnopharmacology* 154 (1):190–197.

Kichu, M., T. Malewska, K. Akter, I. Imchen, D. Harrington, J. Kohen, and J.F. Jamie. 2015. "An ethnobotanical study of medicinal plants of Chungtia village, Nagaland, India." *Journal of Ethnopharmacology* 166:5–17.

Koch, A., P. Tamez, J. Pezzuto, and D. Soejarto. 2005. "Evaluation of plants used for antimalarial treatment by the Maasai of Kenya." *Journal of Ethnopharmacology* 101 (1–3):95–99.

Kosalge, S.B., and R.A. Fursule. 2009. "Investigation of ethnomedicinal claims of some plants used by tribals of Satpuda Hills in India." *Journal of Ethnopharmacology* 121 (3):456–461.

Koudouvo, K., D.S. Karou, K. Kokou, K. Essien, K. Aklikokou, I.A. Glitho, and M. Gbeassor. 2011. "An ethnobotanical study of antimalarial plants in Togo Maritime Region." *Journal of Ethnopharmacology* 134 (1):183–190.

Kültür, S. 2007. "Medicinal plants used in Kirklareli Province (Turkey)." *Journal of Ethnopharmacology* 111 (2):341–364.

Kvist, L.P., S.B. Christensen, H.B. Rasmussen, K. Mejia, and A. Gonzalez. 2006. "Identification and evaluation of Peruvian plants used to treat malaria and leishmaniasis." *Journal of Ethnopharmacology* 106:390–402.

Lacroix, D., S. Prado, D. Kamoga, J. Kasenene, and B. Bodo. 2011. "Structure and *in vitro* antiparasitic activity of constituents of *Citropsis articulata* root bark." *Journal of Natural Products* 74:2286–2289.

Leaman, D.J., J.T. Arnason, R. Yusuf, H. Sangat-Roemantyo, H. Soedjito, C.K. Angerhofer, and J.M. Pezzuto. 1995. "Malaria remedies of the Kenyah of the Apo Kayan, East Kalimantan, Indonesian Borneo: A quantitative assessment of local consensus as an indicator of biological efficacy." *Journal of Ethnopharmacology* 49:1–16.

Leto, C., T. Tuttolomondo, S.L. Bella, and M. Licata. 2013. "Ethnobotanical study in the Madonie Regional Park (Central Sicily, Italy)—Medicinal use of wild shrub and herbaceous plant species." *Journal of Ethnopharmacology* 146 (1):90–112.

Libman, A., S. Bouamanivong, B. Southavong, K. Sydara, and D.D. Soejarto. 2006. "Medicinal plants: An important asset to health care in a region of Central Laos." *Journal of Ethnopharmacology* 106 (3):303–311.

Lingaraju, D.P., M.S. Sudarshana, and N. Rajashekar. 2013. "Ethnopharmacological survey of traditional medicinal plants in tribal areas of Kodagu district, Karnataka, India." *Journal of Pharmacy Research* 6 (2):284–297.

Madge, C. 1998. "Therapeutic landscapes of the Jola, the Gambia, West Africa." *Health & Place* 4 (4):293–311.

Magassouba, F.B., A. Diallo, M. Kouyate, F. Mara, O. Bangoura, A. Camara, and A.M. Balde. 2007. "Ethnobotanical survey and antibacterial activity of some plants used in Guinean traditional medicine." *Journal of Ethnopharmacology* 114:44–53.

Mahishi, P., B.H. Srinivasa, and M.B. Shivanna. 2005. "Medicinal plant wealth of local communities in some villages in Shimoga District of Karnataka, India." *Journal of Ethnopharmacology* 98 (3):307–312.

Mahmood, A., A. Mahmood, R.N. Malik, and Z.K. Shinwari. 2013. "Indigenous knowledge of medicinal plants from Gujranwala district, Pakistan." *Journal of Ethnopharmacology* 148 (2):714–723.

Mahyar, U.W., J.S. Burley, C. Gyllenhaal, and D.D. Soejarto. 1991. "Medicinal plants of Seberida (Riau Province, Sumatra, Indonesia)." *Journal of Ethnopharmacology* 31 (2):217–237.

Malla, B., D.P. Gauchan, and R.B. Chhetri. 2015. "An ethnobotanical study of medicinal plants used by ethnic people in Parbat district of western Nepal." *Journal of Ethnopharmacology* 165:103–117.

Mallik, B.K., T. Panda, and R.N. Padhy. 2012. "Traditional herbal practices by the ethnic people of Kalahandi district of Odisha, India." *Asian Pacific Journal of Tropical Biomedicine* 2 (2):S988–S994.

Maregesi, S.M., O.D. Ngassapa, L. Pieters, and A.J. Vlietinck. 2007. "Ethnopharmacological survey of the Bunda district, Tanzania: Plants used to treat infectious diseases." *Journal of Ethnopharmacology* 113:457–470.

Maxia, A., M.C. Lancioni, N.A. Balia, R. Alborghetti, A. Pieroni, and M.C. Loi. 2008. "Medical ethnobotany of the Tabarkins, a Northern Italian (Ligurian) minority in south-western Sardinia." *Genetic Resources and Crop Evolution* 55:911–924.

Mbatchi, S.F., B. Mbatchi, J.T. Banzouzi, T. Bansimba, G.F. Nsonde Ntandou, J.M. Ouamba, and F. Benoit-Vical. 2006. "*In vitro* antiplasmodial activity of 18 plants used in Congo Brazzaville traditional medicine." *Journal of Ethnopharmacology* 104 (1–2):168–174.

Mesfin, F., S. Demissew, and T. Teklehaymanot. 2009. "An ethnobotanical study of medicinal plants in Wonago Woreda, SNNPR, Ethiopia." *Journal of Ethnobiology and Ethnomedicine* 5:28.

Mesfin, A., M. Giday, A. Animut, and T. Teklehaymanot. 2012. "Ethnobotanical study of antimalarial plants in Shinile District, Somali Region, Ethiopia, and *in vivo* evaluation of selected ones against *Plasmodium berghei*." *Journal of Ethnopharmacology* 139 (1):221–227.

Mesia, G.K., G.L. Tona, T.H. Nanga, R.K. Cimanga, S. Apers, P. Cos, and A.J. Vlietinck. 2008. "Antiprotozoal and cytotoxic screening of 45 plant extracts from Democratic Republic of Congo." *Journal of Ethnopharmacology* 115 (3):409–415.

Milliken, W. 1997a. *Plants for Malaria, Plants for Fever: Medicinal Species in Latin America—A Bibliographic Survey*. United Kingdom: Whitstable Litho.

Milliken, W. 1997b. "Traditional anti-malarial medicine in Roraima, Brazil." *Economic Botany* 51 (3):212–237.

Monigatti, M., R.W. Bussmann, and C.S. Weckerle. 2013. "Medicinal plant use in two Andean communities located at different altitudes in the Bolívar Province, Peru." *Journal of Ethnopharmacology* 145 (2):4504–4564.

Muganza, D.M., B.I. Fruth, J.N. Lami, G.K. Mesia, O.K. Kambu, G.L. Tona, and L. Pieters. 2012. "*In vitro* antiprotozoal and cytotoxic activity of 33 ethonopharmacologically selected medicinal plants from Democratic Republic of Congo." *Journal of Ethnopharmacology* 141 (1):301–308.

Muñoz, V., M. Sauvain, G. Bourdy, S. Arrázola, J. Callapa, G. Ruiz, and E. Deharo. 2000. "A search for natural bioactive compounds in Bolivia through a multidisciplinary approach. Part III. Evaluation of the antimalarial activity of plants used by Alteños Indians." *Journal of Ethnopharmacology* 71 (1–2):123–131.

Muñoz, V., M. Sauvain, G. Bourdy, J. Callapa, S. Bergeron, I. Rojas, and E. Deharo. 2000. "A search for natural bioactive compounds in Bolivia through a multidisciplinary approach. Part I. Evaluation of the antimalarial activity of plants used by the Chacobo Indians." *Journal of Ethnopharmacology* 69:127–137.

Muñoz, V., M. Sauvain, G. Bourdy, J. Callapa, I. Rojas, L. Vargas, and E. Deharo. 2000. "The search for natural bioactive compounds through a multidisciplinary approach in Bolivia. Part II. Antimalarial activity of some plants used by Mosetene Indians." *Journal of Ethnopharmacology* 69 (2):139–155.

Muregi, F.W., A. Ishih, T. Miyase, T. Suzuki, H. Kino, T. Amano, and M. Terada. 2007. "Antimalarial activity of methanolic extracts from plants used in Kenyan ethnomedicine and their interactions with chloroquine (CQ) against a CQ-tolerant rodent parasite, in mice." *Journal of Ethnopharmacology* 111:190–195.

Musa, M.S., F.E. Abdelrasool, E.A. Elsheikh, L.A.M.N. Ahmed, A.L.E. Mahmoud, and S.M. Yagi. 2011. "Ethnobotanical study of medicinal plants in the Blue Nile State, South-eastern Sudan." *Journal of Medicinal Plants Research* 5 (17):4287–4297.

Muthaura, C.N., G.M. Rukunga, S.C. Chhabra, G.M. Mungai, and E.N.M. Njagi. 2007. "Traditional phytotherapy of some remedies used in treatment of malaria in Meru district of Kenya." *South African Journal of Botany* 73 (3):402–411.

Muthee, J.K., D.W. Gakuya, J.M. Mbaria, P.G. Kareru, C.M. Mulei, and F.K. Njonge. 2011. "Ethnobotanical study of anthelmintic and other medicinal plants traditionally used in Loitoktok district of Kenya." *Journal of Ethnopharmacology* 135 (1):15–21.

Nadembega, P., J.I. Boussin, J.B. Nikiema, F. Poli, and F. Antognoni. 2011. "Medicinal plants in Baskoure, Kourittenga Province, Burkina Faso: An ethnobotanical study." *Journal of Ethnopharmacology* 133 (2):378–395.

Nagendrappa, P.B., M.P. Naik, and U. Payyappallimana. 2013. "Ethnobotanical survey of malaria prophylactic remedies in Odisha, India." *Journal of Ethnopharmacology* 146 (3):768–772.

Namsa, N.D., M. Mandal, and S. Tangjang. 2011. "Anti-malarial herbal remedies of northeast India, Assam: An ethnobotanical survey." *Journal of Ethnopharmacology* 133:565–572.

Namukobe, J., J.M. Kasenene, B.T. Kiremire, R. Byamukama, M. Kamatenesi-Mugisha, S. Krief, and J.D. Kabasa. 2011. "Traditional plants used for medicinal purposes by local communities around the northern sector of Kibale National Park, Uganda." *Journal of Ethnopharmacology* 136 (1):236–245.

Nanyingi, M.O., J.M. Mbaria, A.L. Lanyasunya, C.G. Wagate, K.B. Koros, H.F. Kaburia, and W.O. Ogara. 2008. "Ethnopharmacological survey of Samburu district, Kenya." *Journal of Ethnobiology and Ethnomedicine* 4:14.

Nazar, S., S. Ravikumar, and G.P. Williams. 2008. "Ethnopharmacological survey of medicinal plants along the Southwest Coast of India." *Journal of Herbs, Spices & Medicinal Plants* 14 (3–4):219–239.

Neves, J.M., C. Matos, C. Moutinho, G. Queiroz, and L.R. Gomes. 2009. "Ethnopharmacological notes about ancient uses of medicinal plants in Trás-os-Montes (northern of Portugal)." *Journal of Ethnopharmacology* 124:270–283.

Ngarivhume, T., C. van't Klooster, J. de Jong, and V.D.J.H. Westhuizen. 2015. "Medicinal plants used by traditional healers for the treatment of malaria in the Chipinge district in Zimbabwe." *Journal of Ethnopharmacology* 159:224–237.

Nguta, J.M., J.M. Mbaria, D.W. Gakuya, P.K. Gathumbi, and S.G. Kiama. 2010a. "Antimalarial herbal remedies of Msambweni, Kenya." *Journal of Ethnopharmacology* 128 (2): 424–432.

Nguta, J.M., J.M. Mbaria, D.W. Gakuya, P.K. Gathumbi, and S.G. Kiama. 2010b. "Traditional antimalarial phytotherapy remedies used by the South Coast community, Kenya." *Journal of Ethnopharmacology* 131 (2):256–267.

Nguyen-Pouplin, J., H. Tran, H. Tran, T.A. Phan, C. Dolecek, J. Farrar, and P. Grellier. 2007. "Antimalarial and cytotoxic activities of ethnopharmacologically selected medicinal plants from South Vietnam." *Journal of Ethnopharmacology* 109 (3):417–27.

Norscia, I., and S.M. Borgognini-Tarli. 2006. "Ethnobotanical reputation of plant species from two forests of Madagascar: A preliminary investigation." *South African Journal of Botany* 72 (October 2004):656–660.

Novy, J.W. 1997. "Medicinal plants of the eastern region of Madagascar." *Journal of Ethnopharmacology* 55:119–126.

Odonne, G., C. Valadeau, J. Alban-Castillo, D. Stien, and M. Sauvain. 2013. "Medical ethnobotany of the Chayahuita of the Paranapura basin (Peruvian Amazon)." *Journal of Ethnopharmacology* 146:127–153.

Okpekon, T., S. Yolou, C. Gleye, F. Roblot, P. Loiseau, C. Bories, and R. Hocquemiller. 2004. "Antiparasitic activities of medicinal plants used in Ivory Coast." *Journal of Ethnopharmacology* 90 (1):91–97.

Omar, S., J. Zhang, S. MacKinnon, D. Leaman, T. Durst, B.J.R. Philogene, and J.M. Pezzuto. 2003. "Traditionally-used antimalarials from the Meliaceae." *Current Topics in Medicinal Chemistry* 3 (2):133–139.

Ong, H.C., and M. Nordiana. 1999. "Malay ethno-medico botany in Machang, Kelantan, Malaysia." *Fitoterapia* 70 (5):502–513.

Oni, P.I. 2010. "Ethnobotanical survey of a fallow plot for medicinal plants diversity in Idena village Ijebu-Ode, South-western Nigeria." *Journal of Medicinal Plants Research* 4 (7):509–516.

Parveen, B. Upadhyay, S. Roy, and A. Kumar. 2007. "Traditional uses of medicinal plants among the rural communities of Churu district in the Thar Desert, India." *Journal of Ethnopharmacology* 113 (3):387–399.

Pascaline, J., M. Charles, O. George, and C. Lukhoba. 2011. "An inventory of medicinal plants that the people of Nandi use to treat malaria." *Journal of Animal & Plant Sciences* 9 (3):1192–1200.

Passalacqua, N.G., P.M. Guarrera, and G. De Fine. 2007. "Contribution to the knowledge of the folk plant medicine in Calabria region (Southern Italy)." *Fitoterapia* 78 (1):52–68.

Pei, S.-J. 1985. "Preliminary study of ethnobotany in Xishuang Banna, People's Republic of China." *Journal of Ethnopharmacology* 13:121–137.

Pieroni, A., C. Quave, S. Nebel, and M. Heinrich. 2002. "Ethnopharmacy of the ethnic Albanians (Arbereshe) of northern Basilicata, Italy." *Fitoterapia* 73:217–241.

Pieroni, A., C.L. Quave, and R.F. Santoro. 2004. "Folk pharmaceutical knowledge in the territory of the Dolomiti Lucane, inland southern Italy." *Journal of Ethnopharmacology* 95 (2–3):373–384.

Poonam, K., and G.S. Singh. 2009. "Ethnobotanical study of medicinal plants used by the Taungya community in Terai Arc Landscape, India." *Journal of Ethnopharmacology* 123:167–176.

Prabhu, S., S. Vijayakumar, J.E.M. Yabesh, K. Ravichandran, and B. Sakthivel. 2014. "Documentation and quantitative analysis of the local knowledge on medicinal plants in Kalrayan hills of Villupuram district, Tamil Nadu, India." *Journal of Ethnopharmacology* 157:7–20.

Prasad, P.R.C., C.S. Reddy, S.H. Raza, and C.B.S. Dutt. 2008. "Folklore medicinal plants of North Andaman Islands, India." *Fitoterapia* 79 (6):458–464.

Prescott, T.A.K., R. Kiapranis, and S.K. Maciver. 2012. "Comparative ethnobotany and in-the-field antibacterial testing of medicinal plants used by the Bulu and inland Kaulong of Papua New Guinea." *Journal of Ethnopharmacology* 139 (2):497–503.

Pushpangadan, P., and C.K. Atal. 1984. "Ethno-medico-botanical investigations in Kerala I. Some primitive tribals of western ghats and their herbal medicine." *Journal of Ethnopharmacology* 11 (1):59–77.

Qureshi, R., and G.R. Bhatti. 2008. "Ethnobotany of plants used by the Thari people of Nara Desert, Pakistan." *Fitoterapia* 79 (6):468–473.

Rahman, N.N.N.A., T. Furuta, S. Kojima, K. Takane, and M. Ali Mohd. 1999. "Antimalarial activity of extracts of Malaysian medicinal plants." *Journal of Ethnopharmacology* 64 (3):249–254.

Randrianarivelojosia, M., V.T. Rasidimanana, H. Rabarison, P.K. Cheplogoi, M. Ratsimbason, D.A. Mulholland, and P. Mauclère. 2003. "Plants traditionally prescribed to treat tazo (malaria) in the eastern region of Madagascar." *Malaria Journal* 2:25.

Ranganathan, R., R. Vijayalakshmi, and P. Parameswari. 2012. "Ethnomedicinal survey of Jawadhu hills in Tamil Nadu." *Asian Journal of Pharmaceutical and Clinical Research* 5 (2):45–49.

Rasoanaivo, P., A. Petitjean, S. Ratsimamanga-Urverg, and R. Rakoto. 1992. "Medicinal plants used to treat malaria in Madagascar." *Journal of Ethnopharmacology* 37:117–127.

Rehecho, S., I. Uriarte-Pueyo, J. Calvo, L.A. Vivas, and M.I. Calvo. 2011. "Ethnopharmacological survey of medicinal plants in Nor-Yauyos, a part of the Landscape Reserve Nor-Yauyos-Cochas, Peru." *Journal of Ethnopharmacology* 133:75–85.

Roumy, V., G. Garcia-Pizango, A. Gutierrez-Choquevilca, L. Ruiz, V. Jullian, P. Winterton, and A. Valentin. 2007. "Amazonian plants from Peru used by Quechua and Mestizo to treat malaria with evaluation of their activity." *Journal of Ethnopharmacology* 112:482–489.

Ruiz, L., L. Ruiz, M. Maco, M. Cobos, A.-L. Gutierrez-Choquevilca, and V. Roumy. 2011. "Plants used by native Amazonian groups from the Nanay River (Peru) for the treatment of malaria." *Journal of Ethnopharmacology* 133 (2):917–921.

Rukunga, G.M., J.W. Gathirwa, S.A. Omar, F.W. Muregi, C.N. Muthaura, P.G. Kirira, and W.M. Kofi-Tsekpo. 2009. "Anti-plasmodial activity of the extracts of some Kenyan medicinal plants." *Journal of Ethnopharmacology* 121 (2):282–285.

Sadeghi, Z., K. Kuhestani, V. Abdollahi, and A. Mahmood. 2014. "Ethnopharmacological studies of indigenous medicinal plants of Saravan region, Baluchistan, Iran." *Journal of Ethnopharmacology* 153:111–118.

Samuelsson, G., M.H. Farah, P. Claeson, M. Hagos, M. Thulin, O. Hedberg, and M.H. Alin. 1991. "Inventory of plants used in traditional medicine in Somalia. I. Plants of the families Acanthaceae-Chenopodiaceae." *Journal of Ethnopharmacology* 35 (1):25–63.

Samuelsson, G., M.H. Farah, P. Claeson, M. Hagos, M. Thulin, O. Hedberg, and M.H. Alin. 1992. "Inventory of plants used in traditional medicine in Somalia. II. Plants of the families Combretaceae to Labiatae." *Journal of Ethnopharmacology* 37 (1):47–70.

Samuelsson, G., M.H. Farah, P. Claeson, M. Hagos, M. Thulin, O. Hedberg, and M.H. Alin. 1993. "Inventory of plants used in traditional medicine in Somalia. IV. Plants of the families Passifloraceae-Zygophyllaceae." *Journal of Ethnopharmacology* 38 (1):1–29.

Sanon, S., E. Ollivier, N. Azas, V. Mahiou, M. Gasquet, C.T. Ouattara, and F. Fumoux. 2003. "Ethnobotanical survey and *in vitro* antiplasmodial activity of plants used in traditional medicine in Burkina Faso." *Journal of Ethnopharmacology* 86:143–147.

Sanz-Biset, J., J. Campos-de-la-Cruz, M.a. Epiquién-Rivera, and S. Cañigueral. 2009. "A first survey on the medicinal plants of the Chazuta valley (Peruvian Amazon)." *Journal of Ethnopharmacology* 122 (2):333–362.

Scarpa, G.F. 2004. "Medicinal plants used by the Criollos of Northwestern Argentine Chaco." *Journal of Ethnopharmacology* 91 (1):115–135.

Schultes, R.E. 1979. "De plantis toxicariis e mundo novo tropicale commentationes. XIX. Biodynamic apocynaceous plants of the northwest Amazon." *Journal of Ethnopharmacology* 1 (2):165–192.

Seid, M.A., and B.A. Tsegay. 2011. "Ethnobotanical survey of traditional medicinal plants in Tehuledere district, South Wollo, Ethiopia." *Journal of Medicinal Plants Research* 5 (26):6233–6242.

Semenya, S., M. Potgieter, M. Tshisikhawe, S. Shava, and A. Maroyi. 2012. "Medicinal utilization of exotic plants by Bapedi traditional healers to treat human ailments in Limpopo province, South Africa." *Journal of Ethnopharmacology* 144 (3):646–655.

Shankar, R., B.K. Sharma, and S. Deb. 2012. "Antimalarial plants of northeast India: An overview." *Journal of Ayurveda and Integrative Medicine* 3:10.

Sharma, H.K., L. Chhangte, and A.K. Dolui. 2001. "Traditional medicinal plants in Mizoram, India." *Fitoterapia* 72 (2):146–161.

Shemluck, M. 1982. "Medicinal and other uses of the Compositae by Indians in the United States and Canada." *Journal of Ethnopharmacology* 5 (3):303–358.

Shil, S., M. Dutta Choudhury, and S. Das. 2014. "Indigenous knowledge of medicinal plants used by the Reang tribe of Tripura state of India." *Journal of Ethnopharmacology* 152:135–141.

Shinwari, M.I., and M.A. Khan. 2000. "Folk use of medicinal herbs of Margalla Hills National Park, Islamabad." *Journal of Ethnopharmacology* 69 (1):45–56.

Singh, A.K., A.S. Raghubanshi, and J.S. Singh. 2002. "Medical ethnobotany of the tribals of Sonaghati of Sonbhadra district, Uttar Pradesh, India." *Journal of Ethnopharmacology* 81:31–41.

Sivasankari, B., M. Anandharaj, and P. Gunasekaran. 2014. "An ethnobotanical study of indigenous knowledge on medicinal plants used by the village peoples of Thoppampatti, Dindigul district, Tamilnadu, India." *Journal of Ethnopharmacology* 153:408–423.

Ssegawa, P., and J.M. Kasenene. 2007. "Medicinal plant diversity and uses in the Sango bay area, Southern Uganda." *Journal of Ethnopharmacology* 113:521–540.

Stangeland, T., P.E. Alele, E. Katuura, and K.A. Lye. 2011. "Plants used to treat malaria in Nyakayojo sub-county, western Uganda." *Journal of Ethnopharmacology* 137 (1):154–166.

Suleman, S., Z. Mekonnen, G. Tilahun, and S. Chatterjee. 2009. "Utilization of traditional antimalarial ethnophytotherapeutic remedies among Assendabo inhabitants in (South-West) Ethiopia." *Current Drug Therapy* 4:78–91.

Tabuti, J.R.S. 2008. "Herbal medicines used in the treatment of malaria in Budiope county, Uganda." *Journal of Ethnopharmacology* 116:33–42.

Tangjang, S., N.D. Namsa, C. Aran, and A. Litin. 2011. "An ethnobotanical survey of medicinal plants in the Eastern Himalayan zone of Arunachal Pradesh, India." *Journal of Ethnopharmacology* 134 (1):18–25.

Tchacondo, T., S.D. Karou, A. Agban, M. Bako, K. Batawila, M.L. Bawa, and C. de Souza. 2012. "Medicinal plants use in central Togo (Africa) with an emphasis on the timing." *Pharmacognosy Research 4* (2):92–103.

Teklehaymanot, T., and M. Giday. 2010. "Quantitative ethnobotany of medicinal plants used by Kara and Kwego semi-pastoralist people in lower Omo River Valley, Debub Omo Zone, Southern Nations, Nationalities and Peoples Regional State, Ethiopia." *Journal of Ethnopharmacology* 130 (1):76–84.

Teklehaymanot, T., M. Giday, G. Medhin, and Y. Mekonnen. 2007. "Knowledge and use of medicinal plants by people around Debre Libanos monastery in Ethiopia." *Journal of Ethnopharmacology* 111 (2):271–283.

Tene, V., O. Malagon, P.V. Finzi, G. Vidari, C. Armijos, and T. Zaragoza. 2007. "An ethnobotanical survey of medicinal plants used in Loja and Zamora-Chinchipe, Ecuador." *Journal of Ethnopharmacology* 111:63–81.

Tetik, F., S. Civelek, and U. Cakilcioglu. 2013. "Traditional uses of some medicinal plants in Malatya (Turkey)." *Journal of Ethnopharmacology* 146 (1):331–346.

Titanji, V.P.K., D. Zofou, and M.N. Ngemenya. 2008. "The antimalarial potential of medicinal plants used for the treatment of malaria in Cameroonian folk medicine." *African Journal of Traditional, Complementary and Alternative Medicines* 5 (3):302–321.

Togola, A., D. Diallo, S. Dembélé, H. Barsett, and B.S. Paulsen. 2005. "Ethnopharmacological survey of different uses of seven medicinal plants from Mali, (West Africa) in the regions Doila, Kolokani and Siby." *Journal of Ethnobiology and Ethnomedicine* 1:7.

Tor-anyiin, T.A., R. Sha'ato, and H.O.A. Oluma. 2003. "Ethnobotanical survey of antimalarial medicinal plants amongst the Tiv people of Nigeria." *Journal of Herbs, Spices & Medicinal Plants* 10:61–74.

Towns, A.M., D. Quiroz, L. Guinee, H. de Boer, and T. van Andel. 2014. "Volume, value and floristic diversity of Gabon's medicinal plant markets." *Journal of Ehnopharmacology* 155:1184–1193.

Traore, M.S., M.A. Baldé, M.S.T. Diallo, E.S. Baldé, S. Diané, A. Camara, and A.M. Baldé. 2013. "Ethnobotanical survey on medicinal plants used by Guinean traditional healers in the treatment of malaria." *Journal of Ethnopharmacology* 150:1145–1153.

Tsabang, N., P.V.T. Fokou, L.R.A. Tchokouaha, B. Noguem, I. Bakarnga-Via, M.S.D. Nguepi, and F.F. Boyom. 2012. "Ethnopharmacological survey of Annonaceae medicinal plants used to treat malaria in four areas of Cameroon." *Journal of Ethnopharmacology* 139 (1):171–180.

Ullah, M., M.U. Khan, A. Mahmood, R.N. Malik, M. Hussain, S.M. Wazir, and Z.K. Shinwari. 2013. "An ethnobotanical survey of indigenous medicinal plants in Wana district south Waziristan agency, Pakistan." *Journal of Ethnopharmacology* 150:918–924.

Upadhyay, B., Parveen, A.K. Dhaker, and A. Kumar. 2010. "Ethnomedicinal and ethnopharmaco-statistical studies of Eastern Rajasthan, India." *Journal of Ethnopharmacology* 129 (1):64–86.

Uzun, E., G. Sariyar, A. Adsersen, B. Karakoc, G. Otük, E. Oktayoglu, and S. Pirildar. 2004. "Traditional medicine in Sakarya province (Turkey) and antimicrobial activities of selected species." *Journal of Ethnopharmacology* 95 (2–3):287–296.

Valadeau, C., A. Pabon, E. Deharo, J. Albán-Castillo, Y. Estevez, F.A. Lores, and G. Bourdy. 2009. "Medicinal plants from the Yanesha (Peru): Evaluation of the leishmanicidal and antimalarial activity of selected extracts." *Journal of Ethnopharmacology* 123 (3):413–422.

Valadeau, C., J. Alban, M. Sauvain, F.A. Lores, and G. Bourdy. 2010. "The rainbow hurts my skin: Medicinal concepts and plants uses among the Yanesha (Amuesha), an Amazonian Peruvian ethnic group." *Journal of Ethnopharmacology* 127:175–192.

van Andel, T., B. Myren, and S. van Onselen. 2012. "Ghana's herbal market." *Journal of Ethnopharmacology* 140:368–378.

Vigneron, M., X. Deparis, E. Deharo, and G. Bourdy. 2005. "Antimalarial remedies in French Guiana: A knowledge attitudes and practices study." *Journal of Ethnopharmacology* 98:351–360.

Waruruai, J., B. Sipana, M. Koch, L.R. Barrows, T.K. Matainaho, and P.P. Rai. 2011. "An ethnobotanical survey of medicinal plants used in the Siwai and Buin districts of the Autonomous Region of Bougainville." *Journal of Ethnopharmacology* 138 (2):564–577.

Wee, Y.C. 1992. *A guide to medicinal plants*. Singapore: Singapore Science Centre.

Wiart, C. 2000. *Medicinal Plants of Southeast Asia*. Kuala Lumpur: Pelanduk Publications.

Willcox, M. 2011. "Improved traditional phytomedicines in current use for the clinical treatment of malaria." *Planta Medica* 77:662–671.

Wondimu, T., Z. Asfaw, and E. Kelbessa. 2007. "Ethnobotanical study of medicinal plants around 'Dheeraa' town, Arsi Zone, Ethiopia." *Journal of Ethnopharmacology* 112:152–161.

Yaseen, G., M. Ahmad, S. Sultana, A.S. Alharrasi, J. Hussain, M. Zafar, and R. Shafiq Ur. 2015. "Ethnobotany of medicinal plants in the Thar Desert (Sindh) of Pakistan." *Journal of Ethnopharmacology* 163:43–59.

Yetein, M.H., L.G. Houessou, T.O. Lougbe, O. Teka, and B. Tente. 2013. "Ethnobotanical study of medicinal plants used for the treatment of malaria in plateau of Allada, Benin (West Africa)." *Journal of Ethnopharmacology* 146:154–163.

Zerabruk, S., and G. Yirga. 2012. "Traditional knowledge of medicinal plants in Gindeberet district, Western Ethiopia." *South African Journal of Botany* 78:165–169.

Zheng, X.-L., and F.-W. Xing. 2009. "Ethnobotanical study on medicinal plants around Mt. Yinggeling, Hainan Island, China." *Journal of Ethnopharmacology* 124:197–210.

Zirihi, G.N., L. Mambu, F. Guédé-Guina, B. Bodo, and P. Grellier. 2005. "*In vitro* antiplasmodial activity and cytotoxicity of 33 West African plants used for treatment of malaria." *Journal of Ethnopharmacology* 98 (3):281–285.

Index

A

B

D

F

M

P

Q

R

S

W

X

Y

Z

Printed and bound by CPI Group (UK) Ltd, Croydon, CR0 4YY
17/10/2024
01775709-0014